(First row) T. Pollard, R. Goldman, J. Rosenbaum/A. Somlyo, M. Shelanski
(Second row) K. Weber, D. Rifkin/H. Lewis, J. D. Watson, D. Koshland
(Third row) A. Harris/E. W. Taylor/J. Starger
(Fourth row) K. Porter/H. Huxley/B. R. Brinkley/R. Pollack

Cell
Motility

BOOK A

Cell Motility

BOOK A
Motility, Muscle and Non-muscle Cells

edited by

R. Goldman
Carnegie-Mellon University

T. Pollard
Harvard University

J. Rosenbaum
Yale University

COLD SPRING HARBOR CONFERENCES ON CELL PROLIFERATION
VOLUME 3

Cold Spring Harbor Laboratory
1976

Cell Motility

© 1976 by Cold Spring Harbor Laboratory

International Standard Book Number 0–87969–117–4
Library of Congress Catalog Card Number 76–17144

Printed in the United States of America

Cover and book design by Emily Harste

Contents

Section 3. The Association of Microfilaments and Microtubules with Surface Phenomena Related to Cell Motility

BOOK B
Actin, Myosin and Associated Proteins

Section 6. Membranes and Their Association with Contractile Proteins

BOOK C
Microtubules and Related Proteins

Preface

When Jim Watson invited us to organize a "small but comprehensive" symposium on cell motility, we did not realize the enormous decision-making difficulties this would entail. We rapidly found that the terms "small" and "comprehensive" were not compatible when applied to the motility field. The topic list and, consequently, the speaker list grew by leaps and bounds. In retrospect, all of this should have been predictable considering the similar evolution of past Cold Spring Harbor Conferences on Cell Proliferation and the vitality of the motility field.

Since it would be impossible to cover all of cell motility in depth, we decided to emphasize the structural and molecular aspects of motility in non-muscle eukaryotic cells. In developing the sessions, we attempted to select the most rapidly advancing and promising areas of research which stressed the functional aspects of actomyosinlike contractile systems, microtubules and other types of cytoplasmic fibers. For completeness, individual sessions on muscle contraction and on the motility of prokaryotes were also included.

A good index of the growth of the cell motility field can be obtained by comparing this meeting with the last symposium which attempted to achieve similarly broad coverage. The previous symposium, held at Princeton in 1963, was organized by Bob Allen and Noburô Kamiya and was entitled "Symposium on the Mechanisms of Cytoplasmic Streaming, Cell Movement and the Saltatory Motion of Subcellular Particles." Thirty papers were presented, and the meeting was attended by just under one hundred scientists. The majority of the contributors used light microscopic and cell physiological techniques in their attempts to unravel the mysteries of motility. There were only a handful of contributions employing electron microscopy or methods of biochemical analysis. Neither tubulin nor cytoplasmic actin or myosin had been purified at that time. In contrast, these proceedings contain 92 papers, and just under 250 scientists attended the meeting. Looking over the topics covered herein, one can only marvel at the tremendous rate of progress and the increase in interest in motility over the past 12 years. This progress reflects the development of powerful biochemical and morphological techniques with which investigators can probe the rather elusive mechanisms underlying motile phenomena.

The data compiled in this volume demonstrate the complexity of the motility problem and the exciting possibility that a small number of molecular mechanisms may be responsible for most of the diverse motile phenomena seen throughout the animal and plant kingdoms. On the other hand, it is humbling that none of the specific motile mechanisms are understood in detail. We hope that this collection of papers will be a useful resource in designing the new experimental approaches upon which future progress depends.

As organizers of the conference and editors of these proceedings, we would like to express our deep appreciation and gratitude to all of the participants and to the Cold Spring Harbor staff, without whom these volumes would not have been produced. Special thanks are due to Helen Parker, whose patience and common sense allowed the meeting to progress smoothly. We are also grateful to Nancy Ford and Annette Zaninovic for their efforts in assembling and editing the manuscripts. Finally, we wish to thank Jim Watson for providing moral support, for arranging generous financial support from granting institutions, and for the number of times he allowed us to invite "one more speaker."

Bob Goldman
Tom Pollard
Joel Rosenbaum

Introduction:
Motility in Cells

Keith R. Porter

Department of Molecular, Cellular and Developmental Biology
University of Colorado, Boulder, Colorado 80302

Motility in biological systems is expressed in a number of ways, the majority of which will be discussed in this volume. It is probable that such expressions of motility as cytoplasmic streaming, cell shape changes, ciliary motion and muscle contraction have features in common, and that eventually it will be discovered that they all depend on the same or similar structural elements, and that the motive force is a product of similar events at the macromolecular level. At present, however, the available information is insufficient to support any useful discussion of this possibility, and I shall therefore confine this report to a brief review of intracellular motility in selected systems where it is a major product of cell activity. We shall look at these systems to see what they have in common. Hopefully by reviewing some of the better studied systems as an introduction, the reports in the succeeding papers will take on added significance.

We do know enough about motility in cells to recognize that it is frequently an expression of the properties and interactions of several kinds of filaments, of microtubules and of the cytoplasmic ground substance in which they are embedded. The most elaborately evolved of these systems is the myofibril, about which there is a wealth of information. Far less is known about the role of microtubules in motility. They may function primarily as skeletal structures, which are elastic within limits and in some instances subject to distortion. Energy stored in the distortion becomes available for motion when the tubules are released. In other instances, the microtubules by their assembly seem to inflict asymmetry on an otherwise isometric system, the cytoplasmic ground substance. When the microtubules disassemble, a force is released that may transport cytoplasmic structures over substantial distances in the cell. Finally, there is a free flowing of liquid cytoplasm, which, on the basis of the evidence available, involves actin-rich fibers and myosin-associated particles.

I propose to look at certain of these systems and the associated phenomena in some detail and to see where they resemble one another and where they

1

differ. The system will include cyclosis in plant cells, "streaming" in animal cells, the movement of pigment in chromatophores and the motion of chromosomes in the mitotic spindle. Finally, I shall review some of the newer preliminary observations on the cytoplasmic ground substance where the motive force is developed.

Cyclosis (Streaming) in Plant Cells

One of the most remarkable examples of *intracellular* motility is found in plant cells, especially in the giant cells of such green algae as *Nitella* and other Characeae. In this streaming, many particulate components of the cytoplasm and occasionally a few nuclei are transported along pathways in a smooth streaming motion. Velocities as great as 100 μm/sec have been recorded. The motion is O_2-dependent and reaches its greatest velocity when the cell is exposed to sunlight. In *Nitella,* two streams flow in opposite directions along helical paths around the cell. Presumably, the two streams are really a continuous one with the direction reversed at the nodes where a transverse wall ends the coenocyte.

To understand this motion and its relation to other parts of the cell (as well as cytoplasmic streaming in other cells) it is important to look at the structure of these algae cells. The greater part (90%) of the volume of one of these cells is occupied by a vacuole (trophoplast), which is a sink for metabolic wastes. This is surrounded on the outside by a cortical layer of structured cytoplasm (a viscoelastic gel) in which the non-motile chloroplasts are embedded. The whole is limited on the outside by a plasma membrane, subjacent to which there is a single layer of microtubules. A characteristic plant cell wall encloses the whole coenocyte, and it is interesting to note that the fine cellulose filaments in this wall, i.e., in the layer closest to the plasma membrane, are oriented with their long axes parallel to the microtubules. This is a good place to observe also that the streaming we are discussing bears no relation to the microtubules either in position or direction.

The layer of solated cytoplasm (endoplasm) that streams is located between the cortical (parietal) gelled layer (the structured cytoplast) and the surface of the trophoplast or vacuole. The direction of the stream parallels the orientation of a layer of fibers just beneath and closely associated with the chloroplasts. The relation to chloroplasts is so close that when the chloroplasts are isolated, the fibers remain adherent to them (Palevitz and Hepler 1975). In *Nitella,* the fibers (first noted by Pickett-Heaps [1967] and Nagai and Rebhun [1966]) measure about 0.2 μm in diameter and are made up of 50–70-Å filaments (Fig. 1). Palevitz and Hepler (1975) have shown that these filaments decorate with heavy meromyosin (HMM). The location of the streaming layer in relation to other components of the *Nitella* coenocyte is shown in Figure 2.

Fibers similar to those in *Nitella* have been observed in cells of higher plants by O'Brien and Thimann (1966) and O'Brien and McCully (1970). They have noted that streaming is unidirectional with respect to single fibers, and therefore the fibers seem to be polarized. In the cells of higher plants, there is a layer of parietal cytoplasm in which there is no streaming but from which mitochondria and other cytoplasmic components may emerge and enter the

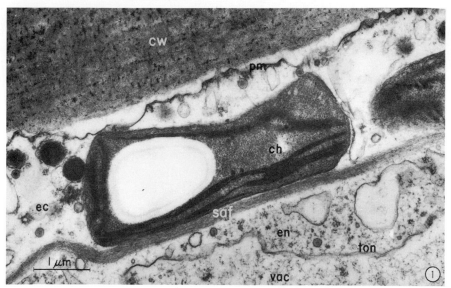

Figure 1

Micrograph showing a portion of a large, vacuolated internodal cell of Chara sp. The image includes a chloroplast (ch), a portion of the cell wall (cw) and the tonoplast (ton) or membrane limiting the vacuole (vac). The subcortical actin fibrils (saf) are prominent at the interface between ectoplasm (ec) and streaming endoplasm (en). A plasma membrane (pm) separates the ectoplasm from the cell wall. It is evident that the fibril is made up of numerous filaments. Magnification, 15,000×. (Micrograph courtesy of J. Pickett-Heaps.)

stream. Thus the organization is much the same as in *Nitella,* with a parietal portion of the cytoplasm a gel and not involved in the motion.

What then is the origin of the motive force? A number of observations suggest that the force derives from some fiber-associated events. For example, in living cells the streaming is most rapid close to the fibers and most vigorous of all in close relation to groups of several fibers similarly oriented. It is also obvious that the fibers are anchored in the structured cytoplasm or ectoplasm. It is assumed that the separation of these zones (ectoplasm and endoplasm) is labile, but of such a nature as to incorporate and contain the actin available for assembly into filaments and fibers.

Direct observations on the filaments with Nomarski optics, using a window technique introduced by Kamitsubo (1972a), has convinced Allen (1974) that the fibers in *Nitella,* or filaments extending from them, execute an undulating motion. The active propagation of such waves is thought to provide the motive force. This interpretation would seem to be supported by the observed undulating motion of fibers in droplets of protoplasm from Characeae (Jarosch 1964) as well as in centrifuged cells (Kamitsubo 1972b).

A different suggestion, however, comes from the studies of Kersey et al. (1976). These investigators have shown that the filaments of the fibers decorate with HMM and that the arrowheads point in a direction opposite to that of the motion. This same relation of polarity to motion is of course evident in striated muscle, where the moving component is myosin. Kersey and co-

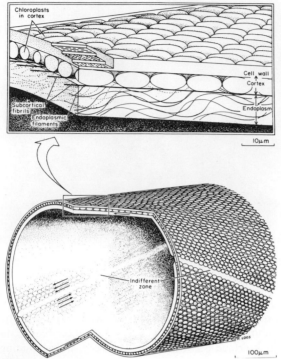

Figure 2

Diagram at the top illustrates the relationship of the subcortical (actin) fibrils (saf in Fig. 1) to the cortical layer, with chloroplasts on the outside and streaming endoplasm on the inside. The vacuole with limiting tonoplast is at the bottom of the image. Unit filaments of the subcortical fibrils are depicted as executing an undulating motion thought to provide the motive force that moves the streaming endoplasm. The lower diagram places the upper image with respect to the whole cell. The arrows indicate the direction of motion in the two streams. (Reprinted, with permission, from Allen 1974.)

workers therefore favored the conclusion that the streaming cytoplasm contains myosin, possibly in aggregates that move along the fixed actin fibers. Kersey et al. (1976) have obtained preliminary evidence that plant cell cytoplasm does, in fact, contain "myosinlike" protein.

Whatever the source of the motive force, the involvement of actin filaments in this cytoplasmic streaming seems incontrovertible. With that conclusion in mind it becomes very interesting to look at similar motility in non-muscle animal cells.

Motility in Animal Cells

"Streaming motion" within most vertebrate cells observed under conditions of in vitro culture is not as a rule so dramatic as in plant cells. The volume of cytoplasm moved is far less. Usually only a few particles move in unison, and the extent of motion may be quite limited. Freed and Lebowitz (1970), for example, describe a lipid granule in HeLa cells as moving approximately 4 μm in about 10 seconds. Such long saltatory motions are of course much less rapid than they appear in the usual time-lapse movie of living cells. The direction of motion is nonrandom, and particles seem to move to and fro in channels within the cytoplasm. The majority of these channels are arranged roughly as radii extending from the cell center (Freed and Lebowitz 1970). The inward motion is observed to be more rapid than the outward. In cells treated with colchicine or vinblastine, these long motions vanish, and this is the strongest evidence in favor of tubule involvement in the motion. Actually, the disassembly of tubules by these drugs may simply occlude channels otherwise dependent on the presence of microtubules.

Streaming motion in nerve cells is more dramatic than in fibroblasts or most other animal cell types. This may be accounted for by their greater size and the therefore greater distances over which components of the cytoplast and metabolites must be transported. In any case, the nerve cell more closely resembles the plant cell in showing channels along which particulate components of the cytoplasm actively stream. These have been observed by many investigators especially in the perikaryon, but perhaps Pomerat and colleagues have made the best recordings (Pomerat et al. 1967). A similar streaming has been visualized in axons and dendrites by Kirkpatrick (1971), McMahan and Kuffler (1971) and probably by others (see review by Wuerker and Kirkpatrick 1972). This relatively rapid streaming in the perikaryon and axons is localized within channels that seem to coincide with the distribution of microtubules and that occupy the clear zones between masses of Golgi cisternae, Nissl substance and included mitochondria (Fig. 3). The components of the stream are of uncertain identity but probably include mitochondria and vesicles derived from the endoplasmic reticulum. Presumably this activity is one expression of a constant restructuring of the perikaryon which facilitates the movement of metabolites between the deeper zones of these large cells and their surfaces.

Other than microtubules, the structural elements involved are not entirely clear. According to such seasoned observers of axoplasm as Palay et al. (1968) and Wuerker (1970), the major elements possibly involved in the motion here are neurotubules (microtubules) and neurofilaments (100 Å) or the "intermediate filament" of Shelanski (Yen et al. 1976). The neurotubules are especially prominent in the central regions of dendrites. In some instances they show "wispy" substances on their surfaces, substances that radiate laterally and are continuous with a similar component of the axoplasmic matrix to form a three-dimensional latticework (Burton and Fernandez 1973) (Fig. 4).

Neurofilaments are not so conspicuous as neurotubules and may be entirely absent, as, for example, in the ventral nerve of the crayfish, a fact which places in some doubt their essentialness to streaming. Like the tubules, they are long, of uniform diameter and unbranched. According to Wuerker (1970), they are made of four subunits arranged as a cylinder around a central space 30 Å in diameter. They also show lateral arms extending into the ground substance.

The endoplasmic reticulum in axoplasm is without attached ribosomes and forms a network of varicose, membrane-bound channels.

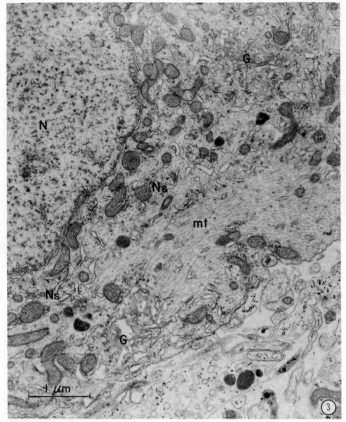

Figure 3

Micrograph of a portion of a neuron from cerebellum of a frog. The nucleus (N) is at the upper left; Nissl substances and Golgi components are indicated by Ns and G, respectively. External to these areas the cytoplasm of the typical neuron appears relatively clear in the light microscope image, and it is in this region that it is common to observe streaming of mitochondria and isolated vesicles of the ER. Microtubules are numerous, but their relation to the motion is not established. Magnification, 16,000×.

Microfilaments, the 50–70-Å variety, have been described in the perikarya of nerve cells, but in no instance do they achieve the prominence typical of neurotubules and neurofilaments.

These filament components of nerve cells have been characterized to varying degrees. *Neurotubules* have been isolated in large quantities from the central nervous system, and much of the biochemical information we have about tubulin has been gleaned from such starting materials. *Neurofilaments* (90–100 Å), on the other hand, have received less attention. It is known, however, that these are made of a unique protein with a molecular weight of about 55,000, which is found in a variety of cells, including smooth muscle and brain tissue (Cooke 1976; Yen et al. 1976). Their composition is not to be confused with either tubulin or actin. *Microfilaments*, on the other hand, have been identified as actin in mouse neuroblastoma cells by HMM-binding

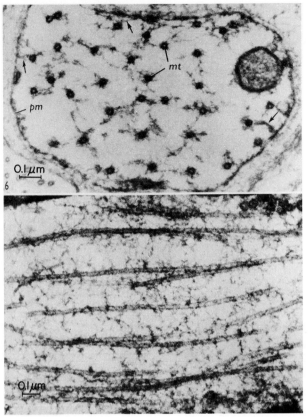

Figure 4
Sections of a crayfish axon stained with lanthanum hydroxide to delineate
selectively the association between tubules and "filamentous material" of the
axoplasmic matrix. The material is thought to be the sectioned image of the
matrix lattice that is evident in electron microscope images of whole cultured
cells (see Fig. 11). Magnifications: (*top*) cross section 57,500×; (*bottom*)
longitudinal section 46,500×. (Reprinted, with permission, from Burton and
Fernandez 1973.)

(Burton and Kirkland 1972) and in neurons of the central nervous system
by similar procedures (LeBeux and Willemot 1975). These findings support and extend the reported isolation of an actinlike protein from mammalian brain tissue (Berl, Puszkin and Nicklas 1973) and from cultured
nerve cells (Fine and Bray 1971), in which instance it was identified as such
by gel electrophoresis. A tropomyosinlike protein has recently been isolated
from brain and growing neurons (Fine et al. 1973). Thus while all the elements essential for a transport mechanism seem to be present in nerve cells,
there is no clear indication of how they work.

It is now generally agreed that transport in nerves falls into two distinct
classes: fast (400 mm/day) and slow (1–2 mm/day), the fast transport being
bidirectional. These two classes are based on well-known measurements of
the movement of radiolabeled proteins down the fibers following injection of

labeled amino acids into the region of nerve cell bodies. The procedure has produced similar results for so many investigators that there seems little reason to question its usefulness or the validity of the results. The same may be said for the validity of experiments demonstrating the inhibition of rapid transport by colchicine and vinblastine. For most investigators, these studies provide incontrovertible evidence that microtubules are essential to this motion. For others, there is still some question, since colchicine inhibition is not complete even where microtubule disassembly is total (Byers 1974). Byers has made the further interesting observation that rapid axoplasmic flow is peripheral to the major column of microtubules, thereby questioning their immediate involvement in the generation of the motive force.

This is doubtless an inadequate summary of available observations on streaming in nerve cells. Even if it were more extensive, however, the origin and location of the motive force would not, I think, be apparent. Microtubules are certainly prevalent wherever streaming occurs, but whether they supply a surface on which a force develops or simply define channels is uncertain. The presence of actin in nerve cells encourages the thought that it is involved in some way in moving the motile portion of the cytoplasm. Needless to say, ideas and models to explain axoplasmic flow (Schmitt 1968) have come in abundance from investigators who have considered this problem.

Among these, Ochs (1966) has suggested that both microtubules and neurofilaments are involved. This possibility is built into what he described as the "transport filament model" on the basis that there is certainly a close connection between the neurofilaments and neurotubules (Yamada, Spooner and Wessells 1971).

This possible interaction of neurofilaments (90–100 Å) and microtubules, with the microtubules providing the directionality and supportive function, has appealed to numerous investigators of this and similar phenomena. One cannot read their collected views without feeling that the truth is being approached, albeit slowly, and that some person of appropriate genius will soon emerge with a totally satisfactory explanation.

If I might attempt a contribution to this process in a field where speculation is rife, it would be to suggest that microtubules provide the frame for a structured, relatively non-motile portion of the cytoplast. This surrounds regions (channels) in the perikaryon and extends out into the dendrites and axons, where it is gradually remodeled in the slow transport mentioned above. There may also be a structured system representing the cell cortex and its limiting plasma membrane. According to this view, at phases within and between these two (or in channels within either one), the structured character breaks down and grades into an unstructured fluid phase in which rapid transport takes place (Byers 1974). Both structured phases of the cytoplast are labile and may contribute to the fluid phase at the boundary between the two. Possibly, as in plant cells, actin filaments populate this boundary and interact with aggregates of myosin in the fluid phase to provide the motive force. In the absence of solid information it is equally possible that the motive force could be provided by propagated waves of structuring or even contractions within the matrix of the structured ground substance. These would be given direction and a base for propagation by the microtubules to which the ground substance meshwork is attached (see below). A somewhat similar view has been published by Samson (1971). Impressed by the evidence that microtubules are surrounded by materials rich in acid mucopolysaccharides and that these materials extend from the tubule surface to form a three-dimensional latticework, Samson makes the point that polyelectrolytes, those highly charged and flexible chains, are responsive in terms of conformation to small changes in their ionic environment and thus suggests that polyelectrolyte gels can transform chemical energy into mechanical work.

Streaming motion within the same rate range as that observed in nerve cells (2–3 μm/sec) and having some of the same characteristics can be found in other cells. For example, it is easily observable and available for experimentation in axopods, those very long, slender arms used to move the cell and to capture prey that reach out in large numbers from the cell body of certain protozoa, called Heliozoa. The flowing motion in axopods is smooth, channeled, bidirectional and involves mitochondria and dense particles thought to be instrumental in narcotizing the prey. The core of each axopod comprises an axoneme constructed of many microtubules in a paracrystallin array (Tilney and Porter 1965; Roth, Pihlaja and Shigenaka 1970). Thus the moving stream is localized between the axoneme, a rigid, form-imparting structure, and the limiting plasma membrane. It is a zone of relative solation. An interesting feature of this stream is that it is not tubule-dependent. The axopods of cells exposed to colchicine rapidly diminish in length and finally exist as mere knobs on the cell's surface (Tilney 1968). This retraction is accompanied by a total disassembly of the axoneme, yet streaming of the cytoplasm continues. As in nerve cells, we are once again confronted with evidence that microtubules influence the location of the stream but not the motive force.

Another interesting observation to emerge from studies on heliozoan axopods is the effect of magnesium ion. At low concentrations (0.1 mM), this ion stops the streaming motion and stabilizes the microtubules as though inducing gel formation in all parts of the axopod (Shigenaka, Watanabe and Kaneda 1974; Shigenaka, Tadokaro and Kaneda 1975).

The Motion of Pigment and Chromosomes

Most students of intracellular motion would agree that the movement of pigment in chromatophores, if understood, might provide an explanation for other, more limited, cytoplasmic motions. Of this one cannot be certain, but pigment motion and chromosome motion have some features in common that deserve attention.

The melanophore of the teleost, *Fundulus heteroclitus* (L.), is easily observed in the light microscope, and the pigment is an obvious marker of any cytoplasmic movements that take place. When the fish pales in response to appropriate stimuli, the melanin granules (melanosomes) aggregate into a tiny, spherical mass around the centrosphere of the cell. On the other hand, when the fish darkens, the pigment disperses.

There are one or two features of this motion that are worth noting. Once the inward flow or aggregation of pigment is initiated, it proceeds in a resolute fashion at a constant velocity until complete, which takes about 2–3 minutes. There is no saltation in this motion (Bikle, Tilney and Porter 1966; Green 1968).

The disaggregation or dispersion is quite different. In this case, the motion is irresolute. The pigment granules move back and forth, with the net movement toward the periphery. Saltation is the rule, and the complete dispersion takes two to three times longer than the aggregation. Aggregation proceeds as though propelled by an elastic component (Green 1968), whereas the dispersion proceeds in a manner suggesting that a hesitant restructuring of the pigment support is involved.

These phenomena are demonstrated equally well by other chromatophores, those from the red squirrel fish, *Holocentrus ascensionis,* commonly found in the Caribbean. These cells contain an extraordinarily large number of microtubules, all oriented radially with respect to the cytocentrum (Fig. 5). The red pigment granules are similarly arranged in files focused on the cell center. Many of the microtubules insert into sickle-shaped bands of dense (fuzzy) material that reside in the centrosphere (where the centrioles also reside). Others of the total population may insert in the cytoplasmic cortex (Porter, Bennett and Junqueira 1970) (Fig. 6). Pigment aggregation into a

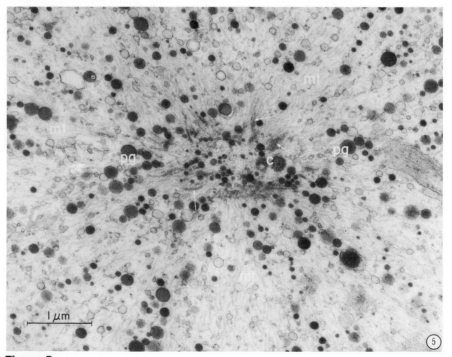

Figure 5
Micrograph of a part of an erythrophore as it appears in a thin section. The centrosphere region containing two centrioles (c) and numerous satellite densities (arrows) into which microtubules insert occupies the center of the image. The pigment granules (pg) are distributed in files among the numerous microtubules radiating from the cell center. The mitochondria and the nucleus reside outside the large central area (see Fig. 6). (Reprinted, with permission, from Porter 1973.)

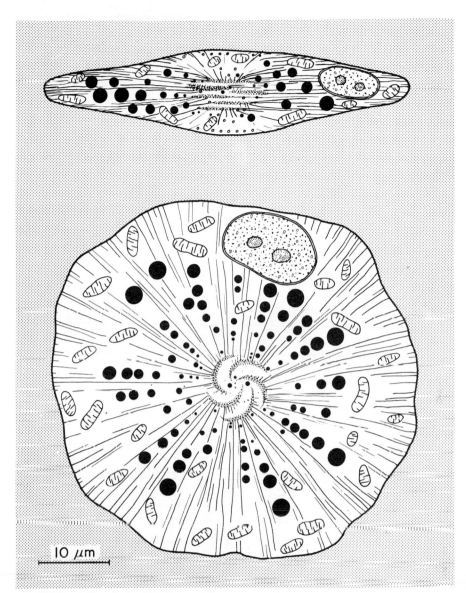

10 μm

Figure 6
Diagram illustrating the disposition and orientation of microtubules as well as the distribution of pigment granules, mitochondria and nucleus in a typical erythrophore. At the cell center, sickle-shaped dense bodies are the insertion points for large numbers of microtubules; other microtubules reside in or near the cell cortex and seem to insert in this part of the cell. The top diagram represents a vertical section; the bottom, a horizontal section. Magnification, ca. 1860×.

spherical mass in the cell center is achieved in 2–3 seconds at uniform velocities of 15–20 μm/sec. Dispersion takes twice as long and involves saltatory motion, as is also seen in the melanophore. Both motions are strictly along radial pathways.

When the fine structure of these cells in the dispersed and aggregated states is compared, a surprising observation emerges. Intact microtubules appear to vanish during aggregation and then to reassemble during dispersion (Fig. 7). The cytoplast pigment changes in form from that of a disc to a sphere and back to a disc (Porter 1973; Schliwa and Bereiter-Hahn 1973).

It should come as no surprise that the cytocentrum of the cell and the microtubules should function, as in this instance, in shaping the cytoplast. The same phenomenon has been observed in a number of other cases. For example, karyoplasts produced by centrifuging CHO cells in the presence of cytochalasin, and thus deprived of cytocentrum and microtubules, are ap-

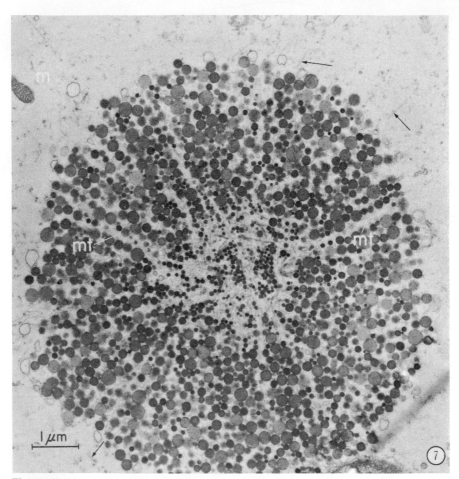

Figure 7
Micrograph of a thin section that includes the center of an erythrophore in which the pigment is just beginning dispersion from the more compact aggregated state. Outside the pigment mass most of the microtubules are disassembled, and only small fragments (arrows) are evident. Inside the mass a few microtubules (mt) have begun to reassemble. The cell center is obvious. A mitochondrion (m) is at the upper left, representing a population that does not move with the pigment. Magnification, ca. 12,200×. (Reprinted, with permission from Porter 1973.)

parently unable to stretch out on the substrate (Shay, Porter and Prescott 1974). On the other hand, the cytoplasts, which contain cytocentrum and microtubules, spread out to resemble the cell of origin (Figs. 8, 9).

Similarly, cells induced with dB-cAMP to assemble a larger than normal number of microtubules show unusual form asymmetries in response to their presence (Porter, Puck and Hsie 1974). However, cells deprived of assembled microtubules by low temperatures or colchicine or in transformation by oncogenic viruses invariably adopt more spherical and pleomorphic forms (Fonte and Porter 1974; Brinkley, Fuller and Highfield 1975).

What is perhaps more remarkable than these form-associated phenomena is the rapidity with which microtubules in the erythrophores break down and then reassemble. This striking behavior serves to remind one that microtubule systems can be extraordinarily dynamic, and that in other kinds of cells, as well, they are probably involved in a constant remodeling of the cyto-

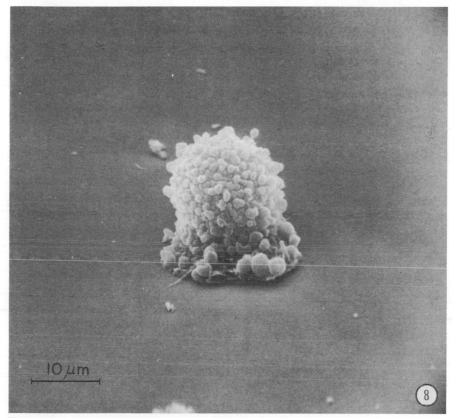

Figure 8
Scanning electron micrograph of a CHO "cell" (a karyoplast) deprived of its centrosphere and associated microtubules. It remains in the more-or-less spherical form for the duration of its life (about 48 hr). Micrographs of thin sections show it to contain an intact nucleus with envelope, mitochondria, ribosome and elements of the ER, as well as a continuous plasma membrane. The significance of the surface blebbing is not understood. Magnification, ca. 1800×. (Reprinted, with permission, from Shay, Porter and Prescott 1974.)

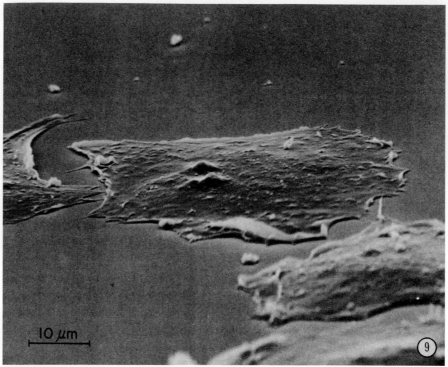

Figure 9
Scanning electron micrograph of a CHO "cell" deprived of its nucleus and a volume of cytoplasm. The cytoplast consists of centrosphere with centrioles and associated microtubules. The cytoplast, unlike the karyoplast, has retained a capacity to spread out on the substrate and adopt a form similar to that of the original CHO cell in S phase. Its surface is usually free of blebs. Magnification, ca. 1600×. (Reprinted, with permission, from Shay, Porter and Prescott 1974.)

plasts, though at a slower rate than in erythrophores. Possibly, disassembly and reassembly are more localized in other cells (e.g., cultured fibroblasts), and more confined, and are reflected only in the occasional movement of lipid granules or lysosomes.

The most available interpretation of these phenomena is that aggregation of pigment is accompanied by a sudden disassembly of microtubules, which releases an elastic force that draws the pigment toward the cell center. Presumably, this elasticity resides in and is one of the properties of the cytoplasmic matrix (a three-dimensional lattice), which will be described in the next section. In the other part of the cycle, dispersion of the pigment is accompanied by a restructuring of the whole tubule complex, including, evidently, the polymerized (or copolymerized) form of the matrix lattice. This restructuring takes longer but is nonetheless remarkably rapid: 4–5 seconds in contrast to 2 seconds for the aggregation. After reassembly, the microtubules are arranged radially around the cytocentrum and occupy space that is essentially discoidal in form. Thus the spherical mass of aggregated pigment is brought into dispersion in the same disc-shaped space it occupied at the beginning of aggregation. It follows that the tubules are involved in reshaping the central cytoplast much as they seem to be involved in shaping other cells (Porter 1973). Information contained in the cell center, where assembly is initiated, then determines the distribution of the tubules in the cytoplast and their orientation in space. We may be misinterpreting the observations, or overinterpreting them, but in either event, the fault would be greater were we not to attempt an interpretation.

The dynamic assembly and disassembly of microtubules and their involvement in motion are conspicuously evident in mitosis, where, in the major events, they are highly orchestrated. Since the microtubules are easily equated with birefringence, their behavior and disassembly can be followed in living material (Inoué and Ritter 1975; Inoué et al. 1975). It should be recognized that the two poles of the spindle, with their centrioles and astral arrays, represent centrospheres (the central apparatus of Wilson 1928), the organization centers of two sister cytoplasts. The independent behavior of these is evident from the time the centrioles separate in later prophase (Wilson 1928).

Certain features of mitosis remind one of the phenomena observed in chromatophores.

1. The saltatory motion of pigment granules, especially at the limits of the dispersion, mimics the saltatory behavior of the chromosomes at the cell plate during the early stages of metaphase. It is as though the chromosomes, residing at the margin between the two sister cytoplasts, are constantly buffeted by changes of these margins (zones of influence) with respect to each other as the microtubules of the cytoplasts assemble and disassemble along with the associated matrix.

2. The movement of the chromosomes at anaphase at a constant velocity is so similar (except in rate) to the motion of the pigment granules that one is compelled to conclude that they are moved by much the same mechanism. Whether in both instances the motion is an expression of the action of an elastic component in the ground substance requires further study, but certainly the possibility exists.

3. In both chromatophores and the mitotic spindle, simultaneously with the motion, there is a disassembly of microtubules, thus presumably releasing the elasticity of the ground substance to yield a kind of syneresis—in one case, moving pigment, and in the other, chromosomes.

4. The dependency of anaphase motion on the disassembly and then reassembly of microtubules has been studied in the case of mitosis (Inoué and Ritter 1975) but not in chromatophores. However, there is good reason to believe that results with the latter system would be similar. We do know that in chromatophores, the use of high concentrations of colchicine (5 mM) at first induces a partial aggregation of pigment, but it then seems quickly to effect another change leading to a release of the pigment into a now solated (depolymerized) ground substance and the consequent random distribution of the pigment granules. The same has been observed to result from sudden shifts to low temperatures (4°C) (K. R. Porter, unpubl. obs.). If the elastic matrix in which the pigment apparently resides, and in which the chromosomes presumably become enmeshed via the chromosome fibers (the kinetochore microtubules), is

depolymerized along with the microtubules by low temperatures and colchicine, one would have an explanation for the failure of the putative contractile component to "work" under these conditions, as has been reported for the spindle by Inoué and Ritter (1975).

The rapid contraction of the cytoplasmic (elastic) matrix apparently involved in pigment aggregation is not unique to the erythrophores of *Holocentrus*. Just how general it is as an accompaniment of microtubule disassembly is not known, but one additional example, in the heliozoan *Actinophrys*, is so remarkable as to warrant mention. When an acceptable food organism touches the tip of an axopod, the axopod shortens as much as 100 μm in a fraction of a second. The available evidence indicates that this is accompanied by tubule breakdown rather than sliding (Ockleford and Tucker 1973). The contractile force would appear to reside in an elastic component of the matrix that is stretched by the axoneme and released by axoneme disassembly. Alternatively, there may be a contractile system that is triggered by the stimulus of contact with the food organism. However, microscopy has thus far not identified anything in the axopods except microtubules and a matrix, including bridges or "links," between them (Roth, Pihlaja and Shigenaka 1970; Ockleford and Tucker 1973).

Structural Correlates in Non-muscle Motion

In this limited review, the discussion has thus far focused on the movement of cytoplasmic components (in non-muscle cells) as a product of interaction between filaments (60 and 100 Å) and their surroundings and also the assembly and disassembly of microtubules in a structured ground substance or matrix assumed to have elastic properties. Clearly, if valid, these concepts, and especially the latter, should be based on evidence that there is, in fact, a structured ground substance or matrix capable of responding by contracting to the removal of struts in the form of microtubules. Pertinent observations have now been reported from several laboratories and they all describe the existence in fixed preparations of a finely constructed, three-dimensional lattice, containing in its continuity microtubules, microfilaments, polysomes and the cisternal elements of the endoplasmic reticulum. That this structured component of the cytoplasm has not received more attention is due to the limited two-dimensional image provided by thin sections and a general, healthy concern shared by most microscopists that structure at these levels may represent little more than an artifact of fixation and/or drying. In this final section I should like to look at some published and unpublished evidence and examine especially some observations made in our Boulder laboratory on whole (unsectioned) cells grown in tissue culture and dried by the critical point method.

As mentioned earlier in this review, students of cell fine structure have been familiar for some time with fiber bundles in fixed cells (Buckley and Porter 1967; Goldman et al. 1973), the so-called stress fibers that decorate with HMM (Ishikawa, Bischoff and Holtzer 1969). There is no question of their relation to real structures because they can be watched in living cells and are observed to change orientation with the changing form of the cell.

They seem designed to exert a unidirectional force parallel to the orientation of their long axes and that of their component filaments (Buckley and Porter 1967). These fiber bundles, or closely associated components of the cytoplasm, are sensitive to the cell's exposure to cytochalasin B (Schroeder 1969; Spooner, Yamada and Wessells 1971; Goldman et al. 1973). For similar reasons, microtubules have acquired acceptance and are generally thought to perform a skeletal role in imparting asymmetries of form to the cytoplast and thence to the whole cell (Roberts 1974). The tubule surfaces in section are usually decorated by a "fuzz," which in some instances, e.g., in cilia, appears in axonemes condensed into discrete spokes (Warner 1970) or bridges extending to adjacent tubules (Tilney and Porter 1965). These tubule surface features are made more convincing in some instances by the periodicity in the disposition (McIntosh, Ogata and Landis 1973). However, where their disposition is less regular, e.g., on tubules in the mitotic spindle (McIntosh 1974) or on neurotubules (Burton and Fernandez 1973), they should probably not be discounted as mere artifacts.

Recently, this awareness of a substructure beyond the more discretely formed elements has come to include an image of the ground substance as comprising a lattice of microtrabeculae (30–60 Å thick) in which most other elements of the cytoplasm appear to be suspended. This is seen in sections of nerves (Burton and Fernandez 1973), in the growth cones of growing nerve fibers (Bunge 1973), in the ruffles and spikes on cells in vitro (Wessells, Spooner and Luduena 1973) and in sections of microvilli, where they establish regular connections with the core of actin filaments (Mooseker and Tilney 1975). In each of these few examples, the existence of such a lattice has been accepted without serious question.

Most recently, the lattice or mesh has been depicted in whole cultured cells that have been fixed and then dried by the critical point method (Buckley 1975; Buckley and Porter 1973, 1975). Though perfectly evident in images taken with the conventional electron microscope operating at 80 or 100 kV, the stereo images are a great improvement due to the better penetration and resolution of the high energy beam provided by the 1 MeV electron microscope.[1]

The thinly spread cultured cell that grows from a variety of tissue explants, and which in many instances is an endothelial cell, is an excellent object for study. It lends itself well to experimental manipulation as well as to light and electron microscopy. Techniques for the latter were described long ago (Porter, Claude and Fullam 1945) and have since been improved (Buckley and Porter 1975) and subsequently simplified to make them available to a wide range of investigations (Gershenbaum, Shay and Porter 1974).

In studies done in collaboration with John Wolosewick, WI38 cells, a strain of human cells derived from fetal lung tissue having a limited life span

---

[1] The high-voltage microscope is installed in the Department for Molecular, Cellular and Developmental Biology at the University of Colorado in Boulder. It is one of two such installations in the country financed by the Division of Research Resources, National Institutes of Health; the other one is at the University of Wisconsin, Madison. Both are available on a part-time basis to investigators from other institutions, and anyone desiring to use the instrument in Boulder should direct their initial inquiry to K. R. Porter at Boulder.

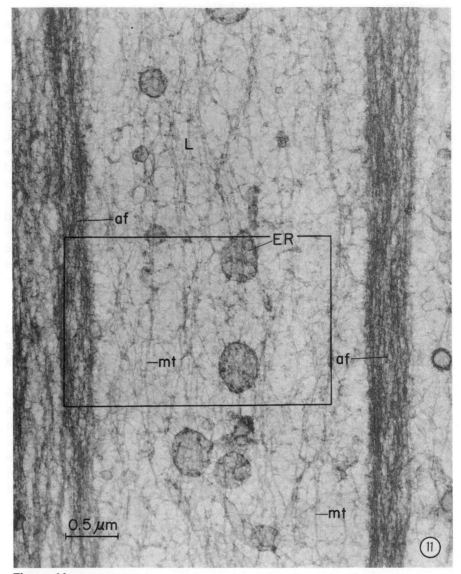

Figure 11

High-voltage TEM of a thin portion of a whole WI38 cell. Two bundles of actin filaments (af) run vertically in the image and represent stress fibers. Vesicles of the ER as well as a few microtubules populate the space between them (see stereos, Fig. 12). The rest of the image comprises a lattice of microtrabeculae which attach to the surfaces of all the other elements in the image. The rectangle outlines the area shown in Fig. 12. Magnification, ca. 29,200×.

sion, can be equated with the lattice of the cytoplast. They have shown that the microtrabeculae of the meshwork decorate with HMM. In the same vein, it is attractive to think of the lattice as the source of the actin-rich mixtures prepared from sea urchin eggs (Kane 1974), *Acanthamoeba* (Pollard 1976) and rabbit macrophages (Stossel and Hartwig 1976). Some coincidence of

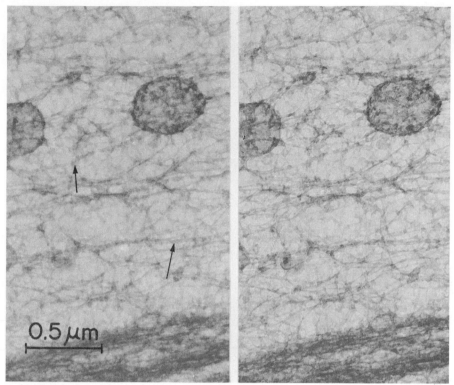

Figure 12

A stereo pair of micrographs of the region outlined in Figure 11. The two images can be fused with the unaided eye (with a little practice) or with a stereo viewer. (Two-lens stereo viewers of satisfactory quality [3×] are available from Abrams Instrument Corporation, Lansing, Michigan.) When this is done, the fine trabeculae of the three-dimensional lattice can be seen to run into upper and lower cortices and to fuse with microtubules (arrows) and the surfaces of ER vesicles. They seem also to be continuous with the material of the fiber (af, see Fig. 11) across the bottom of the micrograph, which is close to the lower surface, the substrate surface, of the cell. Magnification, 40,000×.

content and properties would seem unavoidable considering the procedures used in obtaining these various extracts. Thus, in addition to actin, one may reasonably expect in the final analysis to find myosin and an actin binding protein of high molecular weight. On incubation, these combine to form gels, which, like the cytoplasmic matrix, liquefy on cooling. According to Stossel and Hartwig (1976), myosin is involved in gel contraction. Whether this phenomenon should be regarded as mimicking the behavior of the matrix in chromatophores in pigment aggregation belongs for the present in the realm of rank speculation.

It also seems probable that in these micrographs we are looking at the fixed image of the so-called thixatropic gel that fascinated cell biologists 40 years ago (Marsland and Brown 1936; Freundlich 1942). But the "gel" as it appears in the electron microscope image is more complicated than implied

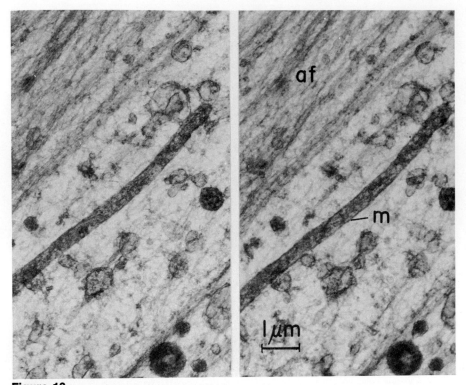

Figure 13

A stereo pair of micrographs depicting a small region in the thin margin of a WI38 cell. The field includes part of a mitochondrion (m) as well as vesicles of the ER and elements of the lattice oriented parallel to constitute a loose fiber bundle (af). Other regions of the cytoplasmic ground substance included in the image show the characteristic random lattice of thin trabeculae. These insert into the surfaces of the ER vesicles but seem not to contain the mitochondrion. Magnification, 10,000×.

by that appellation for it is part of an organized unit, the cytoplast, constructed around the paired centrioles or their equivalent. It increases in density and viscosity toward the cell center. The ER, Golgi and most of the other elements contained in the gel are distributed nonrandomly. It has those elastic or contractile properties that seem to be released or activated when the microtubules are disassembled. So there is more to the "gel" than one visualizes in any one micrograph.

Artifact?

The possibility that we might be viewing artifacts totally unrepresentative of the cytoplasmic gel is becoming more and more remote. The "pros" for its validity seem to outweigh the "cons." It is, for example, preserved in the form depicted when OsO_4 and glutaraldehyde are used separately or together. Dehydration with acetone or alcohol preceding the drying from CO_2 by the critical point route does not alter its appearance. Since the same procedures faithfully preserve some of the most delicate components and surface features known to

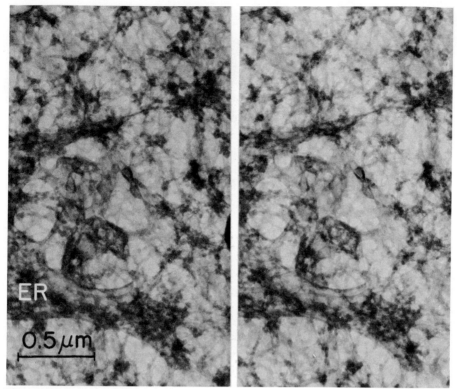

Figure 14

Stereo pair of micrographs of cytoplasmic region including a larger number of polysomes. These are contained in junctional points within the microtrabecular lattice. Some are also associated with the surfaces of an ER cisterna (ER) that comes into the image from the left. The trabeculae of the lattice are severalfold broader (150 Å) here than in Fig. 12, and the intertrabecular spaces are correspondingly smaller (800–1000 Å). This illustrates the typical shift in morphology of the lattice as one moves from the thinner into the thicker, more central regions of the cell. Whether the change is related to other aspects, such as regions with and without ribosomes, has yet to be determined. Magnification, 40,000×.

be part of the cell, it is difficult to suspect them of failing in this instance. Actually, similar trabeculae appear in freeze-dried cells (without a cryoprotectant being used), although with larger dimensions, suggesting that some shrinkage may accompany dehydration with acetone. Similar and additional arguments for the reality of the lattice have been presented by Wessells, Spooner and Luduena (1973) and Buckley (1975). The former find significant the presence of asymmetries in elements of the lattice as it extends into microspikes and ruffles. Buckley finds a similar lattice in negatively stained, unfixed margins of cultured cells. In addition, there is evidence from freeze-fracture preparations for the existence of bridges between the microtubules in the motile axostyles of flagellated protozoa (Bloodgood and Miller 1974).

These observations and arguments in support of the lattice are very compelling, and we have therefore been bold enough to construct a model of it as

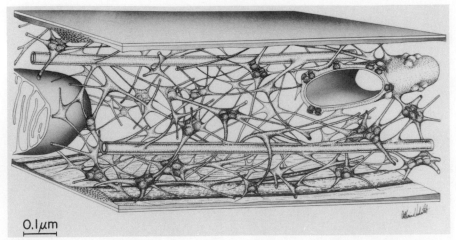

0.1μm

Figure 15

A model of the cytoplasmic ground substance showing lattice and contained micro-tubules, ribosomes and elements of the endoplasmic reticulum. The actin filament bundles are depicted here as part of the cytoplasmic cortex, and it is doubtful that this is uniformly the case. This, then, is our current view of the gel (thixotropic) that occupies the space between the upper and lower surfaces of a thinly spread cell. Its distribution is thought to be dependent on the presence of the highly asymmetric microtubules. Magnification, ca. 87,000×.

it appears in the thin margins of WI38 and other cells (Fig. 15). It is offered as an early edition which may undergo several revisions in time. Already there are details that strike us as not altogether correct, but in the basic features of its relation to the cell cortex, the ER, polysomes and microtubules it is accurate enough for our present purposes, which are, in part, to stimulate further thought on and investigation of this important motile and motility-supporting phase of the cytoplasm.

Acknowledgments

The author is pleased to acknowledge the assistance of Doris De Falco, Karen Anderson and Margaret Isenhart with various aspects of manuscript preparation and John Wolosewick for valuable discussions and high-voltage micrographs. The work reported here was supported by National Institutes of Health Grants PO1 AG00310 and 5 PO7 RR00592, and a National Institutes of Health Fellowship (1 F32 GM00138) to John Wolosewick.

REFERENCES

Allen, N. S. 1974. Endoplasmic filaments generate the motive force for rotational streaming in Nitella. *J. Cell Biol.* **63**:270.

Berl, S., S. Puszkin and W. J. Nicklas. 1973. Actomyosin-like protein in brain. Actomyosin-like protein may function in the release of transmitter material at synaptic endings. *Science* **179**:441.

Bikle, D., L. G. Tilney and K. R. Porter. 1966. Microtubules and pigment migration in the melanophores of *Fundulus heteroclitus* L. *Protoplasma* **61**:322.

Bloodgood, R. A. and K. R. Miller. 1974. Freeze-fracture of microtubules and bridges in motile axostyles. *J. Cell Biol.* **62**:660.

Brinkley, B. R., G. M. Fuller and D. P. Highfield. 1975. Cytoplasmic microtubules in normal and transformed cells in culture: Analysis by tubulin antibody immunofluorescence. *Proc. Nat. Acad. Sci.* **72**:4981.

Buckley, I. K. 1975. Three-dimensional fine structure of cultured cells: Possible implications of subcellular motility. *Tissue and Cell* **7**:51.

Buckley, I. K. and K. R. Porter. 1967. Cytoplasmic fibrils in living cultured cells. *Protoplasma* **64**:349.

———. 1973. Advances in electron microscopy of whole cells grown *in vitro*. *J. Cell Biol.* **59**:37a.

———. 1975. Electron microscopy of critical point dried whole cultured cells. *J. Microsc.* **104**:107.

Bunge, M. B. 1973. Fine structure of nerve fibers and growth cones of isolated sympathetic neurons in culture. *J. Cell Biol.* **56**:713.

Burton, P. R. and H. L. Fernandez. 1973. Delineation by lanthanum staining of filamentous elements associated with the surfaces of axonal microtubules. *J. Cell Sci.* **12**:567.

Burton, P. R. and W. L. Kirkland. 1972. Actin detected in mouse neuroblastoma cells by binding of heavy meromyosin. *Nature New Biol.* **239**:244.

Byers, M. R. 1974. Structural correlates of rapid axonal transport: Evidence that microtubules may not be directly involved. *Brain Res.* **75**:97.

Cooke, P. 1976. A filamentous cytoskeleton in vertebrate smooth muscle fibers. *J. Cell Biol.* **68**:539.

Fine, R. E. and D. Bray. 1971. Actin in growing nerve cells. *Nature New Biol.* **234**:115.

Fine, R. E., A. L. Blitz, S. E. Hitchcock and B. Kaminer. 1973. Tropomyosin in brain and growing neurones. *Nature New Biol.* **245**:182.

Fonte, V. and K. R. Porter. 1974. Topographic changes associated with the viral transformation of normal cells to tumorigenicity. In *Abstracts of the Eighth International Congress on Electron Microscopy*, Canberra, vol. II.

Freed, J. J. and M. M. Lebowitz. 1970. The association of a class of saltatory movements with microtubules in cultured cells. *J. Cell Biol.* **45**:334.

Freundlich, H. 1942. Some mechanical properties of sols and gels and their relation to protoplasmic structure. In *Structure of protoplasm* (ed. W. A. Seifrez), pp. 85–98. Iowa State College Press, Ames.

Gershenbaum, M. R., J. W. Shay and K. R. Porter. 1974. The effects of cytochalasin B on BALB/3T3 mammalian cells cultured *in vitro* as observed by scanning and high voltage electron microscopy. In *Scanning electron microscopy* (ed. O. Johari), part III, p. 589. ITT Research Institute, Chicago.

Goldman, R. D. and D. M. Knipe. 1973. Functions of cytoplasmic fibers in non-muscle cell motility. *Cold Spring Harbor Symp. Quant. Biol.* **37**:523.

Goldman, R. D., G. Berg, A. Bushnell, C-M. Chang, L. Dickerman, N. Hopkins, M. L. Miller, R. Pollack and E. Wang. 1973. Fibrillar systems in cell motility. In *Locomotion of tissue cells. Ciba Foundation Symposium*, vol. 14, p. 83. Associated Scientific Publishers, New York.

Green, L. 1968. Mechanism of movements of granules in melanocytes of *Fundulus heteroclitus*. *Proc. Nat. Acad. Sci.* **59**:1179.

Hayflick, L. 1965. The limited *in vitro* lifetime of human diploid cell strains. *Exp. Cell Res.* **37**:614.

Inoué, S. and H. Ritter, Jr. 1975. Dynamics of mitotic spindle organization and function. In *Molecules and cell movement* (ed. S. Inoué and R. E. Stephens), pp. 3–30. Raven Press, New York.

Inoué, S., J. Fuseler, E. D. Salmon and G. W. Ellis. 1975. Functional organization of mitotic microtubules. *Biophys. J.* **15**:725.

Ishikawa, H., R. Bischoff and H. Holtzer. 1969. Formation of arrowhead complexes with heavy meromyosin in a variety of cell types. *J. Cell Biol.* **43**:312.

Jarosch, R. 1964. Screw mechanical basis of protoplasmic movement. In *Primitive motile systems in cell biology* (ed. R. D. Allen and N. Kamiya), pp. 599–633. Academic Press, New York.

Kamitsubo, E. 1972a. A "window technique" for detailed observation of characean cytoplasmic streaming. *Exp. Cell Res.* **74**:613.

————. 1972b. Motile protoplasmic fibrils in cells of the Characeae. *Protoplasma* **74**:53.

Kane, R. E. 1974. Preparation and purification of polymerized actin from sea urchin egg extracts. *J. Cell Biol.* **66**:305.

Kersey, Y. M., P. K. Hepler, B. A. Palevitz and N. K. Wessels. 1976. Polarity of actin filaments in characean algae. *Proc. Nat. Acad. Sci.* **73**:165.

Kirkpatrick, J. B. 1971. Time-lapse cinematography of axoplasmic flow in peripheral nerves. In *Abstracts of the 11th Annual Meeting of the American Society for Cell Biology,* New Orleans, abstract #671.

Lazarides, E. and K. Burridge. 1975. α-Actinin: Immunofluorescent localization of a muscle structural protein in nonmuscle cells. *Cell* **6**:289.

LeBeux, Y. J. and J. Willemot. 1975. An ultrastructural study of the microfilaments in rat brain by means of E-PTA staining and heavy meromyosin labeling. II. The synapses. *Cell Tissue Res.* **160**:37.

Marsland, D. A. and D. E. S. Brown. 1936. Amoeboid movement at high hydrostatic pressure. *J. Cell. Comp. Physiol.* **8**:167.

McIntosh, J. R. 1974. Bridges between microtubules. *J. Cell Biol.* **61**:166.

McIntosh, J. R., E. S. Ogata and S. C. Landis. 1973. The axostyles of *Saccino aculus. J. Cell Biol.* **56**:304.

McMahan, U. J. and S. W. Kuffler. 1971. Visual identification of synaptic boutons on living ganglion cells and of varicosities in postganglionic axons in the heart of the frog. *Proc. Royal Soc. B* **177**:485.

Mooseker, M. S. and L. G. Tilney. 1975. Organization of an actin filament-membrane complex. *J. Cell Biol.* **67**:725.

Nagai, R. and L. I. Rebhun. 1966. Cytoplasmic microfilaments in streaming Nitella cells. *J. Ultrastruc. Res.* **14**:571.

O'Brien, T. P. and M. E. McCully. 1970. Cytoplasmic fibers associated with streaming and saltatory particle movement in *Heracleum mantegazzianum. Planta* **94**:91.

O'Brien, T. P. and K. V. Thimann. 1966. Intracellular fibers in oat coleoptile cells and their possible significance in cytoplasmic streaming. *Proc. Nat. Acad. Sci.* **56**:888.

Ochs, S. 1966. Axoplasmic flow in neurons. In *Macromolecules and behavior* (ed. J. Gaito), pp. 20–39. Appleton-Century-Crofts, New York.

Ockleford, C. D. and J. B. Tucker. 1973. Growth, breakdown, repair, and rapid contraction of microtubular axopodia in the heliozoan *Actinophrys sol. J. Ultrastruc. Res.* **44**:369.

Palay, S. L., C. Sotelo, A. Peters and P. M. Orkand. 1968. The axon hillock and the initial segment. *J. Cell Biol.* **38**:193.

Palevitz, B. A. and P. K. Hepler. 1975. Identification of actin *in situ* at the ectoplasm-endoplasm interface of Nitella. *J. Cell Biol.* **65**:29.

Pickett-Heaps, J. D. 1967. Ultrastructure and differentiation in Chara sp. I vegetative cells. *Aust. J. Biol. Sci.* **20**:539.

Pollard, T. D. 1976. The role of actin in the temperature-dependent gelation and contraction of extracts of Acanthamoeba. *J. Cell Biol.* **68**:579.

Pomerat, C. M., W. J. Hendelman, C. W. Raiborn, Jr. and J. F. Massey. 1967. In *The neuron* (ed. M. Kyden), pp. 119–178. Elsevier, Amsterdam.

Porter, K. R. 1973. Microtubules in intracellular locomotion. In *Locomotion of tissue cells. Ciba Foundation Symposium,* vol. 14, p. 149. Associated Scientific Publishers, New York.

Porter, K. R., G. S. Bennett and L. C. Junqueira. 1970. Microtubules in intracellular pigment migration. In *Proceedings of the 7th Congres Int'l de Microscopie Electronique,* Grenoble, vol. III, pp. 945–946.

Porter, K. R., A. Claude and E. F. Fullam. 1945. A study of tissue culture cells by electron microscopy. *J. Exp. Med.* **81**:233.

Porter, K. R., T. T. Puck and A. W. Hsie. 1974. An electron microscope study of the effects of dibutyryl cyclic AMP on Chinese hamster ovary cells. *Cell* **2**:145.

Roberts, K. 1974. Cytoplasmic microtubules and their functions. *Prog. Biophys. Mol. Biol.* **28**:273.

Roth, L. E., D. J. Pihlaja and Y. Shigenaka. 1970. Microtubules in the heliozoan axopodium. I. The gradion hypothesis of allosterism in structural proteins. *J. Ultrastruc. Res.* **30**:7.

Samson, F. E., Jr. 1971. Mechanism of axoplasmic transport. *J. Neurobiol.* **2**:347.

Schliwa, M. and J. Bereiter-Hahn. 1973. Pigment movements in fish melanophores: Morphological and physiological studies. II. Cell shape and microtubules. *Z. Zellforsch.* **147**:107.

Schmitt, F. O. 1968. Fibrous proteins-neuronal organelles. *Proc. Nat. Acad. Sci.* **60**:1092.

Schroeder, T. E. 1969. The role of "contractile ring" filaments in dividing Arbacia egg. *Biol. Bull.* **137**:413.

Shay, J. W., K. R. Porter and D. M. Prescott. 1974. The surface morphology and fine structure of CHO (Chinese hamster ovary) cells following enucleation. *Proc. Nat. Acad. Sci.* **71**:3059.

Shigenaka, Y., U. Tadokaro and M. Kaneda. 1975. Microtubules in protozoan cells. I. Effects of light metal ions on the heliozoan microtubule and their kinetic analysis. *Annot. Zool. Japon.* **48**:227.

Shigenaka, Y., K. Watanabe and M. Kaneda. 1974. Degrading and stabilizing efects of Mg^{2+} ions on microtubules-containing axopodia. *Exp. Cell Res.* **85**:391.

Spooner, B. S., K. M. Yamada and N. K. Wessells. 1971. Microfilaments and cell locomotion. *J. Cell Biol.* **49**:595.

Stossel, T. P. and J. H. Hartwig. 1976. Interactions of actin, myosin and a new actin-binding protein of rabbit pulmonary macrophages. II. Role in cytoplasmic movement and phagocytosis. *J. Cell Biol.* **68**:602.

Tilney, L. G. 1968. Studies on the microtubules in heliozoa. IV. The effect of colchicine on the formation and maintenance of the axopodia and the redevelopment of pattern in actinosphaerium nucleofilum (Barrett). *J. Cell Sci.* **3**:549.

Tilney, L. G. and K. R. Porter 1965. Studies on the microtubules in heliozoa. I. The fine structure of actinosphaerium nucleofilum (Barrett) with particular reference to the axial rod structure. *Protoplasma* **60**:317.

Warner, F. D. 1970. New observations on flagellar fine structure. *J. Cell Biol.* **47**:159.

Wessells, N. K., B. S. Spooner and M. A. Luduena. 1973. Surface movements, microfilaments and cell locomotion. In *Locomotion of tissue cells. Ciba Foundation Symposium,* vol. 14, p. 53. Associated Scientific Publishers, New York.

Wilson, E. B. 1928. *The cell in development and heredity,* 3rd ed. The MacMillan Company, New York.

Wolosewick, J. J. and K. R. Porter. 1976. High-voltage electron microscopy of whole cells of the human diploid cell line, WI-38. *Amer. J. Anat.* (in press).

Wuerker, R. B. 1970. Neurofilaments and glial filaments. *Tissue and Cell* **2**:1.

Wuerker, R. B. and J. B. Kirkpatrick. 1972. Neuronal microtubules, neurofilaments and microfilaments. *Int. Rev. Cytol.* **33**:45.

Yamada, K. M., B. S. Spooner and N. K. Wessells. 1971. Ultrastructure and function of growth cones and axons of cultured nerve cells. *J. Cell Biol.* **49**:614.

Yen, S-H., D. Dahl, M. Schachner and M. L. Shelanski. 1976. Biochemistry of the filaments of brain. *Proc. Nat. Acad. Sci.* **73**:529.

Section 1

PRIMITIVE MOTILE SYSTEMS

Introductory Remarks:
Some Aspects of the Structure
and Function of Bacterial Flagella

Julius Adler

Departments of Biochemistry and Genetics
University of Wisconsin, Madison, Wisconsin 53706

At the outset it must be made clear that eukaryotic flagella are very different from bacterial flagella (Fig. 1). The former have the well known 9 + 2 structure and are surrounded by an extension of the cytoplasmic membrane. The latter have but a single, naked filament.

The papers in this section will try to answer the following basic questions:

1. What is the structure of bacterial flagella?
2. How do bacterial flagella function? How is a change in direction achieved? What is the energy source?
3. How does sensory information act on the flagella to bring about attraction or repulsion by stimuli?

As a starting point, presented below is a brief review of recent observations of some aspects of the structure and function of bacterial flagella.

Structure of Bacterial Flagella

It is now possible to isolate "intact" flagella from bacteria, i.e., flagella with the basal structure still attached (DePamphilis and Adler 1971a; Dimmitt and Simon 1971) (Fig. 2). There is the helical filament, a hook and a rod. In the case of *Escherichia coli,* four rings are mounted on the rod (DePamphilis and Adler 1971a), whereas flagella from gram-positive bacteria have only the two inner rings (DePamphilis and Adler 1971a; Dimmitt and Simon 1971). For *E. coli* it has been established that the outer ring is attached to the outer membrane, and the inner ring is attached to the cytoplasmic membrane (Fig. 2) (DePamphilis and Adler 1971b). The basal body thus (a) anchors the flagellum into the cell envelope; (b) provides contact with the cytoplasmic membrane, the place where the energy originates and possibly the place where the sensory information acts (see below); and (c) may serve as the motor (or a part of it) that drives the rotation of the flagellum.

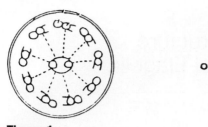

Figure 1
Cross section of flagella: (*left*) eukaryotic; (*right*) bacterial.

Energy Source for Bacterial Motility

Energy for the rotation of the flagellum comes from the intermediate of oxidative phosphorylation (the proton gradient in the Mitchell hypothesis), not from ATP directly (Larsen et al. 1974a; Thipayathasana and Valentine 1974), unlike in the case of eukaryotic flagella or muscle. This is true for both counterclockwise and clockwise rotation (Larsen et al. 1974a).

Chemotaxis, however, requires ATP in addition (Larsen et al. 1974a). The reason is that tumbling, which is essential for chemotaxis (Berg and Brown 1972; Macnab and Koshland 1972; Tsang, Macnab and Koshland 1973; Brown and Berg 1974), requires ATP (see below for the role of ATP in a protein methylation reaction).

Effect of Sensory Information
on Direction of Rotation of Flagella

These bacteria swim in smooth, straight, lines, frequently interrupted by tumbles that bring about a new, randomly chosen, direction (Berg and Brown 1972). The control of tumbling frequency is a central feature of chemotaxis by *E. coli* and *Salmonella typhimurium* (Berg and Brown 1972; Macnab and Koshland 1972; Tsang, Macnab and Koshland 1973; Brown and Berg 1974).

Bacterial flagella work by rotating (Berg and Anderson 1973; Silverman and Simon 1974), and this rotation alternates between counterclockwise and clockwise (Silverman and Simon 1974). This has been demonstrated by tethering the bacteria to a glass slide by means of antibody to the filament (Silverman and Simon 1974); now that the filament is no longer free to rotate, the cell instead rotates, usually counterclockwise but sometimes clockwise (Silverman and Simon 1974).

Addition of attractants to tethered *E. coli* cells causes counterclockwise rotation of the cells, whereas addition of repellents causes clockwise rotation (Larsen et al. 1974b). These responses last for a short time, which is dependent on the strength of the stimulus; the rotation then returns to the unstimulated state, mostly counterclockwise (Larsen et al. 1974b).

Mutants of *E. coli* that swim smoothly and never tumble always rotate counterclockwise, whereas mutants that almost always tumble rotate mostly clockwise (Larsen et al. 1974b).

From these results and from the prior knowledge that increase of attractant

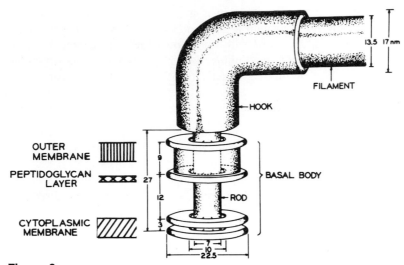

Figure 2
Model of the flagellar base of *E. coli*. Dimensions are in nanometers. (Modified from DePamphilis and Adler 1971a, b.)

concentration causes smooth swimming (i.e., suppressed tumbling) (Berg and Brown 1972; Macnab and Koshland 1972; Brown and Berg 1974), whereas addition of repellent causes tumbling (Tsang, Macnab and Koshland 1973), it was concluded (Larsen et al. 1974b) that smooth swimming results from counterclockwise rotation of flagella and tumbling from clockwise rotation.

Role of Methionine in Chemotaxis and Tumbling

A useful probe for exploring the mechanism of the tumble-regulating mechanism came from the discovery that *E. coli* requires L-methionine for chemotaxis (Adler and Dahl 1967). After removal of methionine, methionine auxotrophs lose the ability to tumble and do not respond to either attractants or repellents (Armstrong 1972a; Adler 1973; Tso and Adler 1974; Aswad and Koshland 1974; Springer et al. 1975), much like mutants that always swim smoothly and fail to carry out any chemotaxis (generally nonchemotactic, or *che* mutants) (Armstrong, Adler and Dahl 1967). Methionine must be continuously present for chemotaxis to occur (Adler and Dahl 1967; Armstrong 1972a). These studies have been extended to indicate that *S*-adenosylmethionine, a metabolite of methionine, is required in chemotaxis, probably as a methyl donor (Armstrong 1972b; Larsen et al. 1974a; Springer et al. 1975; Aswad and Koshland 1975). We (Kort et al. 1975) have now identified a protein in the cytoplasmic membrane of *E. coli* that contains a metabolically labile methyl group donated by methionine. The involvement of this protein methylation reaction in chemotaxis is indicated by four lines of evidence (Kort et al. 1975): (1) The methylation reaction is altered in some generally nonchemotactic mutants and is coreverted with chemotaxis defects.

(2) The methylation level of the protein is affected by chemotactic stimuli. (3) The methyl group of the protein is derived from methionine and is metabolically labile, which is in accord with the known fact that chemotaxis requires a continuous supply of methionine. (4) Methylation is abnormal in mutants having defective or missing flagella (see Kort et al. 1975 for documentation).

The following questions about methylation remain:

1. What is the nature of the methylation system (reactants, products, methylation and demethylation enzymes involved; defects in various types of mutants)?
2. How do sensory receptors bring about a change in methylation?
3. How does a change in methylation bring about a change in the direction of rotation of the flagella?

REFERENCES

Adler, J. 1973. A method for measuring chemotaxis and use of the method to determine optimum conditions for chemotaxis by *Escherichia coli*. *J. Gen. Microbiol.* **74**:77.

Adler, J. and M. M. Dahl. 1967. A method for measuring the motility of bacteria and for comparing random and non-random motility. *J. Gen. Microbiol.* **46**:161.

Armstrong, J. B. 1972a. Chemotaxis and methionine metabolism in *Escherichia coli. Can. J. Microbiol.* **18**:591.

Armstrong, J. B. 1972b. An S-adenosylmethionine requirement for chemotaxis in *Escherichia coli. Can. J. Microbiol.* **18**:1695.

Armstrong, J. B., J. Adler and M. M. Dahl. 1967. Nonchemotatic mutants of *Escherichia coli. J. Bact.* **93**:390.

Aswad, D. and D. E. Koshland, Jr. 1974. Role of methionine in bacterial chemotaxis. *J. Bact.* **118**:640.

———. 1975. Evidence for an S-adenosylmethionine requirement in the chemotactic behavior of *Salmonella typhimurium. J. Mol. Biol.* **97**:207.

Berg, H. C. and R. A. Anderson. 1973. Bacteria swim by rotating their flagellar filaments. *Nature* **245**:380.

Berg, H. C. and D. A. Brown. 1972. Chemotaxis in *Escherichia coli* analyzed by three-dimensional tracking. *Nature* **239**:500.

Brown, D. A. and H. C. Berg. 1974. Temporal stimulation of chemotaxis in *Escherichia coli. Proc. Nat. Acad. Sci.* **71**:1388.

DePamphilis, M. L. and J. Adler. 1971a. Fine structure and isolation of the hook-basal body complex of flagella from *Escherichia coli* and *Bacillus subtilis. J. Bact.* **105**:384.

———. 1971b. Attachment of flagellar basal bodies to the cell envelope: Specific attachment to the outer, lipopolysaccharide membrane and the cytoplasmic membrane. *J. Bact.* **105**:396.

Dimmitt, K. and M. Simon. 1971. Purification and thermal stability of intact *Bacillus subtilis* flagella. *J. Bact.* **105**:369.

Kort, E. N., M. F. Goy, S. H. Larsen and J. Adler. 1975. Methylation of a membrane protein involved in bacterial chemotaxis. *Proc. Nat. Acad. Sci.* **72**:3939.

Larsen, S. H., J. Adler, J. J. Gargus and R. W. Hogg. 1974a. Chemomechanical coupling without ATP: The source of energy for motility and chemotaxis in bacteria. *Proc. Nat. Acad. Sci.* **71**:1239.

Larsen, S. H., R. W. Reader, E. N. Kort, W.-W. Tso and J. Adler. 1974b. Change in direction of flagellar rotation is the basis of the chemotactic response in *Escherichia coli*. *Nature* **249**:74.

Macnab, R. M. and D. E. Koshland, Jr. 1972. The gradient-sensing mechanism in bacterial chemotaxis. *Proc. Nat. Acad. Sci.* **69**:2509.

Silverman, M. and M. Simon. 1974. Flagellar rotation and the mechanism of bacterial motility. *Nature* **249**:73.

Springer, M. S., E. N. Kort, S. H. Larsen, G. W. Ordal, R. W. Reader and J. Adler. 1975. The role of methionine in bacterial chemotaxis: Requirement for tumbling and involvement in information processing. *Proc. Nat. Acad. Sci.* **72**:4640.

Thipayathasana, P. and R. C. Valentine. 1974. The requirement for energy transducing ATPase for anaerobic motility in *Escherichia coli*. *Biochim. Biophys. Acta* **347**:464.

Tsang, N., R. Macnab and D. E. Koshland, Jr. 1973. Common mechanism for repellents and attractants in bacterial chemotaxis. *Science* **181**:60.

Tso, W.-W. and J. Adler. 1974. Negative chemotaxis in *Escherichia coli*. *J. Bact.* **118**:560.

Motility and the Structure of Bacterial Flagella

Marsha Hilmen and Melvin Simon

Department of Biology, University of California, San Diego
La Jolla, California 92093

The flagellar organelles in bacteria are responsible for their motility. These organelles have a relatively simple structure and may provide a model for understanding the mechanisms involved in motility in other cells. Figure 1 shows electron micrographs of the insertion of the flagellar basal structure into the bacterial cell membrane system. Two basal ring structures are associated with the plasma membrane and two other rings are associated with the outer membranes. The four rings are mounted on a shaftlike structure that terminates at the outer surface of the cell in a hooklike structure (Abram, Vatter and Koffler 1966; Dimmitt and Simon 1971; DePamphilis and Adler 1971a,b,c). The hook is relatively short (approx. 70 nm) and ends in the long, helical, flagellar filament. The filament in *E. coli* is 15–18 nm in diameter and is generally from 10–15 microns long (Iino 1974).

The structure, assembly and function of the entire flagellar organelle is controlled by about 20 genes (Fig. 2), which have been defined in *E. coli* (Silverman and Simon 1973a,b; Silverman and Simon 1974a,b) and *Salmonella* (Iino and Enomoto 1966; Yamaguchi et al. 1972). These genes can be thought of as belonging to three overlapping phenotypic categories. These categories, useful for summarizing our understanding of specific flagellar functions, are (1) genes that control the structure of proteins involved in the assembly of the flagellar filament and basal structure; (2) genes that function to regulate the synthesis and the assembly of the flagellar components; and (3) genes that control functions required for flagellar activity and for chemotaxis.

The *hag* gene (Fig. 2) is an example of the first category; it controls the structure of the flagellin subunit that is the major component of the flagellar filament. It is a protein with a molecular weight of about 53,000 daltons (Hilmen, Silverman and Simon 1974; Kondoh and Hotani 1974). On the other hand, the *cfs* locus and the *fluI* gene probably have regulatory functions. Ordinarily, flagellar synthesis is sensitive to catabolite repression, and cyclic AMP is necessary for flagellar synthesis in wild-type cells. However, muta-

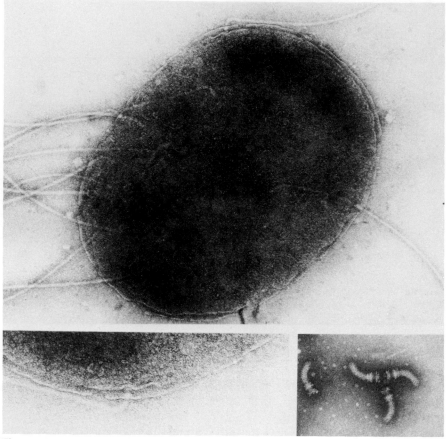

Figure 1
Insertion of the flagellar organelle into the cell membrane. (*Top*) *E. coli* strain
MS1829, which carries the *cfs* mutation (Silverman and Simon 1974b) and over-
produces bacterial flagella, was grown in L-broth medium. The cells were collected
and resuspended in 25% sucrose and incubated at room temperature for 5 min.
The cells were then resuspended in 0.01 M Tris buffer and put on Formvar carbon-
coated grids. They were stained with 0.5% phosphotungstic acid. (*Bottom left*)
Higher magnification of flagellar insertion. (*Bottom right*) Purified isolated hook–
basal structures. The apparent diameter of the hook structure is 17 nm.

tions at the *cfs* locus result in cells that can make flagella in the apparent
absence of cyclic AMP, thus releasing the pathway of flagella synthesis from
control by cyclic AMP. The *cfs* locus is probably part of an operator that controls the rate of synthesis of the *flaI* gene product (Silverman and Simon
1974b). The *tem* locus also appears to have a regulatory function. Wild-type
bacteria generally do not make flagella when the temperature is raised to 42°C.
Mutations at the *tem* locus allow the cells to make flagella at the high temperature (M. Silverman and M. Simon, in prep.). The specific mechanisms
involved in this effect are not clearly understood. The *flaE* gene also seems to
have a regulatory function. Mutations in this gene lead to cells that make long
polyhook structures. The mutant strains lack the ability to terminate the

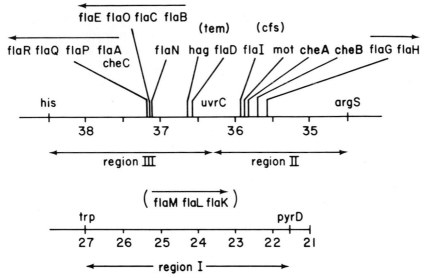

Figure 2
Distribution of flagellar genes on the genetic map of *E. coli.*

assembly of the hook structure and synthesize flagellin slowly (Silverman and Simon 1972). The *mot* gene apparently belongs to the third category; it seems to specifically affect the motility function. *Mot* mutants synthesize flagella apparently identical to those of wild type, except that they are inactive. The *cheA, cheB* and *cheC* genes, defined by Armstrong and Adler (1969), and the other *che* genes, described by Parkinson (1975) in *E. coli* and Koshland et al. (this volume) in *Salmonella,* are examples of genes that specifically affect flagellar function. Bacteria that are mutant at these loci are able to swim but do not respond to gradients of attractants or repellents as do wild-type cells.

We have been interested in the relationship between flagellar structure and motility and the mechanisms involved in flagellar activity. The evidence indicates that bacterial flagella rotate. Rotation has been observed in two ways (Silverman and Simon 1974b; Simon et al. 1975): first, indirectly, by binding either the flagellar filament or the elongated flagellar hook structure to a surface using specific antibodies and observing the subsequent rotation of the cell; and second, more directly, by using polystyrene beads bound to the flagellar filament or polyhook structure and observing the rotation of the beads. The simplest interpretation of these observations is that the entire flagellar filament rotates, and that this rotation is driven at the basal structure. The observation that rotation occurs in polyhook-*hag* double mutants, even in the total absence of a flagellar filament protein, suggests that the filament protein is not at all necessary for rotation. Furthermore, the observation that in strains carrying *mot* mutations, the flagellar apparatus appears to be morphologically the same as in wild-type cells suggests that understanding the mechanism of rotation requires an understanding of the specific nature of the *mot* gene function. The product of the *mot* gene probably serves to transduce energy supplied through the membrane (Larsen et al. 1974) to flagellar rotation. The *mot* gene product could be firmly attached and integrated into the flagellar basal structure. On the other hand, it could be transiently bound to the basal structure and to the cell membrane. Finally, it might be a soluble enzyme in the cell, whose activity leads to the formation of a specific substance required for flagellar rotation. Our approach to distinguishing among these possibilities has been to test the first hypothesis by purifying and isolating the flagellar basal structure and comparing the polypeptides that make up this structure in wild-type cells with those in a variety of *mot* mutants that should entirely lack the *mot* gene product. If the *mot* gene product is part of the flagellar basal structure, it could be identified in this way. The initial results of our isolation procedures suggest that we can consistently prepare structures that have the same morphology as the basal structure and hook; however, no differences in the polypeptide composition of the basal regions isolated from wild-type flagella and from flagella from a variety of *mot* mutants have been found.

THE HOOK–BASAL COMPLEX

Isolation

In wild-type cells, the hook–basal structure complex accounts for approximately 0.001% of the total protein of *E. coli*. Furthermore, the structure has no apparent enzymatic activity, and there is no simple way to assay for it during the purification. Therefore the major criterion used in the purification of the structure is its morphological integrity (Fig. 3). An initial approach has been to isolate the structure in a number of different ways and attempt to purify it until a consistent polypeptide pattern is found on acrylamide gel electrophoresis. Two approaches have been used: The first is to isolate relatively intact flagella, using the procedures described by DePamphilis and Adler (1971a), then to dissociate the flagellar filament, and finally, to purify the hook–basal structure complex, which is relatively more stable than the flagellar filament structure (Abram et al. 1970). The second approach involves using *hag* mutants that are incapable of synthesizing flagellar filaments. These mutants, however, do synthesize hook–basal body structures (Yamaguchi et al. 1972). The structures can be isolated directly from membrane preparations of the *hag* mutants (Simon et al. 1975). Finally in order to track the purification procedure, *hag* cells were occasionally labeled with ^{32}P at high specific activity. Figure 4 shows the results of a typical experiment designed to isolate the hook–basal structure. Relatively intact bacterial flagella were isolated, and the filament was disaggregated by incubation at pH 2.7 for 10 minutes. The resulting mixture was put on a sucrose gradient to separate the hook–basal structures from the residual flagellar protein (Fig. 4, top). The structure was then further purified by isopycnic sucrose gradient centrifugation. The separation can be followed by observing the radioactive phosphorus. It was initially associated with the hook–basal structure; but after the gradients, the radioactivity has completely dissociated from the hook–basal material (Fig. 4, bottom). The gradients can be monitored by electron

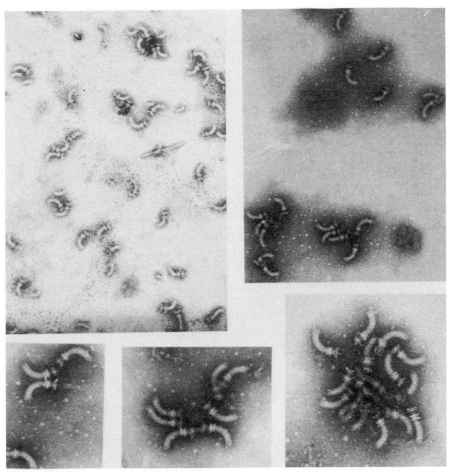

Figure 3
Morphology of purified hook–basal structures. (*Top left*) Structures isolated from
a strain carrying the *hag* mutation. The morphology of the structures isolated from
purified, intact flagella is shown in the other micrographs.

microscopy or, more conveniently, by using antibody to the hook structure
(Hilmen, Silverman and Simon 1974). Using a related procedure, the hook
structure was isolated from *hag* mutants. This isolation did not involve
incubation in acid, nor did it include banding in cesium chloride. It simply
involved cell lysis, solubilization of the cell membranes by Triton-X 100, and
sucrose gradient centrifugation. Figure 3 shows that the hook–basal body
structures isolated by either of these methods appear to be morphologically
identical.

Analysis of Isolated Structures

The isolated structures have been dissociated and analyzed by SDS-acrylamide
gel electrophoresis in a variety of gel systems. Figure 5 shows the gel pattern

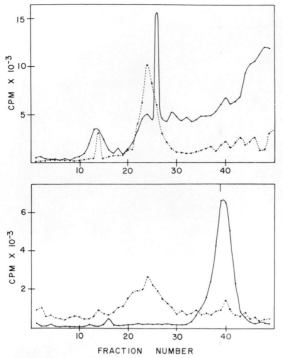

Figure 4

Purification of hook–basal structures. The purification scheme was essentially the same as described previously (Hilmen, Silverman and Simon 1974). Two cultures of MS1350 were grown; one in low-sulfur minimal medium with added $^{35}SO_4$ (the specific activity was 40 $\mu Ci/\mu g$ sulfur), the other in low-phosphate medium with added $^{32}PO_4$ (the specific activity of the phosphorous was about the same as that of the sulfur).

(*Top*) Results of sucrose gradient centrifugation of the hook–basal structures obtained after disaggregating intact flagella by acid treatment. The front peak, including fractions 12–15, contains all of the morphologically distinguishable hook–basal structures. The material in this peak was pooled and run on isopycnic sucrose gradients.

(*Bottom*) Results from run on isopycnic sucrose gradients. Twenty μl of each fraction in the sulfate-labeled gradient and 200 μl of each fraction in the phosphate-labeled gradient were counted. The material containing radioactive phosphorous was collected and extracted with chloroform-methanol. Ninety-nine percent of the counts were found to be soluble in chloroform-methanol.

obtained from material isolated in the experiments described in Figure 4. There are at least 11 clearly distinct bands derived from this structure. In similar experiments with the hook–basal structure isolated from *hag* mutants, the same pattern was found. Table 1 summarizes the apparent molecular weights of the bands that are consistently found. The band that migrates with an apparent molecular weight of 42,000 daltons corresponds to the hook subunit protein. Eleven bands are always found, but this is a minimal estimate.

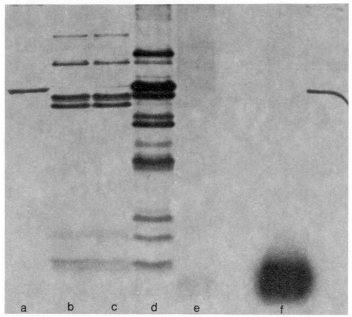

Figure 5

SDS-acrylamide gel electrophoresis of purified hook–basal structures. (*a*) Pure radioactive hook protein, molecular weight 42,000 daltons; (*b*) fraction 24 removed from the gradient shown in Fig. 4 (top); (*c*) fraction 27 removed from the same gradient; (*d*) the purified hook–basal structure (fraction 39, Fig. 4 bottom); (*e*) ^{32}P-labeled material (fraction 25, Fig. 4 bottom); and (*f*) ^{32}P-labeled material (fraction 25, Fig. 4 top).

Table 1

Apparent Molecular Weights of Flagellar Components

Components	MW (daltons)
Flagellin	54,000 ± 3,000
Hook subunit	42,000 ± 2,000
Basal structural components	60,000 ± 3,000
	39,000 ± 2,000
	31,000 ± 3,000
	27,000 ± 3,000
	20,000 ± 2,000
	18,000 ± 2,000
	13,000 ± 1,000
	11,000 ± 1,000
	9,000 ± 1,000

The results shown represent many individual runs (more than 20) using 7, 10, 12.5 and 15% gels.

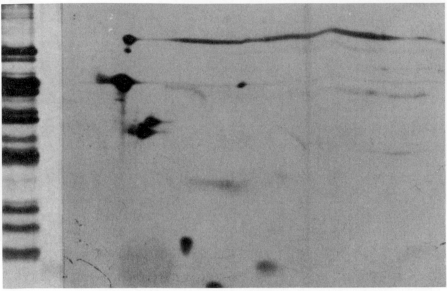

Figure 6

Two-dimensional acrylamide gel of purified hook–basal structures. On the left is the SDS-acrylamide gel pattern obtained with the purified material. This is compared to a two-dimensional separation. The order of the proteins from top to bottom is the same, but the exact distances of migration do not correspond precisely. The pH gradient runs from left to right, starting at about pH 3.0 and going to pH 8.5.

Thus, for example, very often the band that migrates with an apparent molecular weight of 60,000 appears to split into two bands, one migrating at 60,000, the other at 61,000. In order to further examine the distribution of these polypeptides, the two-dimensional gel technique described by O'Farrell (1975) was used. This involves isoelectric focusing in the horizontal direction and SDS electrophoresis in the vertical direction. Figure 6 shows the distribution of these polypeptides on the two-dimensional gels. The pattern is compared with the pattern obtained on the one-dimensional gel. Many of the components of the hook–basal structure complex have relatively acidic isoelectric points. Material with very basic isoelectric points (higher than pI 9.0) would not be seen on this gel. Therefore, for example, the protein that normally forms a band at a molecular weight corresponding to 39,000 daltons does not appear on the two-dimensional gel. Ten spots can be clearly seen. The two bands at 17,000 and 18,000 molecular weight always give a streak rather than a discreet spot on the two-dimensional gel, and the contours of the streaked spot are characteristic of these two polypeptides. Obviously there may be more than 11 polypeptides contained in the hook–basal structure complex. Certainly any polypeptide that lacked sulfur would not appear as a spot on these gels. Furthermore, proteins with molecular weights greater than 160,000 daltons might not enter these gels and therefore would not be observed. Finally, it is possible that a protein with only one sulfur atom per molecule present in only one molecule per flagellar structure might have been

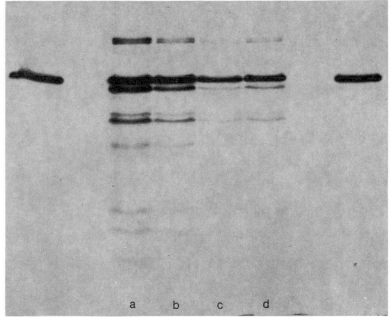

Figure 7
SDS-acrylamide gel patterns of structures isolated from wild-type and *mot*
mutant strains. (*a*) Mot polyhook structures; (*b*) wild type; (*c*) *mot*; (*d*)
wild type. The *mot* mutants were strains carrying amber mutations in the
mot gene.

overlooked. Table 1 summarizes the apparent molecular weights corresponding to the polypeptides consistently observed in these preparations.

Figure 7 shows the comparison between polypeptides derived from the
hook–basal structure isolated from wild-type bacteria and those isolated from
an amber mutant in the *mot* gene. It is clear that all of the polypeptides match
up precisely, and that all of the components present in the wild type are also
present in the *mot* mutant. Similar experiments have been done with bacteriophage *mu*-induced mutants in the *mot* gene. The hook–basal structure isolated from these mutants shows exactly the same polypeptide pattern as the
structure isolated from wild-type cells. Finally, when these polypeptides were
separated by the two-dimensional acrylamide gel technique, the spots found
for wild type and those for the *mot* mutants overlapped precisely. Similar
experiments have been performed with two amber mutants in the *mot* gene
and one strain that carries a missense mutation in the *mot* gene. All of these
strains gave exactly the same peptide pattern.

We can consider the following three interpretations for the data:

1. The *mot* protein is in fact an integral part of the isolated flagellar basal
   structure complex, but that we did not detect it.
2. The *mot* protein is not part of the hook–basal structure complex but is
   only tenuously bound. Thus it is removed from the complex early during
   the purification.

3. The *mot* protein never was part of the structure but is a cell component involved in the formation of a product necessary for motility.

The first possibility seems unlikely. On the basis of the initial specific activity we can calculate that if the *mot* protein represented 0.5% of the protein of the hook–basal structure complex and if it had at least 1 sulfur atom per molecule, then clearly we would have been able to detect it. This raises the possibility that the *mot* protein is either transiently attached to the basal structure, or not associated with the structure at all, but instead acts to synthesize a product (e.g., a polysaccharide or low molecular weight metabolite) or to modify a protein that in turn is necessary for flagellar rotation. In order to distinguish among these possibilities it is necessary to analyze directly the *mot* gene product. Although we have no apparent biochemical assay for the product, we can measure the effects of the *mot* gene. We may resolve the question of the nature of the *mot* gene product by purifying the *mot* gene and studying its transcription and translation in vitro.

CONCLUSIONS

The flagellar hook–basal structure can be purified and isolated using morphological integrity as a criterion for purification. Material prepared in this way shows the presence of 11 polypeptide bands when analyzed by acrylamide gel electrophoresis. These include the flagellin subunit with an apparent molecular weight of 53,000 daltons and the hook protein with an apparent molecular weight of 42,000. Attempts to assign the *mot* function to one of the other polypeptides by comparing the gel patterns obtained with wild-type and mutant structures were unsuccessful. Thus the *mot* gene product may not be a part of the hook–basal structure that was isolated. In fact, the isolated structure only accounts for 11 gene functions. There are at least another eight gene products associated with flagellar function. Therefore there may be another structure associated with the base of the flagellar organelle that we have lost in the purification (Larsen 1975), or there may be some other specialized proteins corresponding to the other gene functions that are not integral parts of the flagellar structure.

Acknowledgment

This work was supported by a grant from the National Science Foundation (BMS 73–01606).

REFERENCES

Abram, D., A. E. Vatter and H. Koffler. 1966. Attachment and structural features of flagella of certain bacilli. *J. Bact.* **91**:2045.

Abram, D., J. R. Mitchen, H. Koffler and A. E. Vatter. 1970. Differentiation within the bacterial flagellum and isolation of proximal hook. *J. Bact.* **101**:250.

Armstrong, J. B. and J. Adler. 1969. Location of genes for motility and chemotaxis on the *Escherichia coli* genetic map. *J. Bact.* **97**:156.

DePamphilis, M. L. and J. Adler. 1971a. Purification of intact flagella from *Escherichia coli* and *Bacillus subtilis*. *J. Bact.* **105**:376.

———. 1971b. Fine structure and isolation of the hook-basal body complex of flagella from *Escherichia coli* and *Bacillus subtilis*. *J. Bact.* **105**:384.

———. 1971c. Attachment of flagellar basal bodies to the cell envelope: Specific attachment to the outer, lipopolysaccharide membrane and the cytoplasmic membrane. *J. Bact.* **105**:396.

Dimmitt, K. and M. I. Simon. 1971. Purification and thermal stability of intact *Escherichia coli* flagella. *J. Bact.* **105**:369.

Hilmen, M., M. Silverman and M. Simon. 1974. The regulation of flagellar formation and function. *J. Supramol. Struc.* **2**:360.

Iino, T. 1974. Assembly of *Salmonella* flagellin *in vitro* and *in vivo*. *J. Supramol. Struc.* **2**:372.

Iino, T. and M. Enomoto. 1966. Genetical studies of non-flagellate mutants of *Salmonella*. *J. Gen. Microbiol.* **43**:315.

Kondoh, H. and H. Hotani. 1974. Flagellin from *Escherichia coli* K12. *Biochim. Biophys. Acta* **336**:119.

Larsen, S. 1975. The structure and function of bacterial flagella. Ph.D. thesis, University of Wisconsin, Madison.

Larsen, S., J. Adler, J. J. Gargus and R. W. Hogg. 1974. Chemomechanical coupling without ATP. *Proc. Nat. Acad. Sci.* **71**:1239.

O'Farrell, P. H. 1975. High resolution two-dimensional electrophoresis of proteins. *J. Biol. Chem.* **250**:4007.

Parkinson, S. 1975. Genetics of chemotactic behavior in bacteria. *Cell* **4**:183.

Silverman, M. and M. I. Simon. 1972. Flagellar assembly mutants in *Escherichia coli*. *J. Bact.* **112**:986.

———. 1973a. Genetic analysis of flagellar mutants in *Escherichia coli*. *J. Bact.* **113**:105.

———. 1973b. Genetic analysis of bacteriophage Mu-induced flagellar mutants in *Escherichia coli*. *J. Bact.* **116**:114.

———. 1974a. Positioning flagellar genes in *Escherichia coli* by deletion analysis. *J. Bact.* **117**:73.

———. 1974b. Characterization of *Escherichia coli* flagella mutants that are insensitive to catabolite repression. *J. Bact.* **120**:196.

Simon, M., M. Hilmen and M. Silverman. 1975. The assembly and function of *E. coli* flagella. In *Proceedings 1st International Congress of IAMS Tokyo*, vol. 1, p. 641. Science Council of Japan, Tokyo.

Yamaguchi, S., T. Iino, T. Moriguchi and K. Ohta. 1972. Genetic analysis of *fla* and *mot* cistrons closely linked to H1 in *Salmonella abortusequi* and its derivatives. *J. Gen. Microbiol.* **70**:59.

Does the Flagellar Rotary Motor Step?

Howard C. Berg

Department of Molecular, Cellular and Developmental Biology
University of Colorado, Boulder, Colorado 80302

A bacterial flagellum is driven at its base by a reversible rotary motor (Berg and Anderson 1973; Silverman and Simon 1974; Berg 1975a,b). The motor is only a few hundred angstroms in diameter (DePamphilis and Adler 1971) and built from a limited number of components (Hilmen, Silverman and Simon 1974). It utilizes as an energy source an intermediate in oxidative phosphorylation (presumably the proton motive force) rather than ATP itself (Larsen et al. 1974a). How does it work? Does it have a ratchetlike drive that moves in steps or a fluid drive that moves continuously?

I have attempted to answer this question by studying the rotation of cells tethered by their flagella to glass cover slips, a technique pioneered by Silverman and Simon (1974) (Fig. 1). In an earlier report (Berg 1974) I described experiments in which the rotation was followed by the tracking microscope. From the smoothness of signals obtained from the tracker (Fig. 2) I inferred that there was little passive slip between the flagellum and the cell wall, and that steps, if any, numbered more than about 25 per revolution. I assumed implicitly that the filament linking the cell to the glass was rigid. Since inertial effects are negligible, this implies that changes in the torque generated by the motor be reflected directly in changes in the rate of rotation of the cell. The assumption was not valid; the filament is elastic. Changes in the torque are masked, in part, by changes in its twist. I try here to set the record straight.

Cells Stop at Nearly Fixed Orientations

One can stop tethered *Escherichia coli* by adding sulfhydryl reagents, such as *p*-chloromercuriphenyl sulfonate, or uncouplers of oxidative phosphorylation, such as 2,4-dinitrophenol (DNP) or carbonylcyanide *m*-chlorophenyl hydrazone (CCCP). (For the effects of such reagents on swimming cells, see, for example, Clayton 1958; Larsen et al. 1974a; Ordal and Goldman 1975.) When cells are exposed to these compounds at concentrations of 2×10^{-3}, 2×10^{-3} and 10^{-5} M, respectively, most stop spinning in 5–10 minutes, some much sooner. The action of the sulfhydryl reagent can be reversed by

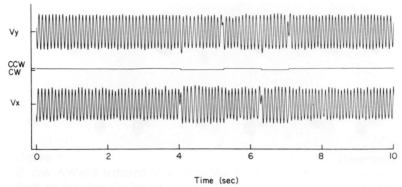

Figure 2

A strip-chart record of the rotation of a tethered *E. coli* obtained with the tracking microscope (Berg 1971, 1974). The signals represent the x and y velocities (V_x, V_y) of the point on the surface of the cell closest to the center of the laboratory reference frame. The cell was spinning at about 11 rps. It changed its direction of rotation four times. The line in the middle is an event marker indicating the direction of rotation, clockwise (CW) or counterclockwise (CCW).

angle M/B. The cell cannot stop when the motor stops because the filament will untwist and drive it in the forward direction. When the motor starts up again, part of the torque will be used to twist the filament back up again. The motion of the body of the cell will be smoother than the motion of the shaft of the motor. The following analysis of this problem is from E. M. Purcell (pers. comm.). Let a cell of frictional drag coefficient f be driven by a fluttery motor that runs at an angular velocity $d\phi/dt = \Omega(1 + \cos n\Omega t)$. The mean angular velocity of the motor is Ω, but it stops n times a revolution. The angular position of the cell relative to the shaft of the motor is

$$\phi = \Omega t + n^{-1} \sin n\Omega t. \tag{4}$$

The angular position of the shaft of the motor relative to the cover slip is the twist of the filament, θ. We would like to know the angular position of the cell relative to the cover slip, $\alpha = \phi - \theta$. In order to twist the filament an angle θ, the cell must generate an equal amount of torque through viscous drag. This implies

$$f(d\alpha/dt) = B\theta. \tag{5}$$

Equations 4 and 5 have the solution

$$\alpha = \Omega t - \theta_o + (n^2 + n^4\theta_o^2)^{-\frac{1}{2}} \sin (n\Omega t - \tan^{-1} n\theta_o), \tag{6}$$

where θ_o is the mean angle of twist of the filament given by Equation 3. In the limit that the filament is infinitely stiff, i.e., $\theta_o = 0$, Equation 6 reduces to Equation 4, and the amplitude of the flutter term is $1/n$. If θ_o is finite, the amplitude of the flutter term is smaller by a factor $1/(1 + n^2\theta_o^2)^{\frac{1}{2}}$. This is the case for the cell spinning at 10 rps (above). Therefore if we are to see the flutter, we must either stiffen the filament or inhibit the motor (reduce its torque).

Since the twist does not change the torsional stiffness of the filament, a spinning cell also is subject to fluctuations in position due to rotational diffusion of mean-square amplitude kT/B (Eq. 1).

A Pinhole and Photomultiplier Outperform the Tracker

The tracking microscope is useful for studying changes in the motion of tethered cells that occur in response to chemotactic stimuli (Berg and Tedesco 1975) but not for looking for high-frequency steps. When the time constants of the servo loops are reduced to the point that such steps might be seen, the output is dominated by oscillations at the resonant frequencies of the transducer (see Berg 1971). Data can be obtained in a much simpler way by imaging a tethered cell on a pinhole and measuring changes in the intensity of the transmitted light (Figs. 3, 4). Cells were studied at room temperature in a flow chamber made by cementing cover slips to a slide (Berg and Tedesco 1975) mounted on a standard mechanical stage. The optical system included a 100-w tungsten-halogen lamp (Sylvania FCR), a heat-transmitting mirror (Optical Industries 03MCS007), a Nikon S-Ke microscope with long focal length phase-contrast condenser, BM 40× objective, HKW 10× ocular and PFM photomicrographic attachment, and an RCA C7164R photomultiplier. The pinhole (0.37 mm diam) and photomultiplier were mounted in place of the 35-mm camera back. The output of the photomultiplier was fed via an

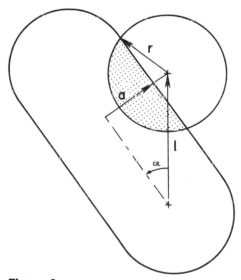

Figure 3
The image of a tethered cell, half-width *a,* sweeping across a pinhole, radius *r.* If the image is of uniform brightness, the intensity of the transmitted light varies as the area of the segment of the circle (stippled). The distance from the center of the pinhole to the edge of the image is $(l \sin \alpha) - a$.

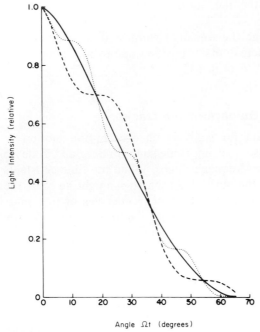

Figure 4

The intensity of the transmitted light (less background) as a function of Ωt for $a = r$, $l = 2.2r$ (Fig. 3). Three cases are shown, each for infinitely stiff filament ($\theta = 0$, $\alpha = \phi$): (————) rotation at a uniform angular velocity; (————) rotation with 10-step flutter, $n = 10$ (Eq. 4); (........) rotation with 20-step flutter, $n = 20$ (Eq. 4). If the filament is twisted a mean angle $\theta_o \simeq 1$, the vertical distances between the second and third curves and the first are smaller by factors 10 and 20, respectively (Eq. 6), and the curves are barely distinguishable.

FET-input operational amplifier (a current-to-voltage converter with 10^8 Ω feedback) to a strip-chart recorder (Brush Mark 220). The frequency response was limited solely by the recorder (3 db down at 125 Hz, ⅕ scale response 100% at 100 Hz).

Steps Are Not Apparent

It is possible to reduce the torque generated by the motor by exposing cells to 10^{-3} M DNP; they gradually slow down and then stop. When the DNP is washed away, they start up again. The changes in speed appear to be continuous. Strip-chart records for two such cells are shown in Figures 5 and 6. Larger cells or cells in media of higher viscosity spin more slowly. This is due to increased drag rather than reduced torque (Berg 1974), so the twist is not affected. Records for a cell of this kind are shown in Figure 7. The bottom trace in each figure was obtained after the cells had stopped. Its mean-square deviation from the mean was used to compute $<\theta^2>$, the conversion from intensity to angle being made from the theoretical curve for constant angular

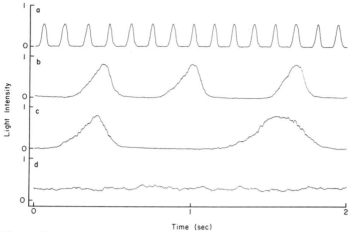

Figure 5
Signals obtained with a cell about 1 μm diam, 3 μm long, tethered near one end. Trace *a,* in tethering medium (0.067 M sodium chloride, 0.01 M potassium phosphate pH 7.0, and 10^{-4} M EDTA). Traces *b–d,* in tethering medium containing 10^{-3} M DNP. The rotation rates were 7.4, 1.6, 0.9–0.7 and 0 rps, respectively. The cell slowed down during trace *c.* (Trace *a* was made at reduced gain.)

velocity (Fig. 4), with a correction for the difference in the position of the rotation axis judged from the fraction of the rotational period required for the image to cross the pinhole. The drag coefficient (*f*) was estimated from the cell's shape and angular velocity and the mean twist (θ_o) from Equation 3. The results are summarized in Table 1. In Figure 6 trace c, 10-step flutter would be attenuated by a factor of 2, 20-step flutter, by a factor of 4. None of

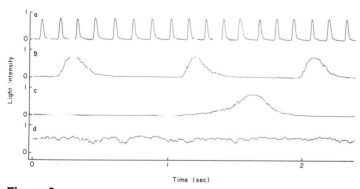

Figure 6
Signals obtained with a cell about 1 μm diam, 3 μm long, tethered near one end. Trace *a,* in tethering medium. Traces *b–d,* in tethering medium containing 10^{-3} M DNP. The rotation rates were 7.5, 0.8, 0.6 or less and 0 rps, respectively. The cell sped up during trace *c;* the preceding transit occurred 4 sec earlier than the one shown, the succeeding one, 1 sec later.

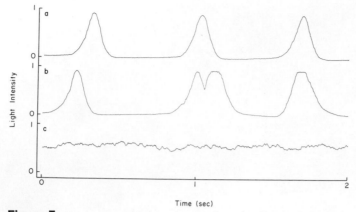

Figure 7

Signals obtained with a double cell (one tethered, another of twice the length stuck to its side, about 2×4 μm overall). Traces *a* and *b*, in tethering medium containing Methocel (Dow Methocel 90 HG 0.14%, $\eta \simeq 2 \ 0.02$ poise). Trace *c*, in Methocel plus 10^{-3} M DNP. The rotation rates were about 1.5, 1.5 and 0 rps, respectively. Note the reversal in trace *b*.

these signals or any that I have seen show a periodic structure. Thus there appear to be more than 10 steps per revolution.

Reversals Are Abrupt

I was struck earlier by the abruptness with which a cell changes its direction of rotation (Berg 1974). The evidence now is more dramatic. A reversal is shown in Figure 7, trace b. The signals on the right half of the trace are flat-topped; those on the left are not. The orientation of the cell differed slightly depending on its direction of rotation, the image completely filling the pinhole for about 0.05 second for one direction but not the other. Note that the

Table 1

Summary of the Data of Figures 5–7

Parameter	Figure 5	Figure 6	Figure 7
$\langle\theta^2\rangle^{1/2}$ (rad)	0.07	0.07	0.04
$\Omega_{\max}$ (rps)	7.4	7.5	1.5
θ_o at $\Omega_{\max}$ (rad)	2.8	2.5	$\simeq 1$
$\Omega_{\min}$ (rps)	0.8	0.6	1.5
θ_o at $\Omega_{\min}$ (rad)	0.3	0.2	$\simeq 1$
$(1 + n^2\theta_o^2)^{1/2}$ at $\Omega_{\min}$ for			
$n = 10$	3	2	10
$n = 20$	6	4	20

When the mean twist is θ_o, the amplitude of the flutter is attenuated by a factor $1/(1 + n^2\theta_o^2)^{1/2}$ (Eq. 6).

reversal was complete in less than 10^{-2} second, even though the filament had to unwind some 60° (Table 1) and wind back up again (in the opposite direction).

The Experiments Can Be Done in Other Ways

Polyhook mutants and mutants with straight filaments (Silverman and Simon 1974) jiggle almost as much as the wild type. It may be possible to tether mutants with normal hooks but no filaments. If not, traces for cells operating at low torque might be averaged over many cycles. The jiggle, being random, should disappear; steps, being periodic, should emerge.

SUMMARY

I conclude, as before (Berg 1974), that there is little passive slip between the flagellum and the cell wall, that the number of steps per revolution is large (more than 10), and that reversals are abrupt.

Acknowledgments

I thank Edward Purcell for his lively interest and Pat Tedesco for her superb technical assistance. This work was supported by a grant from the National Science Foundation (BMS75–05848).

REFERENCES

Berg, H. C. 1971. How to track bacteria. *Rev. Sci. Instrum.* **42**:868.
———, 1974. Dynamic properties of bacterial flagellar motors. *Nature* **249**:77.
———. 1975a. Bacterial behaviour. *Nature* **254**:389.
———. 1975b. How bacteria swim. *Sci. Amer.* **233**:36.
Berg, H. C. and R. A. Anderson. 1973. Bacteria swim by rotating their flagellar filaments. *Nature* **245**:380.
Berg, H. C. and P. M. Tedesco. 1975. Transient response to chemotactic stimuli in *Escherichia coli. Proc. Nat. Acad. Sci.* **72**:3235.
Clayton, R. K. 1958. On the interplay of environmental factors affecting taxis and motility in *Rhodospirillum rubrum. Arch. Mikrobiol.* **29**:189.
DePamphilis, M. L. and J. Adler. 1971. Fine structure and isolation of the hook-basal body complex of flagella from *Escherichia coli* and *Bacillus subtilis. J. Bact.* **105**:384.
Fujime, S., M. Maruyama and S. Asakura. 1972. Flexural rigidity of bacterial flagella studied by quasielastic scattering of laser light. *J. Mol. Biol.* **68**:347.
Hilmen, M., M. Silverman and M. Simon. 1974. The regulation of flagellar formation and function. *J. Supramol. Struc.* **2**:360.
Landau, L. D., A. I. Akhiezer and E. M. Lifshitz. 1967. *General physics* (trans. from Russian by J. B. Sykes, A. D. Petford and C. L. Petford), chap. 13. Pergamon Press, New York.
Larsen, S. H., J. Adler, J. J. Gargus and R. W. Hogg. 1974a. Chemomechanical

coupling without ATP: The source of energy for motility and chemotaxis in bacteria. *Proc. Nat. Acad. Sci.* **71**:1239.

Larsen, S. H., R. W. Reader, E. N. Kort, W.-W. Tso and J. Adler. 1974b. Change in direction of flagellar rotation is the basis of the chemotactic response in *Escherichia coli. Nature* **249**:74.

Ordal, G. W. and D. J. Goldman. 1975. Chemotaxis away from uncouplers of oxidative phosphorylation in *Bacillus subtilis. Science* **189**:802.

Silverman, M. and M. Simon. 1974. Flagellar rotation and the mechanism of bacterial motility. *Nature* **249**:73.

The Control of Flagellar Rotation in Bacterial Behavior

Daniel E. Koshland, Jr., Hans Warrick, Barry Taylor*
and John Spudich

Department of Biochemistry, University of California
Berkeley, California 94720

Bacteria have a simple sensory system that allows them to detect gradients of attractants or repellents and to use this information to direct their migration (Pfeffer 1884; Adler 1969). The mechanism for this process has been shown to be an alteration in the frequency of tumbling (Macnab and Koshland 1972; Berg and Brown 1972). This alteration in the tumbling frequency operates through a memory mechanism that sets the level of a regulator, which in turn controls the frequency of tumbling (Macnab and Koshland 1972).

Since recent reviews (Adler 1969, 1975; Berg 1974; Koshland 1974, 1976) have summarized various aspects of this phenomenon, in this paper three features of its control will be emphasized: first, the effect of light on flagellar rotation; second, the quantitation of the tumbling response; and third, the genetics of the signaling system.

Light Effect

In our initial studies on *Salmonella typhimurium* we used various light sources to record the movements of the bacteria (Macnab and Koshland 1972). To examine the phenomena more carefully we raised the light intensity appreciably using a xenon lamp of high intensity. Two phenomena became apparent: first, it was possible to see individual flagella and not simply the flagella bundle when high intensity light was used; second, the sudden onset of a high-intensity light generated tumbling (Macnab and Koshland 1973). When *Salmonella typhimurium* was examined, the flagellar pattern looked very similar to that in Figure 1. The bacteria in normal motion swam in straight lines, with their flagella coordinated in a bundle. When subjected to high-intensity light, flagella seemed to fly apart briefly and then to resume

* Present address: Department of Biochemistry, John Curtin School of Medical Research, Australian National University, Canberra ACT, Australia 2600.

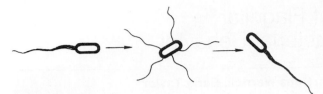

Figure 1
Schematic illustration of tumbling in multiflagellated bacteria. Bacteria swim
with flagella in a bundle. Flagella fly apart when rotation is reversed. Flagella
resume bundle as smooth swimming resumes.

coordinated motion in the bundle, leading to smooth swimming. The flying
apart, therefore, correlated with the tumbling phenomena.

Several possible mechanisms to explain bacterial propulsion were available
in the literature, of which the two most popular were transmission of a conformational wave down the flagella and a propellor type rotation of the flagella. This issue was resolved by the tethering experiments of Silverman and
Simon (1974a,b). The quantitation of the tethered bacteria, as studied by
Berg (1974) and coworkers and Adler (1975) and his coworkers, showed
that the reversal of rotation was correlated with tumbling (Larsen et al.
1974). These studies are extensively discussed in other papers in this volume.

To add further insight on this process we examined a monotrichous organism, *Pseudomonas citronellolis* (Taylor and Koshland 1974). In this case,
the logic was quite simple. The bacterium contained polar flagella, and thus
reversal of rotation should cause a backing up of the bacterium rather than
tumbling, and this could be observed simply by the photographic techniques
developed earlier. In fact, just such observations were made, and a picture
of some data is given in Figure 2. Addition of attractants and repellents to
Pseudomonas gave backing up responses similar to the tumbling in *Salmonella*
and *E. coli*. Moreover, the *Pseudomonas* showed the light phenomenon, i.e.,
in this case, the induction of reversal in strong light.

A simple picture consistent with the actions of the peritrichous and monotrichous organisms therefore emerges. Stimuli that alter the level of tumble
regulator can generate tumbling by causing reversal of flagellar rotation or
suppress tumbling by continued rotation in the clockwise direction. In a
monotrichous organism, reversal causes backing up of the organism. In a
peritrichous organism, the bundle of flagella flies apart, and the independent
action of the individual flagella causes a temporary turning or tumbling in
space which then reverses again to generate smooth swimming. Whether it is
necessary to cause tumbling for all the flagella to reverse or only a few of
them is not known.

Further investigation of the light response revealed properties of the sensing
system. The light response had the action spectrum of a flavin (Macnab and
Koshland 1973) and could be mimicked by adding flavin-type dyes to the
bacteria (Taylor and Koshland 1975). Moreover, careful analysis indicated
that there were two light phenomena affecting the sensory system—an initial
light-generated tumbling phenomenon followed by a light-generated smooth-
swimming response. Thus the light effect acts on the sensory system to perturb

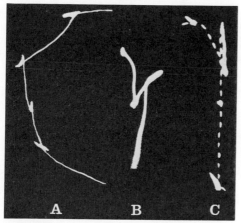

Figure 2

Motility tracks of *Pseudomonas citronellolis*. Photomicrographs were taken in dark field with a stroboscopic lamp operating at 60 (*A, B*) or 10 (*C*) pulses per sec. (*A*) The bacterium entered at the lower right and backed up four times during the exposure. The change of direction is sometimes 180°, but frequently it is a smaller angle, resulting in a random, walk-type motion. (*B*) Reorientation of *P. citronellolis* during reversal. The tracks are broader than those in *A* because the bacterium was above the plane of focus. (*C*) A slower strobing frequency allows comparison of the velocities in the forward and reverse directions. *P. citronellolis* apparently swims more slowly in the reverse direction. Cells were grown in minimal medium with glycerol as sole carbon source.

the flow of information and must intervene at two different points since it is impossible to both increase and decrease the tumbling by action at a single locus. The primary photochemical event can be the same in the two cases but it must act on two different chemical compounds. In fact, the action spectrum is that of a flavin for the tumbling as well as for the smooth response. On prolonged exposure to light, bacteria are paralyzed, but this could be a complex phenomenon affecting many systems besides the signaling system.

That this light response is a perturbation of the sensing system can be shown in several ways (Taylor and Koshland 1975). Mutants that are incapable of sensing gradients did not give the light response. Bacteria subjected to large stimuli, which tend to saturate their receptors and suppress tumbling, did not give light-induced tumbling. Moreover, the light response was not observed in the absence of electron transport. Hence a reasonable hypothesis is that the light response generates an alteration in electron flow, thus perturbing the sensory response. The perturbation can occur at two loci, one of which increases and the other decreases the level of the tumble regulator.

Quantitation of the Signal

Adler (1969) has demonstrated that the chemotactic signal is initiated by receptors that have specificities similar to receptors in other sensory systems.

The ribose receptor was isolated in our laboratory and shown to have high but not absolute specificity (Aksamit and Koshland 1972, 1974). The availability of both this purified receptor and a quantitative procedure (Spudich and Koshland 1975) made it possible to correlate the properties of the purified protein with the biological response in the whole cell.

The quantitative method developed followed from earlier studies on the temporal gradient method (Macnab and Koshland 1972) in which a sudden increase in concentration of attractant generates smooth swimming in a bacterium for a prolonged period of time. The stimulus can be considerably larger than the normal stimulus a bacterium sees in swimming up a gradient, and hence the response time is exaggerated and more easy to measure. By varying the degree of stimulation it can be shown that the mechanism does not change with the amount of stimulation; only the intensity of the response changes. The method is illustrated in Figure 3, in which a constantly tumbling mutant is subjected to a sudden increase in serine concentration. The bacterial motion is observed by opening the shutter of a camera and recording the tracks from four stroboscopic flashes within a period of 0.8 second. Subsequent observations are made at various times after exposure to the stimulus.

For convenience in quantitation, a constant tumbling mutant is used. Such a mutant shows a single light spot because successive stroboscopic flashes catch the bacterium in roughly the same position in space. After a sudden serine stimulus, the bacteria are all seen swimming smoothly at 0.4 minute after stimulus (Fig. 3a). At 0.5 minute, some bacteria have resumed tumbling, at 0.6 minute, more, and at 0.9 minute, all bacteria are tumbling constantly (Fig. 3b,c,d).

The number of smooth-swimming and tumbling bacteria in any of these pictures can be easily quantitated. When the results are plotted as shown in Figure 4, an exponential decay can be observed. To show that the constantly tumbling mutant is not aberrant in other ways, wild-type data are shown on the same graph. The constantly tumbling mutant is easier to quantitate because the background level is zero, but the same principles apply to a wild type, and checks against wild type can always be made if there is any doubt that a mutant may have an altered metabolism.

These quantitative studies have allowed us to reach a number of conclusions. In the first place, the number of tumbles suppressed is a function of the change in receptor occupancy and not the rate of change of receptor occupancy (Spudich and Koshland 1975). By receptor occupancy we mean the amount of receptor (R) and chemoeffector (C) combined in a complex (RC) as shown in Equation 1:

$$R + C \rightleftarrows RC. \tag{1}$$

A priori it could not be known whether the response depended on $\Delta(RC)$, $\Delta RC/\Delta t$, or other more complex functions, and various ideas were suggested. The quantitative test allowed us to say that the tumbling suppression depended on $\Delta(RC)$. At first this might seem anomalous since the memory mechanism argument showed that the bacteria were recording stimuli over time (Macnab and Koshland 1972). But one must be careful to be precise when defining a "response," and examination of the total tumbles suppressed shows that they are proportional to the change in receptor occupancy. This

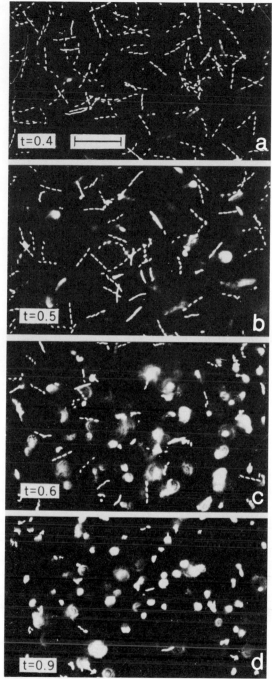

Figure 3

Tumble frequency assay. The disappearance of tracks made by constantly tumbling mutant after an L-serine temporal gradient stimulus ($0 \rightarrow 0.02$ mM) is shown. A 0.9-ml aliquot of a suspension of ST171 (a constantly tumbling mutant) was rapidly mixed with 0.1 ml of attractant to yield a final concentration of 0.02 mM L-serine. Photographs, at the indicated number of minutes after mixing ($t = 0$), were taken by 0.8-sec open-shutter exposures to stroboscopic illumination. The length of the bar in the photograph at 0.4 min is 50 μm.

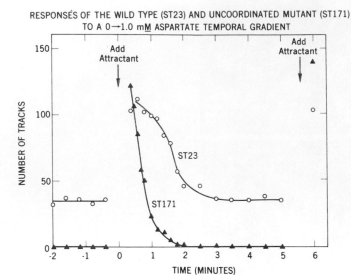

RESPONSES OF THE WILD TYPE (ST23) AND UNCOORDINATED MUTANT (ST171) TO A 0→1.0 m$\underline{M}$ ASPARTATE TEMPORAL GRADIENT

Figure 4
Responses of wild-type and a constantly tumbling mutant to the tumble frequency assay. Number of tracks measured in control prior to stimulus is zero for constantly tumbling mutant (▲) and finite for wild-type (○). Plot of tracks as a function of time after stimulus leads to return to normal motility after a few minutes in both cases.

means that the bacterium integrates the information of receptor occupancy so that whether the stimulus is supplied over 2 seconds or 20 seconds, the number of tumbles suppressed will be the same. The number of tumbles suppressed *per unit of time* will vary, however, if the stimuli are generated more slowly or more rapidly.

A second conclusion derived from these quantitative studies was that the stimuli were additive. A particularly strong stimulus, such as a change from 0–0.5 mM serine, can be broken into a series of smaller stimuli, 0–0.02, 0.02–0.4, 0.4–0.5, and the sum of the areas under the curves for these three stimuli add up to the area under the curve for the large stimulus. Thus there is additivity in the stimuli and recovery times.

Finally, a correlation can be obtained with the purified receptor protein and the bacterial response of the whole system. This is shown in Figure 5. The solid curves represent the calculated binding curve for the bindings of ribose and allose to the purified ribose receptor in *Salmonella typhimurium*. The points represent the observations of a tumbling assay of the type shown in Figure 3. The method is completely objective since it is quite easy to quantitate smooth-swimming or tumbling tracks without subjectivity. The results of Figure 5 indicate that the protein in vitro has the same dissociation constant as in vivo. Since the in vitro study is made in aqueous solution, this suggests that the sugar binding region of the ribose receptor is at least jutting out into the aqueous part of the environment and is not buried in the lipid layer of the membrane.

Additive relationships are found to apply to repellents as well as to attract-

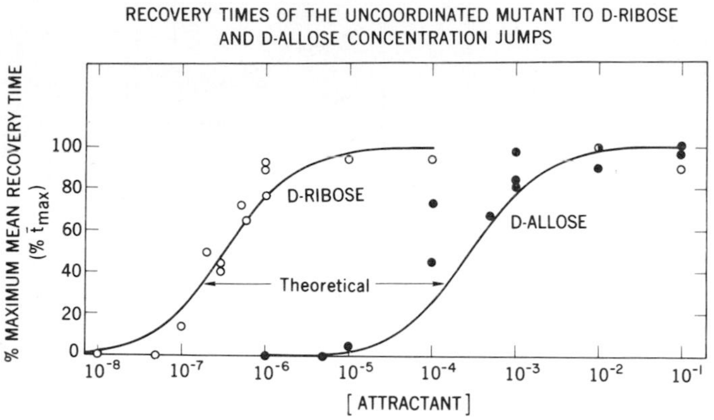

RECOVERY TIMES OF THE UNCOORDINATED MUTANT TO D-RIBOSE
AND D-ALLOSE CONCENTRATION JUMPS

Figure 5
Comparison of response with receptor occupancy. Points are percent maximum mean recovery times ($\bar{t}_R$) for D-ribose and D-allose concentration increases from 0 to the concentration shown on the abscissa. Each point represents the average of three consecutive assays. Theoretical curves were calculated assuming (1) noncooperative chemoreceptor binding constants of 3.3×10^{-7} for ribose and 3.0×10^{-4} for allose (as determined in vitro), and (2) that the response $\bar{t}_R$ is proportional to the change in the fraction of binding protein occupied. The symbol ⊙ represents coincidence of the points ○ and ●.

ants, and a rough additivity prevails between attractants and repellents as well as between two attractants or two repellents. However, the exact quantitation is not precise, and this provides significant clues regarding the chemotactic process. These are being pursued at the moment.

A mechanism to explain the response of the bacteria is shown in Figure 6. A parameter, which for convenience we call the tumble regulator, is varying in a Poissonian manner in the absence of a gradient, thus generating tumbles periodically in the wild type. The level of tumble regulator is elevated by increases in attractants or decreases in repellents, and this suppresses tumbling for an interval. As the level of tumble regulator returns to normal, the tumbling returns to normal. The constantly tumbling mutant behaves similarly, but in this case, the steady-state level of the tumble regulator is below threshold levels so that the bacterium tumbles constantly when a gradient is absent.

If the picture of Figure 6 is correct, it would seem that the inverse experiment to Figure 3, i.e., the generation of tumbling in a smooth-swimming mutant by addition of repellent, would be possible, and this has been observed (R. Zukin and D. E. Koshland, Jr., unpubl.). Thus the relation between the tumble generator and the attractants and repellents can be explained in a simple scheme, as shown in Figure 6.

Selection of Mutants

Preparation of mutants is a classic bacterial tool, and such mutants were studied by Adler and coworkers (Armstrong, Adler and Dahl 1967) who dis-

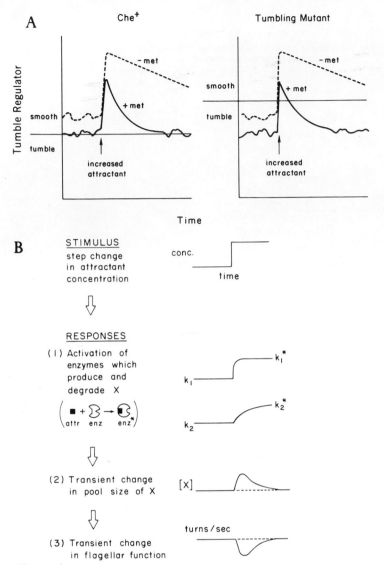

Figure 6

A model for the regulation of tumbling frequency by gradients. (*A*) When the tumble regulator (wavy line) rises above the threshold value (horizontal solid line), tumbling is suppressed; when it falls below, tumbling is increased. In the absence of a gradient, the regulator level fluctuates randomly near the threshold level, but a rapid temporal increase in attractant makes it rise far above the threshold level, suppressing tumbling completely until attractant returns to normal. The regulator is maintained at a steady-state level by a delicate balance between its synthesis and degradation. *S*-adenosyl methionine (SAM) may be necessary for the degradation reaction. Thus during methionine starvation, the level of regulator is higher than normal and tumbling is suppressed, and when the level is further perturbed by a temporal gradient, it takes longer than normal to return to its steady level. By assuming that the uncoordinated mutant has an abnormally high threshold, one can easily explain its behavior in both the presence and absence of methionine. (*B*) A model for the way in which the attractant can alter the rate of formation and degradation of the tumble regulator so that its level rises and falls on ascending and descending gradients.

covered three genes in *E. coli,* which they labeled *cheA, cheB* and *cheC*. Parkinson discovered a fourth, which he called *cheD* (Parkinson 1975). These mutants mapped in the general region of flagella and motility on the *E. coli* map. It seemed to us that a further search for genes involving the general sensory system would be desirable. In part, the individual genes could be used to draw knowledge regarding the sensing system, but it also might be possible to obtain all of the genes responsible for the bacteria behavior and hence gain knowledge as to the total complexity of the sensing system.

A rapid screening method was developed (Aswad and Koshland 1975), and the rationale for its selection system is shown in Figure 7. The procedure involves placing the bacterial culture in the middle of a test tube containing a preformed gradient, with attractant in the upper part of the tube. The nonmotile bacteria stay in the center. Bacteria that sense gradients swim into the upper region, and bacteria that cannot sense a gradient but are motile expand in all directions. Hence the nonchemotactic but motile bacteria concentrate in the lower part of the vessel. After an appropriate interval of about 15 minutes, the upper volume of fluid is removed from the test tube. The test tube is then washed carefully and the mutants in the lower region of the test tube separated and analyzed. Utilizing such a separation we first obtained

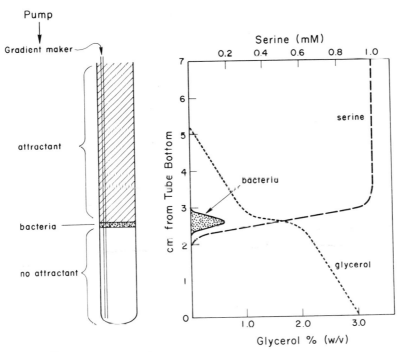

Figure 7
Preformed "liquid" gradient method for isolation of generally nonchemotactic mutants. The drawing on the left shows how the gradient was constructed, and the drawing on the right, the distribution of gradients at the time zero. Wild-type bacteria swim up to attractant. Nonmotiles stay at the origin. Nonchemotactic mutants, which cannot sense a gradient but are motile, wander into the lower region and are collected.

68 generally nonchemotactic mutants, which were separated into six complementation groups (Aswad and Koshland 1975). Recently, the procedure was
repeated to obtain 104 additional mutants, which have now been separated
into nine complementation groups. The identification of the regions of these
groups is shown in Figure 8, and their properties are given in Table 1. Five
of them are identified in a cluster adjacent to each other. Identification was
obtained by transducing phage, episomes and deletion mutants. Two of the
complementation groups have not been mapped, but they lie outside this
region.

Two of the mutants have been located in precisely the gene positions at
which *fla* genes have previously been identified (*cheU* and *cheV*). This is not
the first time that the latter phenomenon has been observed. Silverman and
Simon (1974b) found a constant tumbling mutant that mapped in a *fla* gene
region of *E. coli*, and M. L. Collins and B. Stocker (pers. comm.) have observed the same thing in *Salmonella*. What does this mean? *Fla* genes are
identified because mutant bacteria are found to lack flagella. In our selection
test, however, the bacteria selected for motility possess flagella yet map at
a *fla* locus. The logical explanation is that both processes require the same
protein. Some mutations of the protein damage the flagellar machinery, causing
a loss of flagella formation. Other protein changes allow assembly of the
flagellum but prevent its receiving messages from the sensory system. Either
a regulatory site on the protein that receives the signal has been damaged,

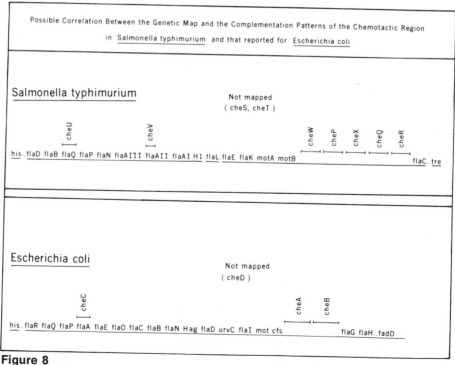

Figure 8
Genetic map of generally nonchemotactic mutants in *E. coli* and *Salmonella
typhimurium.*

Table 1

The Properties of the Motile but Nonchemotactic Mutant Classes

Class	Motility pattern[a]	Complementation pattern	Response to positive gradient of serine	Blue light response	Response to positive gradient of phenol
cheP	smooth	recessive	0	0	0
cheQ	smooth	recessive	0	0	0
cheR	smooth	recessive	0	smooth to tumbling	smooth to tumbling
cheS	smooth	dominant	0	0	smooth to tumbling
cheT	tumbling	recessive	tumbling to smooth	tumbling to smooth	0
cheU	tumbling to smooth	dominant	0	0	0
cheV	smooth	recessive	0	0	0
cheW	smooth	recessive	0	0	0
cheX	tumbling	recessive	0	0	0
Wild type	wild type		wild type to smooth	wild type to tumbling	wild type to tumbling

[a] Smooth indicates no observed tumbles; tumbling indicates constantly tumbling; wild type indicates smooth runs with Poissonian distribution of tumbles.

or some site that allows the organization of the flagella in such a way as to receive signals has been altered.

The total complexity of the signaling system controlling flagellar rotation has not yet been revealed. In the first selection of about 100 mutants, six groups were identified. In the second selection, three additional groups were located (in addition, of course, to duplicates of the original complementation groups). Further selections are being performed, and until no more groups are found, the total number cannot be estimated. Although we tentatively identify these groups with separate genes, this identification cannot be considered final until all possibility of intragenic complementation is eliminated. Of course, these are the genes common to all sensing systems, since genes for individual receptors are not included in this selection. If this is correct, a sensing system of moderate complexity is indicated.

That these genes are interesting subjects for study in the sensing system is clear. Some of these mutants do not show the "light effect," but others do (Table 1). This would suggest that the level of tumble regulator is somewhat different in different complementation groups and that it can be perturbed somewhat differently in the several types of mutants. This will be a handy tool in future studies. As discussed above, some smooth-swimming mutants have been treated with repellents and shown to tumble. The results are consistent with the general picture in which the sensing system provides a level of tumble regulator important to proper tumbling frequency and its proper alteration under the influence of stimuli.

CONCLUSION

Bacterial behavior is controlled by a sensing system that delivers information to bacterial flagella, the motor apparatus of the bacteria. The signal involves a relatively simple switch mechanism in which reversal of flagellar rotation causes backing up in the case of monotrichous organisms and tumbling in the case of peritrichous organisms. The signal can be perturbed by a light signal that can either increase tumbling or suppress it and which has the action spectrum of a flavin.

The response to chemicals can be quantitated and shown to be explained by a change in receptor occupancy, i.e., $\Delta(RC)$, where R (receptor) + C (chemoeffector) $\rightleftarrows$ RC (the complex). The receptor occupancy can be calculated from the properties of a purified receptor protein, and hence the biological response of the whole animal is quantitatively related to the purified receptor properties. Thus the quantitative assay has, for this system, similarities to the action potential for the nerve axon.

The development of a quick genetic screening method has allowed the delineation of nine complementation groups in the *Salmonella* general sensing system. These fall into a cluster of five groups mapping near the flagellar region and two groups mapping at the specific loci of flagellar genes. Two other groups have yet to be mapped. The results give a rough general picture of the complexity of the system controlling flagellar motility.

Acknowledgments

This work was supported by research grants from the U.S. Public Health Service (AM-GM-10765) and the National Science Foundation (GB-7057). We wish to acknowledge the suggestions and helpfulness of Drs. Bruce Ames, Giovanna Ames and Bruce Stocker in the genetic analyses.

REFERENCES

Adler, J. 1969. Chemoreceptors in bacteria. *Science* **166**:1588.

————. 1975. Chemotaxis in bacteria. *Annu. Rev. Biochem.* **44**:341.

Aksamit, R. and D. E. Koshland, Jr. 1972. A ribose binding protein in *Salmonella typhimurium. Biochem. Biophys. Res. Comm.* **48**:1348.

————. 1974. Identification of the ribose binding protein as the receptor for ribose chemotaxis in *Salmonella typhimurium. Biochemistry* **13**:4473.

Armstrong, J. B., J. Adler and M. M. Dahl. 1967. Nonchemotactic mutants of *Escherichi coli. J. Bact.* **93**:390.

Aswad, D. and D. E. Koshland, Jr. 1975. Isolation, characterization and complementation of *Salmonella typhimurium* chemotaxis mutants. *J. Mol. Biol.* **97**:225.

Berg, H. C. 1974. Dynamic properties of bacterial flagellar motors. *Nature* **249**:77.

Berg, H. C. and D. A. Brown. 1972. Chemotaxis in *Escherichia coli* analyzed by three dimensional tracking. *Nature* **239**:500.

Koshland, D. E., Jr. 1974. Chemotaxis as a model for sensory systems. *FEBS Letters* (Suppl.) **40**:S3.

————. 1976. Sensory response in bacteria. *Adv. Neurochem.* (in press).

Larsen, S. H., R. W. Reader, E. N. Kort, W.-W. Tso and J. Adler. 1974. Change in direction of flagellar rotation is the basis of chemotactic response. *Nature* **249**: 74.

Macnab, R. M. and D. E. Koshland, Jr. 1972. The gradient sensing mechanism in bacterial chemotaxis. *Proc. Nat. Acad. Sci.* **69**:2509.

————. 1973. Persistence as a concept in the motility of chemotactic bacteria. *J. Mechanochem. Cell Motil.* **2**:141.

Parkinson, J. F. 1975. Data processing by the chemotaxis machinery of *Escherichia coli. Nature* **252**:317.

Pfeffer, W. 1884. Locomotorische Richtungsbewegunge durch chemisch Reize. *Untersuch Botan. Inst. Tubingen* **1**:363.

Silverman, M. R. and M. I. Simon. 1974a. Flagellar rotation and the mechanism of bacterial motility. *Nature* **249**:73.

————. 1974b. Positioning flagellar genes in *Escherichia coli* by deletion analysis. *J. Bact.* **117**:37.

Spudich, J. L. and D. E. Koshland, Jr. 1975. Quantitation of the sensory response in bacterial chemotaxis. *Proc. Nat. Acad. Sci.* **72**:710.

Taylor, B. and D. E. Koshland, Jr. 1974. Reversal of flagellar rotation in monotrichous and peritrichous bacteria: Generation and changes in direction. *J. Bact.* **119**:640.

————. 1975. The intrinsic and extrinsic light responses of *Salmonella typhimurium* and *Escherichia coli. J. Bact.* **123**:557.

Assembly of the Contractile Tail of Bacteriophage T4

Yoshiko Kikuchi and Jonathan King

Department of Biology, Massachusetts Institute of Technology
Cambridge, Massachusetts 02139

When the long tail fibers of phage T4 bind to their appropriate cell surface receptor sites, they set in motion a series of structural rearrangements of the phage tail (Simon and Anderson 1967a; Arscott and Goldberg 1976). The first of these is the activation of the baseplate; baseplate fiberlets bind to another set of receptors, and the baseplate rearranges from a hexagonal to a star-shaped structure (Simon and Anderson 1967b). This apparently releases the baseplate–tail tube bonds and initiates the contraction of the sheath (Fig. 1). Sheath contraction is then thought to force the released tail tube tip through the cell surface. Further steps are needed for the actual release of the DNA (Benz and Goldberg 1973; Male and Kozloff 1973; Yamamoto and Uchida 1975). The sheath was once thought of as a model muscle, but it is now clear that it represents a different kind of contractile device.

The extended sheath is composed of 24 annuli of six subunits each (Moody 1967a,b). The subunits are 80,000 MW monomers and are the product of T4 gene 18 (King 1968; King and Mykolajewycz 1973). The hollow tail tube, which the sheath subunits must slide past, is also composed of 24 annuli of six subunits each (Moody 1971). The tube subunits are 20,000 MW monomers and are the product of T4 gene 19 (King 1971). DeRosier and Klug (1968) obtained a three-dimensional reconstruction of the extended sheath, showing many details of the subunit arrangements, and this has been subsequently refined (Amos and Klug 1975; Smith et al. 1976). The contracted sheath is thought also to represent 24 annuli of six subunits, but with the annuli closer together and the subunits moved out to higher radius (Moody 1976b; Amos and Klug 1975).

Before contraction, sheath subunits are only weakly bonded to each other (King 1968). After contraction, however, they are extremely tightly bound and can only be dissociated by very strong denaturing agents (Brenner et al. 1959; To, Kellenberger and Eisenstark 1969; Poglazov 1973). No one has yet succeeded in obtaining native subunits from contracted sheaths. A third form of sheath, polysheath, is an aberrant aggregation product of precursor

71

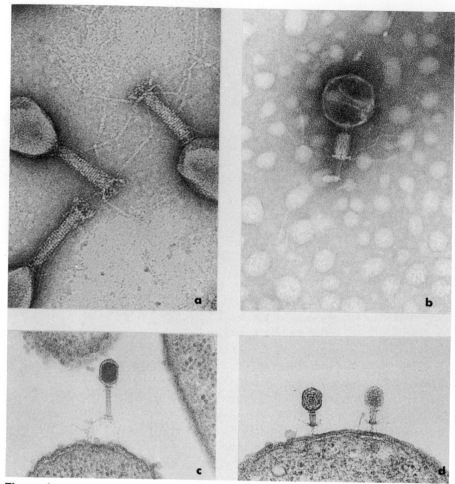

Figure 1

T4 phage particles with extended and contracted tails. (*a*) Purified, negatively stained phage particles. Spikes project from the bottom of the baseplate. The slender tail fibers are attached to the apices of the baseplate. 164,000×. (*b*) Negatively stained ghost, prepared by heat inactivation. Note the altered appearance of sheath and baseplate and the exposed tail tube. The DNA has been ejected from the head. 164,000×. (*c*) Thin section of phage particle attached to host cell by long tail fibers. The head is full of the positively staining condensed DNA. (*d*) Thin sections of T4 particles which have injected their DNA into the host cell. The baseplates have triggered, and the sheaths are contracted. Embedding and staining according to Simon and Anderson (1967a).

subunits that forms at late times within infected cells (Kellenberger and Boy de la Tour 1964; King 1968). The subunits in polysheath are less strongly joined together than in contracted sheath (To, Kellenberger and Eisenstark 1969).

There is no evidence for the requirement of any exogenous energy source for contraction. In fact, there are incomplete particles of T4, lacking a baseplate protein (gene 9 product), that are unstable and contract spontaneously (Simon, Swan and Flatgaard 1970).

Detailed structural studies of the sheath have enabled Moody (1973) and Amos and Klug (1975) to describe precisely the rearrangement of the extended to the contracted state. However, the irreversible nature of contraction, the independence from an exogenous energy source, and the complexity of the triggering process—baseplate expansion—have made it difficult to study the contraction process directly. Nonetheless, it is clear that to some extent sheath contractility represents an assembly problem; a metastable high-energy structure must be formed that will subsequently transform to the more stable contracted state. Aspects of contractile function may in general reflect constraints on the kinds of structures that can be assembled rather than the kinds of mechanisms that transduce chemical to mechanical energy.

Because of the extensive genetic characterization of bacteriophage (Epstein et al. 1964; Edgar 1969) and their general experimental tractability, a great deal is known of bacteriophage structure and assembly. Below we review our studies on the assembly of the contractile tail, including the baseplate, which may itself be a transducing device. The experiments focus on the properties of the precursor proteins—those which have not yet assembled—rather than on proteins isolated from the mature structure.

Genes for Proteins Involved in Tail Assembly

The distinctive character of studies on bacterial virus assembly stems from the availability of mutants that block assembly, thereby identifying the genes involved in the assembly process (Epstein et al. 1964; Edgar 1969). Cells infected with mutants blocked in assembly provide a unique source of precursor structures and precursor proteins for further characterization (Edgar and Wood 1966; Edgar and Lielausis 1968; Kikuchi and King 1975a).

The products of 22 T4 genes are known to be needed for tail formation. Their relative chromosomal positions are as follows:

. . 57 . . . 3 . . . . 53–5–6–7–8–9–10–11–12 . . . 15 . . . 18–19 . . . 25–26–51–27–28–29–48–54 . .

The dashes separate adjacent genes. Two of the dispersed genes, 3 and 15, specify proteins involved in the termination of the sheath and the formation of the site for head attachment. The adjacent pair, 18 and 19, specify the sheath subunit and the tube subunit, respectively (King 1968, 1971). The 17 genes in the two large clusters all specify proteins needed for assembly of a functional baseplate. The product of the leftmost gene, 57, is pleiotropic; it is necessary for the folding or maturation of two tail fiber proteins and one baseplate protein—the gene 12 product (King and Laemmli 1971; Ward and Dickson 1971). All three of these proteins are fibrous in nature (Kells and Haselkorn 1974; Bishop, Conley and Wood 1974).

All of the above genes are essential: if their products are defective due to the introduction of an amber or temperature-sensitive mutation, no viable phage are formed, although incomplete structures may assemble. Two other phage gene products have been implicated in tail assembly—the enzymes dihydrofolate reductase and thymidylate synthetase (Kozloff et al. 1970a;

Capco and Mathews 1973). These proteins appear to be structurally associated with the phage baseplate and may play a critical role in triggering the contraction process (Male and Kozloff 1973). However, cells infected with amber mutants in the genes thought to specify these proteins still produce viable phage (Kozloff, Lute and Crosby 1975), and we have not studied them in our experiments.

Tail Proteins

Since T4 infection shuts off host protein synthesis, incubation of phage-infected cells with radioactive amino acids results in labeling of only phage-specified proteins. The presence of an amber mutation results in the synthesis (in the restrictive host) of only a fragment of the protein of the mutant gene. By comparing the SDS gel patterns of the labeled proteins from wild-type-infected cells with the gel patterns of the proteins from amber mutant-infected cells, it is possible to identify bands in the wild-type gel pattern with the genes that specify them (Hosoda and Levinthal 1968; Laemmli 1970). The products of 19 of the 22 tail genes have been so identified (King and Laemmli 1973; Vanderslice and Yegian 1974; Kikuchi and King 1975b).

Complete phage tails can be readily isolated from cells infected with mutants blocked in head assembly. SDS gel electrophoresis of purified precursor tails reveals 19 constituent polypeptide chains, ranging in molecular weight from 140,000 down to 15,000. Eighteen of them are among the gene products just described; one is not the product of a known gene (Kikuchi and King 1975a).

Thus of the 22 gene products needed for making a functional tail, 18 are actually incorporated into the mature structure, whereas the remaining four—the products of genes 26, 28, 51 and 57—are not. As noted above, the gene 57 product acts in the maturation of two fiber proteins and one baseplate protein. As will be described below, the products of genes 26, 28 and 51 all act in the maturation of a protein that is an early protein in the baseplate assembly pathway. Gene dosage experiments (Snustad 1968) suggest that they function catalytically. The experiments of Kozloff, Lute and Baugh (1973) suggest that the gene 28 product may be involved in the synthesis of the folic acid conjugate found in the tail (Kozloff, Lute and Crosby 1975).

In contrast to ribosomes, the various tail proteins are present in very different molar ratios. Our best estimate of the numbers of the various molecules per tail is given in Table 1.

Most of the tail proteins are in fact part of the baseplate (King and Mykolajewycz 1973). We have determined the stoichiometry of these proteins by preparing phage tails from cells continuously exposed to radioactive amino acids. The proteins of isolated tails have also been stained with Coomassie brilliant blue and the amount of dye bound to the major bands quantitated with a microdensitometer. As a standard we used the core subunit, assuming 144 subunits per tail. Since the baseplate has sixfold rotational symmetry, we have also assumed that most of the major proteins would be present in multiples of six. Thus the calculated number has been rounded off to the closest multiple of six (Y. Kikuchi and J. King, in prep.).

There are no bands present in six or more copies per tail that correspond

Table 1

Stoichiometry of Baseplate Proteins

Proteins	Molecular weight	Copies/baseplate
gp19	20,000	144
gp7	140,000	6
gp10	88,000	12
gp6	85,000	12
gp29	80,000	$\leqq 6$
gp12	55,000	18
pX (gp27)	48,000	$\leqq 6$
gp8	46,000	$\geqq 6$
gp5	44,000	$\leqq 6$
gp48	44,000	$\leqq 6$
gp54	36,000	$\leqq 6$
gp9	34,000	24
gp11	24,000	24
gp53	23,000	$\geqq 6$
pY	22,000	$\leqq 6$
gp25	<15,000	$\leqq 6$

to dihydrofolate reductase or thymidylate synthetase (Kozloff, Lute and Crosby 1975). These proteins may be present in less than six copies, be hidden under other bands, or be lost during purification.

The contracted sheath, containing 144 protein subunits, is readily purified because of its resistance to disaggregation. Studies on the amino acid composition, terminal amino acids and optical properties are summarized in Poglazov (1973).

Tail Assembly

Most of our knowledge of tail assembly derives from the characterization of the structural intermediates accumulating in cells infected with mutants blocked in assembly (King 1968, 1971; King and Mykolajewycz 1973). These have been examined by electron microscopy, sedimentation analyses and SDS gel electrophoresis. A fourth and singularly important approach, assembly activity, depends on the ability to assemble the intermediate structures into complete tails (and viable phage) in vitro, as first described by Edgar and Wood (1966). Cells infected with mutants blocked prior to the formation of the structure of interest serve as a source of protein subunits for the completion of the assembly reaction in vitro. This reflects the important fact that when tail assembly is blocked, proteins downstream of the block accumulate in functional unaggregated form. The use of the in vitro complementation assay to determine the sequence of a series of assembly reactions, as shown in Figure 2, is described by Edgar and Lielausis (1968).

Sheath Assembly

Figure 2 shows the pathway for the assembly of the tail tube and tail sheath on the phage baseplate. All of these reactions proceed in vitro. Initiation of

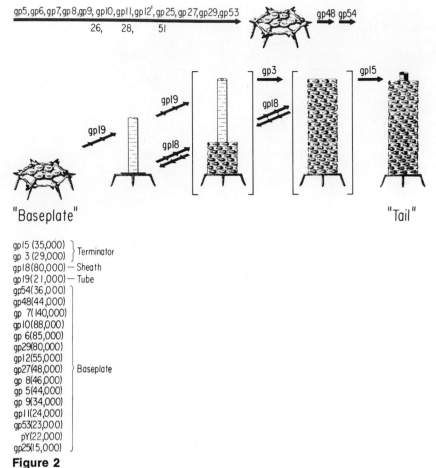

Figure 2

Pathway for T4 tail tube and tail sheath assembly. The structures shown are intermediates in the assembly process. The gene products taking part in each step are shown above the arrows. Brackets around a structure indicate that the structure is unstable outside the cell. At the bottom is a list of the polypeptide chains found in the complete tail, giving molecular weights and genetic identification (gp = gene product). All the reactions shown proceed in vitro (Edgar and Lielausis 1968; King 1971). The figure summarizes data from King and Mykolajewycz (1973) and Kikuchi and King (1975b).

sheath subunit polymerization depends on the presence of a baseplate with tail tube. If tube assembly is blocked by mutation, the sheath subunits accumulate as functional monomers (King 1971; Kikuchi and King 1975a). Addition of tube-baseplates to an extract containing unassembled sheath subunits results in the rapid assembly of the subunits onto the substrate tube-baseplates. Polymerization of the sheath subunits is reversible, indicating that the sheath subunits are only weakly bonded to each other and to the tail tube, in the extended form. Stabilization of the sheath depends on the formation of a special bond between the gene 3 product, located at the tip of the tail tube, and the top annulus of the sheath. This joint is formed of the gene 15 protein and serves as the site for head attachment. Cells infected with mutants defective in either gene 3 or 15 accumulate within the cell tails with polymerized sheath subunits. Upon lysis and dilution of the pool of free subunits, the subunits fall off sequentially from the neck end of these incomplete tails (King 1968). Thus assembly of the sheath can be dissected into three stages: specific initiation dependent on a tube-baseplate; elongation via incorporation of free subunits into the growing structure; and termination via the formation of a special sheath–tube bond.

Note that the injection process depends on the maintenance of the bonds between the bottom annulus of the sheath and the baseplate and between the top annulus of the sheath and the tip of the tail tube. All other sheath annuli must be able to slide past the tail tube. It is unclear whether there is any functional interaction between sheath and tail tube during contraction. Clearly, however, some interaction, probably weak, does take place between them during assembly; sheath subunits do not interact with baseplates lacking a tail tube.

At late times after phage infection, especially in cells with a very large pool of free sheath subunits (for example, infected with a mutant defective in tube assembly), some of the subunits polymerize into an aberrant structure called polysheath (Kellenberger and Boy de la Tour 1964; King 1968). The subunits in polysheath are not tightly bonded to each other (To, Kellenberger and Eisenstark 1969). It is not clear whether they are arranged as in extended sheath, contracted sheath, or some third mode.

Tail Tube Assembly

The polymerization of tail tube subunits depends absolutely on the presence of the baseplate as an initiation structure (King 1971). Cells infected with mutants blocked in baseplate assembly accumulate the normal amount of tail tube protein (King and Laemmli 1973), but this accumulates as free subunits (Fig. 3) rather than as aggregates. In contrast to the sheath polymerization process, tube polymerization is irreversible; incomplete-length tail tubes are stable and can be completed by adding additional subunits (Y. Kikuchi and H. Uchida, unpubl.). The mechanism for determining the length of the tail tube is unclear. We favor a model in which extended protein chains extending from the baseplate specify the extent of tube subunit polymerization. The gene 3 product would then recognize the unique site formed by the end of the length-determiner molecules and the (last) tube annulus.

Precursor and Mature Forms of Tube and Sheath Subunits

Both tail tube and tail sheath subunits share the property of not spontaneously polymerizing into the structures they are designed to form. Both require an initiation structure, and then they polymerize only via incorporation into that structure. This implies that they are synthesized in an essentially nonaggregating form. Incorporation into a substrate structure activates that set of subunits with respect to the soluble pool. As further subunits are incorporated, they are sequentially activated. Such a mechanism—incorporation into a seed structure followed by subunit activation—has been clearly demonstrated by Asakura and colleagues for *Salmonella* flagellin (Uratani, Asakura and Imahori 1972).

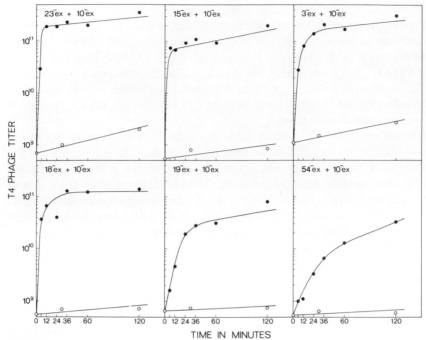

Figure 3

Assembly of the tail tube and tail sheath in vitro in mixtures of extracts of mutant-infected cells. Concentrated extracts were prepared of cells infected with mutants defective in genes for tail assembly. These were incubated together at 30°C and samples taken and plated for viable phage at various times (——•——). The extracts were also incubated alone and these background values averaged (——o——). Gene 23 specifies the major head protein. Thus, 23⁻ extracts accumulate complete phage tails. The gene 10 protein is needed for baseplate assembly. 10⁻ extracts lack tails but contain heads. Thus the upper left panel shows the head-to-tail joining reaction (Edgar and Wood 1966). The other five mixtures require various steps in tail completion to proceed in vitro. In all cases, the 10⁻ extract serves as a source of precursor proteins, and the other member of the pair donates incomplete structures. (Reprinted, with permission, from King 1971.)

Tail tube protein isolated by degradation from mature tail tubes behaves differently from the precursor subunits. This material aggregates spontaneously into tubelike structures (Poglazov and Nikoskaya 1969; To, Kellenberger and Eisenstark 1969; Poglazov 1973). Presumably these subunits are isolated in the "sticky" state rather than in the precursor "nonsticky" state. As noted above, it is extremely difficult to obtain soluble sheath subunits by dissociation of the sheath. In all known degradation processes, the sheath contracts and is then very difficult to dissociate (To, Kellenberger and Eisenstark 1969; Roslansky and Baylor 1970; Poglazov 1973).

The experimental evidence suggests three stable states for sheath subunits: the precursor, soluble form; the assembled, "sticky," but uncontracted form; and the contracted, tightly bound, insoluble form. The state in the aberrant aggregates, polysheath, is unclear. The protein in all four forms has the same molecular weight, as near as can be determined by SDS gel electrophoresis (King and Laemmli 1973). It is interesting to note that early investigators reported that the N termini of the sheath subunits were blocked and that the protein contained sugar (Brenner et al. 1959).

Baseplate Morphogenesis

As noted above, the contraction of the sheath appears to be triggered or generated by a rearrangement of the baseplate. This structural change from hexagonal to star-shaped must represent a mechanism different from the sliding filament type. We have studied the structure and assembly of this wheel-like organelle in detail.

Baseplate Structure

Electron micrographs of negatively stained baseplates are shown in Figure 4a. Their diameters are about 400 Å. These complete baseplates, isolated from cells defective in tail tube formation, have a distinct plug in the center, surrounded by structure that appears to connect the outer rim with the central plug. In computer-filtered images of such baseplates (Fig. 5), additional detail is revealed, including bridges between the outer rim and the inner material (Crowther and Amos 1971). Baseplates occasionally sit on the grid on edge; these images display a shallow, 50×200-Å platform projecting from the face where the tube will form, opposite the spikes.

Baseplates lacking certain proteins—for example, the gene 12 protein associated with the spike face—are unstable if their assembly into phage is blocked. These structures transform to the star shape during preparation for microscopy (Fig. 4b). We presume this mimics the change that occurs in infectious phage during absorption. The star structures generally lack the central plug and have expanded to a diameter of about 600 Å (A. Crowther, Y. Kikuchi, E. Lenk and J. King, unpubl.). A number of groups have studied phage mutants that affect not assembly, but the triggering process in mature phage (Dawes and Goldberg 1973; Kozloff, Lute and Baugh 1973; Yamamoto and Uchida 1975). All these studies indicate that the triggering process involves many proteins of the baseplate and is a multistep process. No mutations are known that affect the ability of the assembled sheath protein to contract. However, these have not been explicitly looked for.

Baseplate Assembly

The mature baseplate consists of at least 15 species of proteins (Fig. 6). Since 14 of these are the products of known phage genes, we can effectively remove them one at a time from infected cells, by mutation, and then examine the state of the remaining proteins. We presume that there are intermediate complexes in the formation of the baseplate, and that they may accumulate in at least some of the mutant-infected cells. Regardless of their organizational state, we can ask whether the proteins accumulate as functional precursors for assembly in vitro. In fact, we find that when concentrated extracts of different baseplate mutant-infected cells are mixed together and incubated, viable phage are formed (Kikuchi and King 1975a). In these mixtures of extracts, baseplates must be assembling from the accumulating proteins or protein

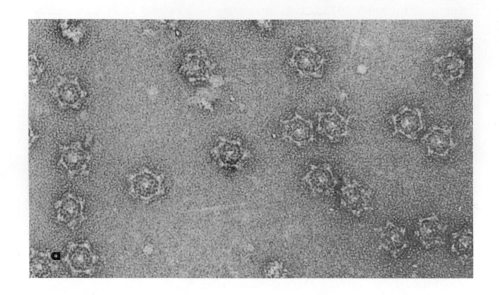

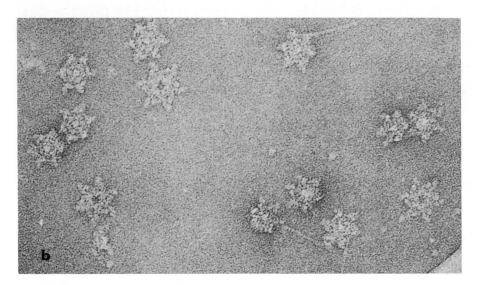

Figure 4

Hexagonal and star-shaped baseplates. (*a*) Complete baseplates, purified by sucrose gradient centrifugation from the lysate of cells infected with an amber mutant of the tail tube gene. The structures are suspended in a film of uranyl acetate over a hole in the carbon film. 200,000×. (*b*) Baseplates isolated from cells lacking both the tail tube protein and the gene 12 product. The gene 12 protein normally adds to the outer face of the baseplate and interacts with the host cell (Kells and Haselkorn 1975). Baseplates lacking gp12 convert to star-shaped structures during preparation for microscopy. Note the hole in the center of the star. 200,000×. A detailed analysis of these structures has been performed using computer filtering procedures (A. Crowther, Y. Kikuchi, E. Lenk and J. King, in prep.).

complexes. We have fractionated these active species by sucrose gradient centrifugation, using their ability to complement mutant extracts in vitro as the assay. That is, extracts prepared from amber mutants defective in each of the baseplate genes have been fractionated by gradient centrifugation. The fractions are then tested for their ability to stimulate phage formation in crude tester extracts prepared from the appropriate amber mutants. These experiments (described in detail in Kikuchi and King 1975a,b,c) have enabled us to define the complete pathway for baseplate assembly (Fig. 7). Here we review the findings.

The baseplate is assembled via two subassembly pathways, one for the formation of outer arms and the other for the formation of the central plug or hub. The sixfold structure of the baseplate represents the polymerization of six arm structures around a central hub.

FORMATION OF THE BASEPLATE ARMS. The major structural precursor of the baseplate is a 15S arm complex formed of the proteins of six clustered genes: 53, 6, 7, 8, 10 and 11. The gene 53 protein is the last protein to be incor-

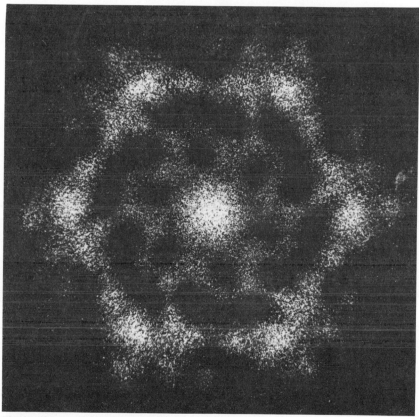

Figure 5
Rotationally filtered T4 baseplate. Negatively stained image of a 19⁻ base-plate that has had noise removed by the computer filtering process described by Crowther and Amos (1971). (Photo courtesy of R. A. Crowther.)

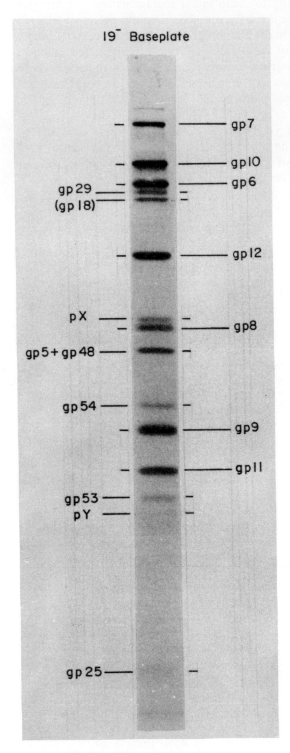

Figure 6 (*See facing page for legend*)

porated into the complex and is required for aggregation of the complexes into a hexagonal array. In its absence, a complex of the remaining proteins accumulates, still sedimenting at 15S. This is shown in Figure 8a. Cells infected with an amber mutant defective in gene 53 have been concentrated, lysed, and centrifuged through a sucrose gradient so as to band structures in the sedimentation range of 15S. Each fraction has been incubated with a 6$^-$ extract, a 7$^-$ extract, an 8$^-$ extract and a 10$^-$ extract. Viable phage are formed in all cases, and the active species sediment together as a complex. In contrast, if the gene 10 protein is removed by mutation, a 15S complex is not found. The 6$^+$, 7$^+$ and 8$^+$ assembly activities can still be recovered (Fig. 8b) but they are now sedimenting much more slowly, not complexed together. Many such experiments reveal the pathway shown in the bottom left of Figure 7. The six proteins interact strictly sequentially. If any protein is removed by mutation, the ones before the block assemble into a precursor complex, whereas the ones downstream remain soluble and functional.

One might question whether the active species recovered from extracts is representative of the state of the total population of that species. For example, Figure 8 shows that the gene 6 product can be recovered as a slowly sedimenting species from a 10$^-$ extract. Perhaps this represents only 10% of the gene 6 protein, and the remainder is in various size aggregates and not active as a precursor for in vitro assembly. By preparing radioactively labeled extracts, fractionating by sedimentation, and assaying the protein in each fraction by SDS gel electrophoresis, we can characterize the behavior of all the molecules of each species, active or not. Such an experiment is shown in Figure 9. The total protein population behaves just like the active species determined by in vitro complementation.

FORMATION OF THE CENTRAL PLUG.   The mature baseplate contains proteins that are not part of the 15S arm precursor. These proteins—for example, gp5, gp27 and gp29—make up a much smaller fraction of the mass of the baseplate than do the arm proteins. Cells infected with mutants defective in these genes accumulate the 15S arm complexes but they also accumulate small amounts of 70S hexagonal structures. When examined in the electron microscope, these appear similar to mature baseplates, except that they lack the central plug (Fig. 10). Such structures are found in cells infected with mutants in the clustered genes 25, 26, 51, 27, 28 and 29. We have not been able to convert these to viable phage in vitro (Kikuchi and King 1975b), and they appear to be aberrant aggregates. This is consistent with a model in which the proteins specified by these second cluster genes are needed for the formation of the central plug, which is needed for proper assembly of the arms.

If extracts prepared from cells infected with mutants of the plug genes are incubated together, viable phage are formed. Thus despite the formation of

Figure 6

Protein composition of complete baseplates. Autoradiogram of ^{14}C-labeled proteins from 19$^-$ baseplates separated by electrophoresis through an SDS-acrylamide gel. Bands are identified according to the phage gene that specifies them (Kikuchi and King 1975a,b). The band pX is in fact the product of gene 27 (Berget and Warner 1975). Molecular weights and stoichiometry are summarized in Table 1.

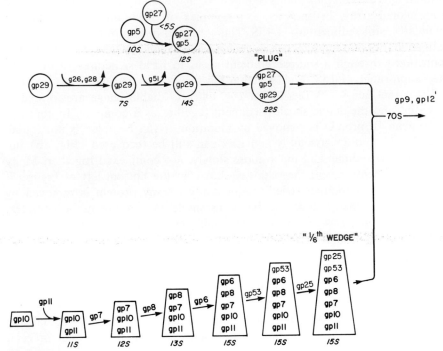

Figure 7

Overall pathway for T4 tail morphogenesis. This figure incorporates all the data on
in vitro assembly activity and protein composition of intermediate structures in tail
assembly. All reactions along the lower pathway proceed in vitro. Thus in cells
infected with an amber mutant of gene 10, all proteins downstream accumulate as
soluble precursors. Incubation with an extract containing gp10 results in the entire
series of reactions proceeding in vitro yielding viable phage. Since the pathway is
branched, the plug proteins assemble in 10⁻-infected cells and accumulate as a
complex. The first set of reactions in the plug pathway (involving the 29, 26 and
28 products) does not proceed in vitro, but all subsequent ones do. Cells infected
with a mutant of gene 51 accumulate all proteins downstream in soluble form. The
arm proteins assemble into 15S complexes, which accumulate. (However, some
fraction of these do aggregate in the absence of the plug to form inactive, dead-end
structures.) Thus the overall character of the assembly process is the synthesis of
the proteins in a nonaggregating form. The sequential nature of the assembly reaction proceeds via stepwise activation of each species as it is incorporated into the
growing structure. (Reprinted, with permission, from Kikuchi and King 1975c.)

some aberrant aggregates, most of the baseplate proteins accumulate in such
cells as functional precursors. We can ask, however, in what state are the
minor (plug) proteins if their incorporation into a baseplate is blocked by a
mutation in arm assembly? Fractionation of the assembly activities shows that
they are associated together as a 22S complex, which we presume represents the
plug itself. Formation of this complex requires the products of genes 26, 51,
27, 28 and 29. Proceeding in the general fashion described above, we fractionated the active species from plug⁻ extracts by centrifugation, to identify
precursor complexes to the plug itself.

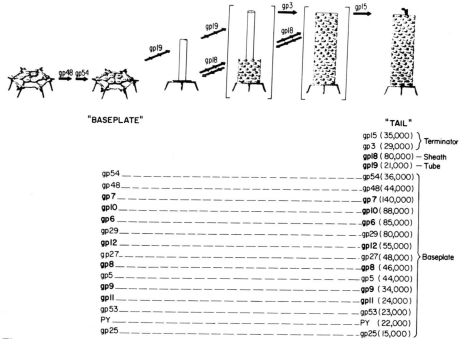

"BASEPLATE"
"TAIL"

gpl5 (35,000) ⎫
gp3 (29,000) ⎬ Terminator
gpl8 (80,000) — Sheath
gpl9 (21,000) — Tube

gp54 ———————————————————————— gp54(36,000)
gp48 ———————————————————————— gp48(44,000)
gp7 ———————————————————————— gp7 (140,000)
gpl0 ———————————————————————— gpl0(88,000)
gp6 ———————————————————————— gp6 (85,000)
gp29 ———————————————————————— gp29(80,000)
gpl2 ———————————————————————— gpl2 (55,000)
gp27 ———————————————————————— gp27(48,000) ⎫
gp8 ———————————————————————— gp8 (46,000) ⎬ Baseplate
gp5 ———————————————————————— gp5 (44,000)
gp9 ———————————————————————— gp9 (34,000)
gpll ———————————————————————— gpll (24,000)
gp53 ———————————————————————— gp53(23,000)
PY ———————————————————————— PY (22,000)
gp25 ———————————————————————— gp25(15,000) ⎭

Figure 7 (*continued*)

The results of many such experiments are summarized in Figure 11. The 22S complex is formed from two smaller complexes, one requiring the products of genes 29, 28, 26 and 51, the other requiring the products of genes 5 and 27. The products of three of the genes of the first group are *not* found in complete baseplates; they presumably function catalytically in assembly (Snustad 1968). These experiments show that the catalytic functions have to do with the maturation of the gene 29 protein. Kozloff, Lute and Baugh (1973) have shown that cells infected with mutants of gene 28 are altered in their synthesis of folic acid. Pteroylhexaglutamate is a structural component of the baseplate (Kozloff et al. 1970a) and is absent from uninfected cells. Taken together, these experiments indicate that folic acid is associated with the gene 29 protein and is required for its maturation.

Since the baseplate has overall sixfold symmetry, we assume that the plug has sixfold symmetry. The conversion of the 7S form of gp29 to a 14S form, under the control of the gene 51 product, might therefore represent polymerization to a hexameric form. The gene 5 and gene 27 proteins may also be forming a hexameric 12S complex. The final reaction in the plug pathway would then be the joining of two hexamers. Alternatively, the last step may be the formation of a mixed hexamer from the 14S and 12S complexes. The gene 25 product is not needed for plug assembly; rather, it must be incorporated into the arm complex if these are to bind to the plug.

The bonds between the tail tube and baseplate must be broken upon infection. Since this is a quite stable joint, special machinery must be built into the baseplate to do the job. We suspect that the disappearance of the plug during the hexagon–star transition of the baseplate reflects this process.

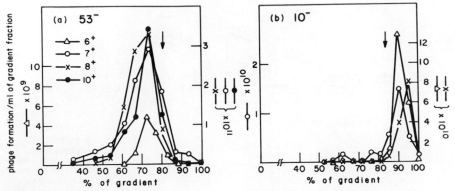

Figure 8

Assay of baseplate precursor proteins by in vitro complementation. Concentrated extracts of mutant-infected cells were centrifuged through 15–30% sucrose gradients for 3 hr at 45,000 rpm at 15°C. Aliquots of each fraction were then incubated with crude tester extracts to determine the sedimentation behavior of the precursor proteins.

(a) Gradient analysis of 53⁻ extract. Gradient fractions were incubated with four different tester extracts: 6⁻ extract to assay for 6⁺ activity; 7⁻ extract to assay for 7⁺ activity; 8⁻ extract to assay for 8⁺ activity; and 10⁻ extract to assay for 10⁺ activity. All four activities were recovered from 53⁻ extract and cosedimented as a 15S complex.

(b) Gradient analysis of 10⁻ extract. The fractions were assayed as described above. In this case, 6⁺, 7⁺ and 8⁺ activities can still be recovered but they are no longer sedimenting together as a complex. Arrows show the position of 34⁺ activity, a tail fiber precursor (King and Wood 1969) which serves as an internal marker. (Reprinted, with permission, from Kikuchi and King 1975a.)

The complexity of the plug, and in fact of the whole baseplate, presumably reflects the complex functions it performs during infection. We suspect that the transformation of the baseplate in fact alters the conformation of the six sheath subunits bound to the baseplate. This conformational change is then propagated and is the motive force of contraction. The studies of Moody (1973), Amos and Klug (1975) and Smith et al. (1976) all support the idea of a major conformational change during contraction.

CONCLUSIONS

These studies reveal that unassembled precursor structural proteins have properties different from those found in the mature structure, despite the absence of detectable protein processing. The proteins within the phage tail are tightly complexed. They are synthesized, however, in a form in which they do not spontaneously aggregate. The activation of these precursor proteins to a reactive form results from their incorporation into a substrate structure. The proteins are not activated in their soluble forms but only during the assembly reactions. These reactions generate new substrate structures capable of incorporating the next protein in the reaction sequence. Thus the assembly

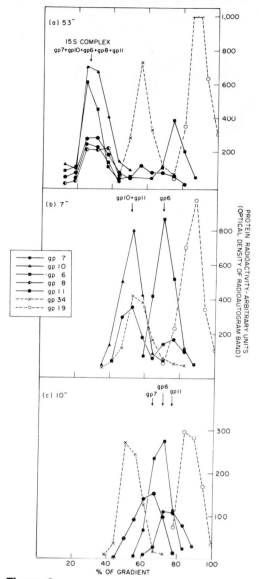

Figure 9

Sedimentation analysis of ^{14}C-labeled baseplate precursor proteins. Radio-actively labeled lysates prepared from mutants defective in genes 53, 7 and 10 were centrifuged through 5–20% sucrose gradients for 7 hr at 45,000 rpm. Each fraction was then heated in SDS and electrophoresed through an SDS-10% acrylamide gel to separate the proteins. Autoradiograms of the gels were then scanned with a microdensitometer to quantitate the distribution of the various tail proteins throughout the gradients. Data for major baseplate proteins are shown in the figure. The dotted lines represent non-baseplate proteins used as internal references. Note, however, that the slowly sedimenting marker (– – –o– – –) is the tail tube protein, showing its nonaggregating character.

The distribution of protein determined by labeling is the same as that determined by in vitro complementation. Thus in the absence of gp10, gp7, gp6 and gp11 all sediment independently of each other. If gp10 is present, as in the 53$^-$ lysate, assembly proceeds, and the proteins accumulate as the 15S arm complex. (Data from Kikuchi and King 1975a.)

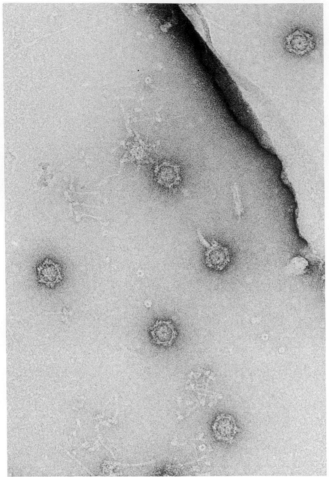

Figure 10
Aberrant, plugless baseplate structures from 27⁻ lysate. Cells infected with
mutants defective in the genes specifying the minor and catalytic baseplate
proteins accumulate small amounts of organized structures. These sediment
similarly to complete baseplates but can be seen to be lacking the central
plug. SDS gel electrophoresis shows that these structures are aggregation
products of the 15S arm complexes. They lack the minor gene products gp5,
gp27, gp29, gp48 and gp54. These structures cannot be converted to viable
phage by incubating with the missing proteins.

pathway is a series of stepwise incorporation/activation steps, with reactive
sites effectively limited to growing structures.

Since the reactivity of the structural subunits is altered by the assembly
reaction, proteins isolated from mature structures cannot a priori be expected
to exhibit the same properties as unassembled precursors. Whether this will
also turn out to be true for proteins that take part in reversible contractile
interactions will depend upon studies of the precursor forms of these proteins.

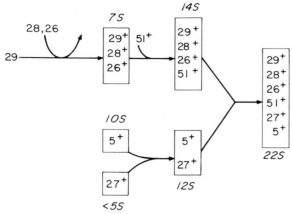

Figure 11

Pathway for plug formation. The figure shows the order of interaction of gene products involved in plug assembly. It was determined by experiments similar to that shown in Figure 8. From such experiments one cannot distinguish whether a particular gene product is actually incorporated into one of the complexes or instead acts catalytically. Since the products of genes 26, 28 and 51 are not found in the mature baseplate, we assume they act catalytically. This is consistent with the gene dosage experiments of Snustad (1968). Kozloff, Lute, and Baugh (1973) presented evidence that the gene 28 product is necessary for the synthesis of phage-specific folic acid conjugate, which is found in the baseplate (Kozloff et al. 1970b). Perhaps the folic acid is necessary for the proper folding of the gene 29 protein or for later steps in its assembly.

Acknowledgments

We thank Elaine Lenk for her expert assistance with electron microscopy. This research was supported by Grant 17,980 from the Institute of General Medical Sciences of the National Institutes of Health.

REFERENCES

Amos, L. and A. Klug. 1975. Three dimensional image reconstructions of the contractile tail of T4 bacteriophage. *J. Mol. Biol.* **99**:51.

Arscott, P. G. and E. B. Goldberg. 1976. Co-operative action of the T4 tail fibers and baseplate in triggering conformational change and in determining host range. *Virology* **69**:15.

Benz, W. C. and E. B. Goldberg. 1973. Interactions between phage T4 adsorption intermediates and the bacterial envelope. *Virology* **53**:225.

Berget, P. and H. Warner. 1975. Identification of P48 and P54 structural components of bacteriophage T4 baseplates. *J. Virol.* **16**:1669.

Bishop, R., M. Conley and W. B. Wood. 1974. Assembly and attachment of bacteriophage T4 tail fibers. *J. Supramol. Struc.* **2**:196.

Brenner, S., G. Streisinger, R. W. Horne, S. P. Champe, S. Benzer and M. W. Rees. 1959. Structural components of bacteriophage. *J. Mol. Biol.* **1**:281.

Capco, G. R. and C. K. Mathews. 1973. Bacteriophage-coded thymidylate synthetase. Evidence that the T4 enzyme is a capsid protein. *Arch. Biochem. Biophys.* **158**:380.

Crowther, A. R. and L. A. Amos. 1971. Harmonic analysis of electron microscope images with rotational symmetry. *J. Mol. Biol.* **60**:123.

Dawes, J. and E. B. Goldberg. 1973. Functions of baseplate components in bacteriophage T4 infection. II. Products of genes 5, 6, 7, 8 and 10. *Virology* **55**:391.

DeRosier, D. J. and A. Klug. 1968. Reconstruction of three-dimensional structure from electron micrographs. *Nature* **217**:130.

Edgar, R. S. 1969. The genome of bacteriophage T4. In *Harvey Lectures*, series 63, p. 263. Academic Press, New York.

Edgar, R. S. and I. Lielausis. 1968. Some steps in the assembly of bacteriophage T4. *J. Mol. Biol.* **32**:263.

Edgar, R. S. and W. B. Wood. 1966. Morphogenesis of bacteriophage T4 in extracts of mutant infected cells. *Proc. Nat. Acad. Sci.* **55**:498.

Epstein, R. H., A. Bolle, E. M. Steinberg, E. Kellenberger, E. Boy de la Tour, R. Chevalley, R. S. Edgar, M. Susman, G. H. Denhardt and A. Lielausis. 1964. Physiological studies of conditional lethal mutants of bacteriophage T4D. *Cold Spring Harbor Symp. Quant. Biol.* **28**:375.

Hosoda, J. and C. Levinthal. 1968. Protein synthesis in *Escherichia coli* infected with bacteriophage T4D. *Virology* **34**:709.

Kellenberger, E. and E. Boy de la Tour. 1964. On the fine structure of normal and "polymerized" tail sheath of phage T4. *J. Ultrastruc. Res.* **11**:545.

Kells, S. and R. Haselkorn. 1974. Bacteriophage T4 short tail fibers are the products of gene 12. *J. Mol. Biol.* **83**:473.

Kikuchi, Y. and J. King. 1975a. Genetic control of bacteriophage T4 baseplate morphogenesis. I. Sequential assembly of the major precursor, *in vivo* and *in vitro*. *J. Mol. Biol.* **99**:645.

———. 1975b. Genetic control of bacteriophage T4 baseplate morphogenesis. II. Mutants unable to form the central part of the baseplate. *J. Mol. Biol.* **99**:673.

———. 1975c. Genetic control of bacteriophage T4 baseplate morphogenesis. III. Formation of the central plug and overall assembly pathway. *J. Mol. Biol.* **99**:695.

King, J. 1968. Assembly of the tail of bacteriophage T4. *J. Mol. Biol.* **32**:231.

———. 1971. Bacteriophage T4 tail assembly: Four steps in core formation. *J. Mol. Biol.* **58**:693.

King, J. and U. K. Laemmli. 1971. Polypeptides of the tail fibers of bacteriophage T4. *J. Mol. Biol.* **62**:465.

———. 1973. Bacteriophage T4 tail assembly: Structural proteins and their genetic identification. *J. Mol. Biol.* **75**:315.

King, J. and N. Mykolajewycz. 1973. Bacteriophage T4 tail assembly: Proteins of the sheath, core and baseplate. *J. Mol. Biol.* **75**:339.

King, J. and W. B. Wood. 1969. Assembly of bacteriophage T4 tail fibers: The sequence of gene product interaction. *J. Mol. Biol.* **39**:583.

Kozloff, L. M., M. Lute and C. M. Baugh. 1973. Bacteriophage tail components. V. Complementation of T4D gene 28⁻ infected bacterial extracts with pteroylhexaglutamate. *J. Virol.* **4**:637.

Kozloff, L. M., M. Lute and L. K. Crosby. 1975. Bacteriophage baseplate components. I. Binding and location of the folic acid. *J. Virol.* **16**:1391.

Kozloff, L. M., C. Verses, M. Lute and L. K. Crosby. 1970a. Bacteriophage tail components. II. Dihydrofolate reductase in T4D bacteriophage. *J. Virol.* **5**:740.

Kozloff, L. M., M. Lute, L. K. Crosby, N. Rao, V. A. Chapman and S. S. DeLong.

1970b. Bacteriophage tail components. I. Pteroyl polyglutamates in T-even bacteriophages. *J. Virol.* **5**:726.

Laemmli, U. K. 1970. Cleavage of structural proteins during the assembly of the head of bacteriophage T4. *Nature* **227**:680.

Male, C. J. and L. M. Kozloff. 1973. Function of T4D structural dihydrofolate reductase in bacteriophage infection. *J. Virol.* **11**:840.

Moody, M. F. 1967a. Structure of the sheath of bacteriophage T4. I. Structure of the contracted sheaths and polysheaths. *J. Mol. Biol.* **25**:167.

————. 1967b. Structure of the sheath of bacteriophage T4. II. Rearrangement of the sheath subunits during contraction. *J. Mol. Biol.* **25**:201.

————. 1971. Application of optical diffraction to helical structures in the bacteriophage tail. *Phil. Trans. Roy. Soc. London B* **261**:181.

————. 1973. Sheath of bacteriophage T4. III. Contraction mechanism deduced from partially contracted sheaths. *J. Mol. Biol.* **80**:613.

Poglazov, B. F., ed. 1973. *Morphogenesis of T-even bacteriophage. Monographs in developmental biology,* vol. 7. S. Karger, Basel.

Poglazov, B. F. and T. J. Nikolskaya. 1969. Self assembly of the proteins of bacteriophage T2 tail cores. *J. Mol. Biol.* **43**:231.

Roslansky, P. F. and M. B. Baylor. 1970. Studies of the proteins of contracted sheath of T-even coliphage. *Virology* **40**:260.

Simon, L. D. and T. F. Anderson. 1967a. The infection of *Escherichia coli* by T2 and T4 bacteriophages seen in the electron microscope. I. Attachment and penetration. *Virology* **32**:279.

————. 1967b. The infection of *Escherichia coli* by T2 and T4 bacteriophages seen in the electron microscope. II. Structure and function of the baseplate. *Virology* **32**:298.

Simon, L. D., J. Swan and J. Flatgaard. 1970. Functional defects in T4 bacteriophages lacking the gene 11 and gene 12 products. *Virology* **41**:77.

Smith, P. R., U. Aebi, R. Josephs and M. Kessel. 1976. Concerning the recovery of structural information from electron micrographs of the extended tail sheath of bacteriophage T4. *J. Mol. Biol.* (in press).

Snustad, D. P. 1968. Dominance interactions in *Escherichia coli* cell mixedly infected with bacteriophage T4D wild-type and amber mutants and their possible implications as to type of gene-product function: Catalytic vs. stoichiometric. *Virology* **35**:550.

To, C. M., E. Kellenberger and A. Eisenstark. 1969. Disassembly of T-even bacteriophage into structural parts and subunits. *J. Mol. Biol.* **46**:493.

Uratani, Y., S. Asakura and K. Imahori. 1972. A circular dichroism study of *Salmonella* flagellin: Evidence for conformational change on polymerization. *J. Mol. Biol.* **67**:85.

Vanderslice, R. W. and C. A. Yegian. 1974. Identification of late bacteriophage T4 proteins on sodium dodecyl sulfate polyacrylamide gels. *Virology* **60**:265.

Yamamoto, M. and H. Uchida. 1975. Organization and function of the tail of bacteriophage T4: II. Structural control of tail contraction. *J. Mol. Biol.* **92**:207.

Ward, S. and R. C. Dickson. 1971. Assembly of T4 tail fibers. III. Genetic control of the major tail fiber polypeptides. *J. Mol. Biol.* **62**:479.

New Calcium-binding Contractile Proteins

**Lewis M. Routledge, William B. Amos, Foch F. Yew
and Torkel Weis-Fogh**

Department of Zoology, University of Cambridge
Cambridge CB2 3EJ, England

The extremely rapid contraction of the vorticellid ciliates has been a curiosity for several hundred years (van Leeuwenhoek 1676; Engelmann 1875). High-speed films of *Vorticella, Carchesium* and *Zoothamnium* give values of between 2 msec and 10 msec for the duration of the contraction (Jones, Jahn and Fonseca 1970; Amos 1975). This is equivalent to a maximum contraction rate of 200 lengths/second, in contrast to 22 lengths/second of the very rapidly contracting mouse finger muscle (Close 1965).

The contractile organelle of the vorticellid ciliates is the spasmoneme (Fig. 1), which has a birefringence similar in magnitude to that of the A band in striated muscle. However, the spasmoneme's birefringence falls close to zero during contraction (Schmidt 1940; Amos 1971), whereas that of the A band remains high. In the electron microscope, the spasmoneme consists largely of filaments, 2–4 nm in diameter (Fig. 2b), which presumably give rise to the birefringence (Amos 1972, 1975; Allen 1973), and membraneous tubules (Fig. 2a), which are thought to store calcium (Carasso and Favard 1966).

Physiological studies showed that the contraction and reextension were dependent only on the calcium ion concentration and could not be poisoned by a wide range of metabolic inhibitors (Hoffmann-Berling 1958; Amos 1971). Also in contrast to muscle, magnesium ions and ATP are not required for the contraction. In the spasmoneme, the change in calcium ion concentration appears to drive the contraction rather than to merely initiate it (Amos 1971). Recently it has been possible to measure directly the amount of calcium bound by the spasmoneme during the contraction (Routledge et al. 1975). In this paper we wish to present our current analysis of the spasmoneme proteins from *Vorticella, Carchesium* and *Zoothamnium*. These proteins show a remarkable consistency in composition, and their properties suggest how this new mechanism of motility may function.

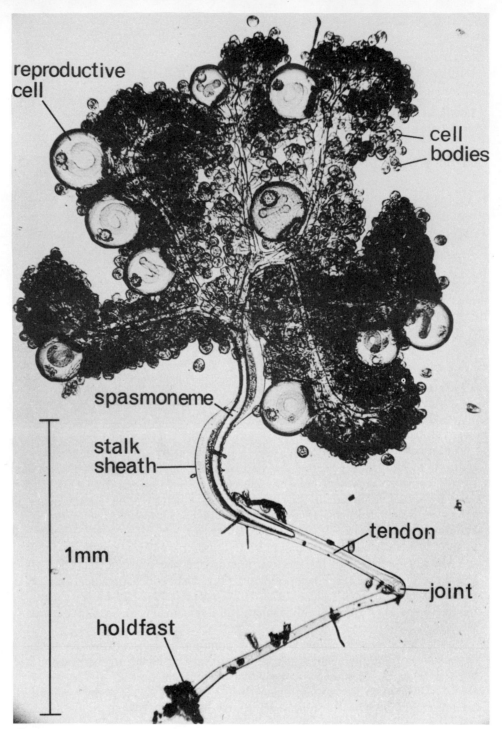

Figure 1

A partly contracted colony of *Zoothamnium geniculatum* after treatment with 2% glutaraldehyde. Several hundred cell bodies (zooids) are connected to contractile branches that unite into the common contractile stem in which the refractile spasmoneme is clearly visible. The cytoplasmic extension surrounding the spasmoneme is seen as a gray envelope. Other longitudinal myonemes can be seen as striations within the reproductive cells. (Photomicrograph taken by W. B. Amos, reprinted, with permission, from Weis-Fogh 1975).

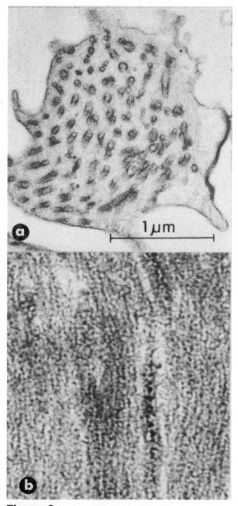

Figure 2
(*a*) Electron micrograph of a nearly transverse thick section of the spasmoneme of *Vorticella*, showing dispersed membraneous tubules, which are believed to store calcium. (*b*) Electron micrograph of a longitudinal section of the spasmoneme of *Zoothamnium*, showing longitudinal filaments that have a beaded appearance similar to the glycerinated and negatively stained material. 56,000×.

EXPERIMENTAL PROCEDURES

Specimen Preparation

Samples of *Vorticella convallaria, Carchesium polypinum and Zoothamnium geniculatum* Ayrton were collected in East Anglian rivers, cultured in sterile river water with *Aerobacter aerogenes* as food for several generations, or glycerinated directly after collection.

The vorticellids to be glycerinated were washed briefly in distilled water and allowed to settle slowly in vials, through distilled water, into a deeper layer of solution containing 50% v/v glycerol, 100 mM potassium chloride, 2 mM EDTA, and 10 mM histidine buffer, adjusted to pH 7.0 with potassium hydroxide. Glycerination was performed for 2 hours at 0–2°C, and the vials were then stored in a freezer at $-20°C$ for 1 to 3 months.

Electron Microprobe Measurements

The giant spasmoneme of the main stalk of each glycerinated colony of *Z. geniculatum* was removed by dissection in the glycerol medium (Weis-Fogh and Amos 1972). The organelle was washed with two changes of a solution containing 50 mM KCl, 20 mM Tris-HCl buffer pH 7.0 and an EGTA buffer adjusted to give a free calcium ion concentration of 10^{-8} M. The ratio of calcium chloride to EGTA was adjusted to give the required free calcium ion concentration using the constants given by Sillén and Martell (1964) and tabulated by Amos et al. (1976). At a free calcium ion concentration of 10^{-6} M, the spasmoneme contracts. Contracted or extended organelles, containing the same total calcium concentration (including the Ca-EGTA complex) but with free Ca^{++} of 10^{-6} and 10^{-8} M, respectively, were placed on collets bearing an aluminized nylon film (Hall, Roeckert and Saunders 1972). Excess solvent was removed and the specimen coated with a film of aluminum under vacuum.

Analysis was carried out in a JEOL JXA X-ray microanalyzer. The calcium $K\alpha$-radiation was selected by means of a diffracting crystal. The continuum X-ray radiation was used as a measure of local mass per unit area, and quantitative analysis was performed as described elsewhere (Hall, Anderson and Appleton 1973). The calcium standards were ultrathin sections of mineral apatite embedded in methacrylate.

Extraction of Spasmoneme Proteins

For biochemical experiments in which the spasmoneme alone was required, the organelles were dissected individually from glycerinated colonies of *Zoothamnium*. A higher yield was obtained by removing the zooids (cell bodies) from the colonies and then extracting the spasmonemal material chemically from the remaining stalks. To remove the zooids, the colonies were subjected to a shearing process in 0.02% SDS by repeated pipetting. The spasmoneme proteins were then solubilized either in 2% SDS with 0.06 M Tris buffer at pH 6.8 or in 3 M guanidine hydrochloride.

Before electrophoresis, the guanidine hydrochloride extract was dialyzed against distilled water or buffer in a Colover ultramicrodialysis cell (Electrothermal Engineering, London). The dialyzed protein was then concentrated to a volume of 5–25 μl in the same apparatus by absorbing the water into 30% polyethylene glycol (MW 20,000). Both operations were performed at 4°C.

Electrophoresis

Polyacrylamide gels were run using a microslab gel apparatus (Amos 1976). Small samples, from as few as two colonies of *Zoothamnium,* each containing 0.7 μg of spasmonemal protein, could easily be analyzed. Discontinuous SDS gels were made according to the formulae of Laemmli and Favre (1973). Discontinuous non-SDS gels were run according to Amos, Routledge and Yew

(1975). In these gels, the calcium ion concentration was controlled by the addition of EGTA.

The gels were stained with either 0.2% Coomassie blue or fast green in 50% methanol with 7% acetic acid and destained in 5% methanol with 7% acetic acid. They were scanned directly with a Joyce-Loebl microdensitometer, and the relative amounts of protein were estimated from the areas under the peaks.

Amino Acid Analysis

In order to obtain proteins for amino acid analysis, 15% polyacrylamide gels were frozen and cut into 0.3-mm thick sections in a Joyce-Loebl gel slicer. The position of the proteins was determined either by staining with Coomassie blue before sectioning or by staining a parallel gel. Microgram quantities of protein were eluted from the gel slices with two changes of 1% SDS. The resulting solution was combined with nine times its volume of cold acetone and the precipitate collected.

The precipitate was hydrolyzed in 6 N HCl containing 0.001 M thioglycollic acid, in sealed tubes under nitrogen, for 20 hours at 110°C. The thioglycollic acid concentration was increased to 2% and the tubes sealed in vacuum ($<$50 μm mercury) for tryptophane determinations. Cysteine was oxidized to cysteic acid by performic acid for determination.

Because only microgram quantities of purified protein were available, a sensitive fluorescence technique (Roth and Hampai 1973) was used to detect amino acids being eluted from the continuous, buffer gradient, chromatography column (Amos, Routledge and Yew 1975).

Birefringence

The glycerinated and dissected spasmonemes were placed in a simple stretching apparatus constructed on a microscope slide. The stretched spasmoneme was infiltrated with solutions of different refractive index and the birefringence measured using a λ/30 compensator of the Brace-Kohler type. In order to avoid the possibility of protein denaturation, organic solvents were not used as imbibition media; instead, aqueous sucrose solutions and water/glycerol mixtures (up to absolute glycerol) were used. The free calcium ion concentration was controlled in all solutions (except 100% glycerol) by the use of EGTA buffers.

RESULTS

Calcium Binding to the Spasmoneme

When the concentration of free calcium ion was changed from 10^{-8} to 10^{-6} M, the glycerinated spasmoneme from *Zoothamnium* contracted to at least 60% of the extended length. This contraction was completely reversible. The organelle from *Z. geniculatum* is large (1 mm $\times$ 30–40 μm) when compared to that of a large *Vorticella* (1 mm $\times$ 1 μm). It can be handled with fine forceps, and its form was well preserved during preparation for microprobe analysis. The organelle could be seen clearly in both transmission and back-

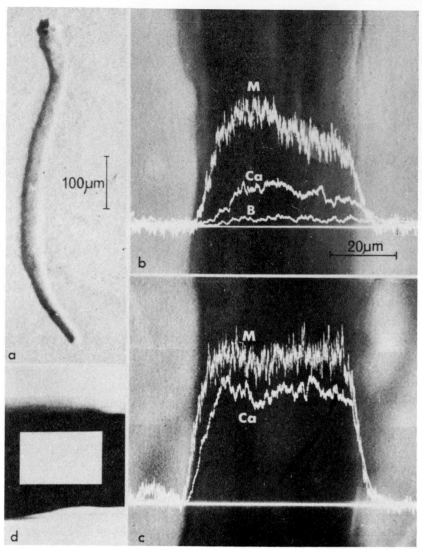

Figure 3

(*a*) Secondary electron image of a spasmoneme formed in the JEOL JXA-50A microanalyzer. The specimen was dried down from a solution containing 10^{-8} M free calcium ions and is extended. (*b*) Counting rates recorded during a linear scan across an extended spasmoneme and the supporting film. M is the total dry mass fraction; Ca, the calcium mass fraction; B, the background component (see text for details). The total concentration of calcium, including Ca-EGTA, was 0.2 mM. (*c*) As above, but from a contracted spasmoneme. It is evident that the calcium mass fraction is much greater in this case. (*d*) Transmission image of a spasmoneme with a superimposed image of the raster covering a rectangular median area of the specimen. The area is $40 \times 26 \ \mu\text{m}^2$. (Reprinted with permission, from Routledge et al. 1975.)

98

scattered electron images (Fig. 3a). It was easy to position the organelle so that a rectangular median area could be selected for analysis (Fig. 3d).

Line scans were also made across the organelle. In Figure 3b, the horizontal line is the line of scan and is also the baseline for the other traces, after subtraction of the background from the supporting film. In the extended organelle (Fig. 3b), the upper trace M is proportional to the total mass present in the line of scan measured by an energy-dispersive silicon detector. Trace Ca is the signal obtained by a diffracting spectrometer set at peak Kα-radiation for calcium. Trace B is the signal from a diffracting spectrometer offset from the Ca Kα peak to record the background component. The amount of calcium present in the spasmoneme is given by the area between the traces Ca and B. In the contracted spasmoneme (Fig. 3c), the background trace B was not significantly different from that obtained with the extended organelle and is omitted. Though the scales for the calcium and mass fraction counts are arbitrary, it can clearly be seen that the ratio of calcium to dry mass in the contracted organelle is much higher than in the extended organelle, even though both organelles were dried down in a calcium buffer containing the same amount of total calcium.

It is difficult to quantitate the calcium mass fraction from line scans because of the low level of X-ray counts from the small area involved. For quantitative measurements, small rectangular areas were selected along the organelle (Fig. 3d) and scanned by an electron beam moving on a raster.

The amount of calcium in the contracted and extended spasmoneme at varying total calcium concentrations (including Ca-EGTA) is shown in Figure 4. Each point shows the mean and standard error of measurements for five or six spasmonemes, with four separate areas assayed within each spasmoneme.

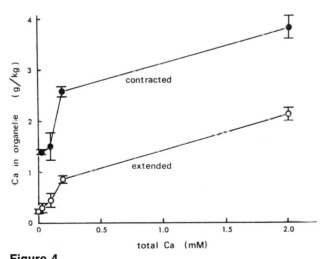

Figure 4
Variation in calcium mass fraction of spasmonemes with total calcium concentration of the bathing solution. The free calcium ion concentration was adjusted to 10^{-6} M to produce contraction or to 10^{-8} M to produce extension. The standard error limits are indicated. (Reprinted, with permission, from Routledge et al. 1975.)

Table 1

Microprobe Measurements of Calcium in the Spasmoneme of
Zoothamnium

Total Ca (μM)	Ca in organelle (g/kg dry mass)		Ca bound during contraction (g/kg dry mass)
	contracted	extended	
0	—	0.21(0.04)	—
21	1.39(0.06)	0.27(0.07)	1.12(0.09)
104	1.58(0.27)	0.46(0.13)	1.12(0.30)
208	2.59(0.09)	0.89(0.06)	1.70(0.11)
2075	3.81(0.22)	2.12(0.12)	1.69(0.25)

Data from Routledge et al. (1975). The figures in parentheses are standard errors.

The calcium was presumed to arise from two sources: binding of calcium to sites within the organelle and accretion principally of Ca-EGTA during drying down. The effect of accretion should be directly proportional to the total calcium concentration in the buffer. The near linear decrease in calcium content of the extended spasmoneme with decreasing total calcium concentration in the buffer (Fig. 4) suggests that the calcium content of the extended organelle is indeed an accretion artifact. The difference in calcium content between contracted and extended organelles remains approximately constant and must represent a true calcium-binding. The amount of calcium bound (Table 1) over a 100-fold total variation in total calcium (20 μM–2 mM) was between 1.12 and 1.70 g calcium per kilogram dry mass of the spasmoneme. In order to relate these results to the hydrated state of the organelle, the dry mass concentration of a hydrated, glycerinated spasmoneme was measured in an interference microscope and was found to be 21 g/100 ml. The calcium binding therefore corresponds to an upper limit of 0.36 g/kg spasmoneme wet mass.

SDS Electrophoresis of Spasmoneme Proteins

Spasmonemes from *Zoothamnium geniculatum* were dissected free from other parts of the colonies and their composition examined. The isolated organelle showed only a feeble periodic acid-Schiff reaction and probably consists almost entirely of protein. It is insoluble in KCl solutions but dissolves completely in 1% SDS and almost entirely in 9 M urea or 3 M guanidine hydrochloride. After extraction in 9 M urea or 3 M guanidine hydrochloride, no further protein could be extracted by SDS; thus the "ghost" after nondetergent extraction may be traces of carbohydrate or lipid.

The solubilized spasmoneme extracts were subjected to electrophoresis on 15% polyacrylamide-SDS gels in parallel with chick brain tubulin and rabbit skeletal muscle actin in order to determine the molecular weights of the proteins present and their possible relationship to other contractile systems (Fig. 5). No significant differences in the gel pattern were observed between

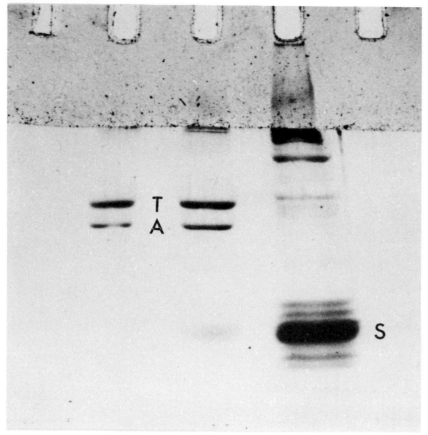

Figure 5

A 15% polyacrylamide microslab gel in which electrophoresis has been carried out on three samples. The right-hand one is a solution obtained by dissolving ten *Zoothamnium* spasmonemes in SDS; on the left is a sample containing 0.2 μg each of actin (A) and tubulin (T); in the center is mixture containing actin, tubulin and a small quantity of the spasmoneme sample. No band from the spasmoneme corresponds precisely to either tubulin or actin. The spasmin band (MW 20,000) is prominent (S). (Reprinted with permission, from Amos, Routledge and Yew 1975.)

the detergent-solubilized and the dialyzed, nondetergent-solubilized spasmoneme material. The resulting pattern (Figs. 5, 6) showed a prominent band, corresponding to a molecular weight of 20,000, that contained 50–60% of the total stainable material. Adding mercaptoethanol to reduce labile disulfide bonds or preheating the sample did not affect this prominent band and affected the remaining proteins only in minor details. This prominent band consists of two components (A and B) and several minor satellite peaks (Figs. 5, 6). No peaks corresponding to actin or tubulin were present. A minor band (3%) was present at 58,000, although this is probably not tubulin but instead an aggregate of an unknown protein since the peak was diminished by preheating. Most of the remaining protein had a molecular weight above 90,000 and remained close to the origin on 15% polyacrylamide gels. The proteins extracted from the complete stalk or the whole zooid-free *Zoothamnium* colony (the stalk sheaths being insoluble) showed no additional major components. This showed that it was not necessary to dissect out the organelle to obtain the spasmoneme proteins, and the method was extended to *Carchesium* and *Vorticella,* in which the spasmonemes are too small to dissect. The proteins from several glycerinated, zooid-free colonies of *Carchesium* were

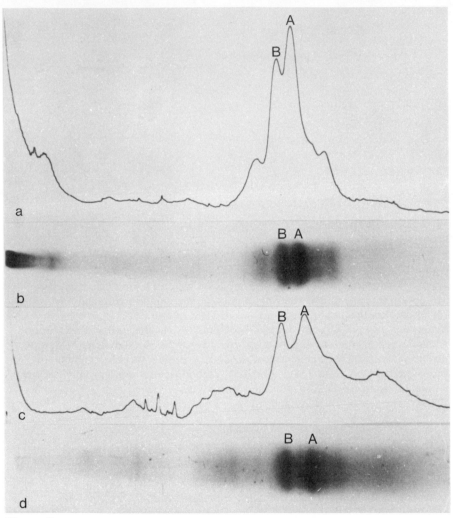

Figure 6

(*a,c*) Densitometer traces of the polyacrylamide gels *b* and *d*. (*b*) Proteins from five glycerinated spasmonemes of *Zoothamnium*. (*d*) Pattern from ten glycerinated colonies of *Carchesium* from which the cell bodies had been removed. The two samples were run in parallel in the same microslab. The 20,000 MW components, spasmins A and B, are labeled. Migration during electrophoresis was from left to right (anodal).

coelectrophoresed with dissected spasmonemes from *Zoothamnium* (Fig. 6). Again 50–60% of the proteins from *Carchesium* were in the 20,000 molecular weight range and consisted of two major bands with minor satellite peaks. The other proteins present in *Carchesium* showed a similar distribution of molecular weights but were not obviously identical to any in *Zoothamnium*. Several hundred stalks from glycerinated *Vorticella* were extracted with 2 μl SDS (2%) and electrophoresed. The major spasmoneme proteins in *Vorticella* are similar to those of *Carchesium* and *Zoothamnium* and comprise 50–60% of the total extractable protein.

In both *Zoothamnium* and *Carchesium,* the ratio of the two major 20,000 MW components did not vary with different extraction techniques. They are probably nonidentical proteins and not breakdown products of a single protein. To test for autolysis, colonies of *Carchesium* were glycerinated in the presence of several protease inhibitors, including *p*-tosyl arginine methyl ester, diisopropylfluorophosphate and sodium mersalyl at a concentration of 10 mM, after which they were still able to contract and extend normally in calcium buffers. The proteins were then extracted in SDS and run on 15% polyacrylamide-SDS gels. There were no changes in the ratio of the two major 20,000 MW components and only small changes in some of the high molecular weight proteins. Therefore the presence of the two 20,000 MW components (A and B) is characteristic of the organelle and is probably essential to the contractile function.

Calcium Binding to the Spasmoneme Proteins

When electrophoresed on a non-SDS 15% polyacrylamide gel at pH 8.8, the nondetergent-extracted spasmoneme proteins gave a pattern similar to those shown in Figure 7, with a very fast migrating component A and a slightly slower component B. When electrophoresed in a second dimension in a 15% polyacrylamide-SDS gel, components A and B corresponded to the two major 20,000 MW proteins, with A having a lower molecular weight than B.

On a non-SDS gel at pH 8.8, the 20,000 MW components migrated faster (anodally) than myoglobin (pI 6.9) and α-casein (pI 4.9), which are in the same molecular weight range. This indicates that the spasmoneme proteins A and B are acidic. This was confirmed by isoelectric focusing, which showed two major bands between pH 4.7 and pH 4.8. The amino acid composition (Table 2) of both A and B indicates a high percentage of possible acidic residues, as opposed to basic residues. The absolute number of net negative charges on the two components is difficult to estimate for several reasons. There is a large standard error in the analysis because of the small quantity of protein available; also, the technique did not distinguish between aspartic acid and asparagine or glutamic acid and glutamine.

Again because of the small quantities of material available, a direct assay of calcium binding could not be made. In order to test whether the 20,000 MW proteins bind calcium, their electrophoretic mobilities were compared at different calcium levels. It was convenient to have the separation gel at pH 8 in order to use the discontinuous buffer system, and at this pH, the glycerinated spasmoneme is still able to bind calcium and contract. At or below

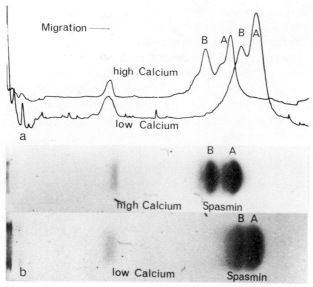

Figure 7

The effect of free calcium ion concentration on the electrophoretic mobility of spasmoneme proteins. (*a*) Superimposed densitometer tracings of stained polyacrylamide gels at high (10^{-6} M free calcium) and low (10^{-8} M free calcium) calcium. The total calcium ($Ca^{++} + Ca\text{-}EGTA$) in both gels was 1 mM. At high calcium, the migration of spasmins A and B was reduced, and both show prominent trailing shoulders.

10^{-8} M free calcium in the polyacrylamide gel, the proteins A and B migrate close to the dye front as two separate bands (Fig. 7, low calcium). When the free calcium concentration was increased to 10^{-6} M, the mobility of proteins A and B was decreased, and both proteins showed a prominent secondary band as a shoulder on the main band (Fig. 7, high calcium). The mobility of a more slowly migrating minor component (58,000 MW) was not altered by the change in the free calcium concentration.

The change in free calcium from 10^{-8} to 10^{-6} M is in the same range as that in which the calcium-binding regulatory protein of skeletal muscle, troponin C, is known to function (Hartshorne and Pyun 1971). Rabbit muscle troponin C was examined for calcium-binding in the calcium-buffered gel system. At 10^{-8} M calcium, troponin C (18,500 MW) migrated with a similar mobility to the 20,000 MW proteins from the spasmoneme. When the free calcium concentration was increased to 10^{-6} M, the electrophoretic mobility of troponin C increased, in contrast to the decrease in electrophoretic mobility of the spasmoneme proteins. To test further the specificity of calcium-binding, experiments were conducted in which gels buffered for Mg^{++} by means of EDTA were used. No change in the electrophoretic mobility of the 20,000 MW spasmoneme proteins or of troponin C occurred with up to 10^{-5} M Mg. A decrease in mobility of the spasmoneme proteins, however, was observed with Mg^{++}, added to the gel as a simple solution of magnesium chloride, at 5×10^{-4} M.

Table 2

Amino Acid Composition of the 20,000 MW Components from the Spasmoneme of *Zoothamnium*

Amino acid	Fast band (A)		Slow band (B)	
		S.E.		S.E.
Aspartic acid + asparagine	12.48	1.24	11.30	2.03
Threonine	7.95	0.17	6.20	0.94
Serine	15.05	4.15	12.58	2.25
Glutamic acid + glutamine	8.57	1.09	7.01	0.50
Proline	6.33	1.47	5.83	1.38
Glycine	9.32	2.07	8.90	2.08
Alanine	6.87	0.89	6.99	1.11
Valine	3.98	0.47	4.15	0.74
Cystine	0		0	
Methionine	0		0	
Isoleucine	2.82	0.26	4.08	0.53
Leucine	4.34	0.41	4.92	1.11
Tyrosine	3.89	0.80	3.30	0.76
Phenylalanine	1.95	0.37	2.20	0.74
Tryptophane	1.15	0.23	1.77	0.36
Lysine	6.95	1.10	7.85	1.35
Histidine	3.74	0.71	7.63	1.64
Arginine	4.59	0.71	6.33	2.42

Data from Amos, Routledge and Yew 1975. The results are expressed as residues per 100 residues. Proline was determined separately from the other amino acids. Each value is the average of six determinations. Standard errors are given.

Birefringence

When the extended glycerinated spasmoneme was stretched, the birefringence increased (Weis-Fogh and Amos 1972). Figure 8 shows the effect on the birefringence of the spasmoneme, at constant extension and 10^{-8} M free calcium, of changing the refractive index of the solvent. At 70% glycerol, the spasmoneme was still able to contract and reextend, in 10^{-6} and 10^{-8} M free calcium, respectively. Even after exposure to 100% glycerol, the spasmoneme behaved normally, and the birefringence curve was fully reversible.

The decrease in birefringence with increasing refractive index indicates that part of the optical anisotropy is due to *form birefringence*. Form birefringence is defined as that due to the discontinuities in refractive index between the solvent and oriented, but optically isotropic, structures. Where birefringence is due to polarizable bonds it is termed *intrinsic birefringence* and is not altered by immersion in high refractive index media. The birefringence of oriented isoprene units in rubber or of oriented regions of polypeptide chain in proteins is of the intrinsic type.

These results indicate that the major contribution to the birefringence in the spasmoneme is due to oriented structures. These may be the 2–4-nm diameter

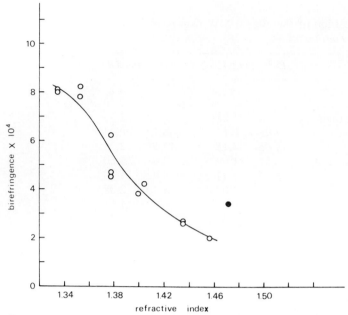

Figure 8

Variation of birefringence with the refractive index of the solution in which it is placed of an isolated, glycerinated spasmoneme of *Zoothamnium*. The solutions contained 0.1 mole/liter KCl and a Ca-EGTA buffer at 10^{-8} M Ca^{++}, together with various concentrations of glycerol up to 86% v/v. The specimen was clamped at a constant, moderately stretched length, and the width remained almost constant. (•) Measurement made in pure glycerol, in which the specimen shrank to 68% of its normal width (W. B. Amos, unpubl.).

filaments observed in the electron microscope, although elongated subunits in the filaments could also contribute to form birefringence.

DISCUSSION

Biochemical Composition in Relation to Other Contractile Systems

The fact that actin could not be detected (Fig. 5) and, if present, is less than 1% of the total proteins as measured by densitometry of the SDS-polyacrylamide gels is clear evidence that the mechanism of contraction is different from that of muscle. The power output of the spasmoneme during contraction exceeds that of muscle (Amos 1971; Weis-Fogh 1975), which makes it unlikely that trace amounts of actin or of any other proteins could be responsible for contraction. In the myonemes of *Stentor,* which are ultrastructurally similar to the spasmoneme of vorticellids (Bannister and Tatchell 1968), the microfilaments are unable to bind heavy meromyosin and are unlikely to be actin (Kristensen, Engdahl Nielsen and Rostgaard 1974).

Tubulin is absent from the electrophoretic pattern, and few microtubules have been found in the stalks of vorticellids (Amos 1972; Allen 1973). Some microtubules in the cytoplasm around the spasmoneme may be lost during glycerination in the cold, and others are probably lost when the spasmonemes are isolated by dissection.

As neither tubulin nor actin were sufficiently abundant to be responsible for contraction, our attention turned to the major protein components: the 20,000 MW bands. These 20,000 MW components are a constant feature of the spasmonemes of the three vorticellid genera examined, *Vorticella, Carchesium* and *Zoothamnium,* which are morphologically quite distinct from one another. In all three vorticellids, there are two major 20,000 MW bands, A and B, and a number of minor satellite bands.

The high molecular weight components, which comprise 30% of the spasmoneme proteins, are present in all three vorticellids, though the absolute amount can vary with different methods of glycerination. Their function is unknown. They are not acidic and do not appear to bind calcium. In *Vorticella* glycerinated for less than 1 month, the addition of calcium and equimolar magnesium and ATP causes cyclic contraction and extension of the stalks. This effect is abolished by ATPase inhibitors, which do not, however, affect the contraction and extension due to changes in calcium concentration (Hoffmann-Berling 1958; Amos 1971). Thus proteins with Mg-ATPase activity, which differ from the contractile proteins, are present in the spasmoneme. Histochemical studies suggest that calcium is stored in membraneous tubules dispersed within the spasmoneme (Carasso and Favard 1966), and these are probably the site of an ATP-dependent calcium pump. Most of the ATPase proteins characterized so far have high molecular weights, and the ATPase from the spasmoneme may be expected to be present in the material remaining close to the origin in the 15% polyacrylamide gels.

Energetics

Amos (1971) calculated the amount of work performed when *Vorticella* contracts. This he did by assuming that the cell body is a sphere that moves through a viscous medium, water, with an average velocity as measured from high-speed films. The viscous drag on the body is given by Stokes' formula and was calculated as a tension per unit of cross-sectional area of spasmoneme of 1.1×10^4 Nm^{-2}. This is a minimum value since it does not take into account the energy lost in overcoming the resistance of stiffening fibers and internal viscous forces in the stalk. The tension developed during contraction of a living specimen of *Carchesium* was later measured (Rahat, Pri-Paz and Parnas 1973) and found to be $4-8 \times 10^4$ Nm^{-2}, showing that the calculated value is of the correct order. The total work done by the spasmoneme in a single twitch was calculated to be 11 joules per kilogram of wet weight.

As contraction and extension of the spasmoneme can occur repeatedly, whereas metabolic activity and ATP splitting are abolished by inhibitors, Amos (1971) suggested a different energy source for the contraction. Contraction can be elicited experimentally by changing the calcium concentration from 10^{-8} to 10^{-6} M. It is reasonable to assume that the cell can vary its internal free calcium ion concentration over the same range, and Ettienne (1970) has detected large and rapid fluctuations in calcium ion level during contraction in the ciliate *Spirostomum*. The change in chemical potential in such a situation is approximately 10^4 J/mole of calcium ions. This is calculated from the equation

$$\Delta\mu_{Ca} = RT \log_e \frac{[Ca^{++}] \text{ higher}}{[Ca^{++}] \text{ lower}},$$

where $\Delta\mu$ is the change in chemical potential of calcium, R is the gas constant, and T is the absolute temperature. Thus for every mole of calcium bound, 10^4 J of energy is available to do work. Thus in a single contraction, at least $11/10^4$ moles of calcium, or 0.044 g calcium/kg wet weight, must be bound to the spasmoneme for this mechanism to be possible.

Measured directly, the glycerinated spasmoneme during contraction bound up to 1.7 g calcium/kg of dry mass. In terms of calcium bound to the hydrated organelle, this is 0.36 g/kg wet weight, since the organelle contains 21% w/v dry mass. Binding of 0.36 g calcium/kg wet weight is more than eight times the amount needed to supply the work done against viscous forces in the chemical potential theory. This amount of calcium is also greater than that required to generate the maximum tension measured in *Carchesium* (Rahat, Pri-Paz and Parnas 1973). The chemical potential theory is the only mechanism, at present, that can fully explain the contraction of the spasmoneme, and the recent measurements fully substantiate the original theory.

Calcium-binding in Relation to Other Systems

The electrophoresis gels indicate that it is the 20,000 MW proteins that bind calcium. This binding is likely to be relevant to contraction since it occurs with a similar calcium threshold which is not altered by micromolar concentrations of magnesium. Because these proteins are intimately involved in the contractile process in the spasmoneme, we have given them the general name of *spasmins* (Amos, Routledge and Yew 1975).

In the presence of calcium, the electrophoretic mobility of troponin C increased. It has been reported previously that troponin C is an asymmetric molecule that becomes more compact on binding calcium (Murray and Kay 1972; Head and Perry 1974), which would explain its increase in mobility on the gel. The mammalian intestinal calcium-binding protein, on the other hand, shows a decrease in mobility on electrophoresis in the presence of calcium (Hitchman, Kerr and Harrison 1973). The elution of this protein from an anionic exchange column indicated a decrease in the net negative charge of the protein on binding calcium, but sedimentation experiments indicate no large conformational changes (Dorrington et al. 1974). Thus a change in net negative charge is the dominant feature of calcium-binding in the intestinal calcium-binding protein. The same may be true for the spasmins, a decrease in the net negative charge predominating over a shape change, though this cannot be affirmed with certainty at present.

The number of calcium atoms bound by each spasmin molecule can be calculated from the knowledge that they are the major components of the spasmoneme (40–60%) and that they bind calcium between 10^{-8} and 10^{-6} M calcium. The microprobe results indicated that up to 1.7 g calcium/kg dry weight are bound to the spasmoneme between 10^{-8} and 10^{-6} M calcium. On this basis, the number of calcium atoms bound to each spasmin molecule (20,000 MW) is between 1.4 and 2.1.

The calcium-buffered gels suggest the binding of more than one calcium atom to each spasmin molecule. At 10^{-8} M, spasmins A and B migrate as two single bands (Fig. 7). At 10^{-6} M, mobility of both spasmins is reduced, and both show doubling of the bands, perhaps indicating the formation of more than one calcium-bound species (Fig. 7). Determining the exact number of calcium atoms bound to each spasmin and the affinity at each site will require the purification of both components in sufficient quantity to calculate binding curves by a direct method. All that can be said at this time is that each spasmin binds a small number of calcium atoms with a high affinity (between 10^{-8} and 10^{-6} M).

Several proteins with high affinities for calcium have recently been characterized. These include parvalbumin (Kretsinger and Nockolds 1973), troponin C (Hartshorne and Pyun 1971), intestinal calcium-binding protein (Dorrington et al. 1974), sarcoplasmic reticulum high-affinity protein (Ostwald and MacLennan 1974) and neuronal calcium-binding protein (Wolff and Siegel 1972; Alema et al. 1973). They are all acidic and bind a few calcium atoms with high affinity. In parvalbumin and troponin C, homologies in the amino acid sequence and tertiary structure occur especially in the region of the calcium binding site. The calcium atom is coordinated by six oxygen atoms arranged at the vertices of an octahedron. The spasmins show some similarity in amino acid composition to parvalbumin and troponin C. They have a large percentage of acidic residues and few aromatic residues. The spasmins have no sulfur-containing amino acids and a high serine content, in contrast to parvalbumin and troponin C. Determination of whether the spasmin calcium-binding sites are homologous to those of parvalbumin and troponin C will have to await sequencing of the protein.

Towards a Molecular Mechanism of Contraction

The ultrastructure of the spasmoneme is that of 2–4-nm diameter filaments (Fig. 2b) lying roughly in the long axis of the organelle. Dispersed among these filaments are membraneous tubules running longitudinally (Fig. 2a) and some mitochondria. Calcium is probably released from the tubules on stimulation and bound to the filaments to cause contraction. During the slow relaxation phase (1 sec, in contrast with 4–10 msec for contraction), calcium is presumably pumped back into the tubules by a Mg-ATPase system. We have shown that two proteins of 20,000 MW (spasmins) comprise the bulk of the organelle and that they bind a small number of calcium atoms with the same high affinity as required to produce contraction.

We have not shown directly that the filaments are composed of spasmin, though this seems likely. The subunit spacing along the filament measured by Amos (1975) was 3.5 nm, and the diameter of a 20,000 MW subunit, calculated on the basis of a protein density of 810 daltons $\cdot$ nm^{-3} (Lake and Leonard 1974), would be 3.6 nm.

If the filaments are linear polymers of spasmin molecules, how do they function? Three basic mechanisms for spasmoneme function have been considered: an electrostatic model (Hoffmann-Berling 1958), an entropic rubber model (Weis-Fogh and Amos 1972) and, most recently, a helical coiled filament model (Huang and Pitelka 1973; Kristensen, Engdahl Nielsen and Rostgaard 1974).

If the electrical charges on polyacrylic fibers are neutralized, the fibers shorten thermokinetically (Katchalsky 1949). This is quoted by Hoffmann-Berling (1958) as a possible mechanism for spasmoneme contraction. The extension at low Ca^{++} would be due to electrostatic repulsion between fixed negative charges on the filaments. The addition of calcium would neutralize these charges and cause the network to collapse. Hoffmann-Berling realized that the electrostatic model did not fully explain the contraction, due to the specificity for calcium ion. Strontium is almost as effective in producing contraction, but other divalent cations, such as beryllium and magnesium, are totally ineffective. Magnesium did not alter the threshold for the calcium response even in 1000-fold excess over calcium. An electrostatic model should also be dependent on ionic strength of the media. The spasmoneme can, however, contract and extend over a wide range of salt concentrations. Only a small decrease in the calcium threshold occurs even at high ionic strength (0.6 M KCl; Amos, unpubl.). The finding that spasmin is an acidic protein (pI 4.7–4.8) does not necessarily strengthen the electrostatic mechanism. Other filamentous proteins, such as actin, flagellin and neurofilament protein, are also acidic, but they are not thought to be involved in electrostatic contractile mechanisms.

The length/tension and birefringence measurements on the spasmoneme in the high calcium state indicate a rubberlike behavior (Weis-Fogh and Amos 1972). The spasmoneme shows a high degree of passive extensibility in both the high and low calcium states. It can be stretched to four times the rest length and, when released, it returns to its rest length. This rubber- or spring-like property may be related to the contractile mechanism. Weis-Fogh and Amos (1972) suggested that the protein backbone in the isotropic, high calcium state spasmoneme consisted largely of thermally agitated, kinetically free, polypeptide chains. When calcium is removed, the spasmoneme actively extends but it is still rubberlike, although now optically somewhat anisotropic. This, they suggested, was due to the introduction of additional Ca-sensitive cross-links in a nonrandom fashion. This could occur in such a way that the structure is stretched due to internal strains, and elastic energy is stored in the network to an extent corresponding to the decrease in configurational entropy of the chains. When calcium is available, the additional bonds would be broken, and the structure would return to a random network, releasing stored energy if allowed to shorten. This theory would produce a high rate of shortening. Hawkes and Holberton (1975) have examined the myonemal contraction in *Spirostomum* against different viscous loads. They found that the force of contraction does not vary with the velocity of shortening. Thus the contractile fibers behave like a rubber band or a metal spring that has been stretched and released. In contrast, a muscle produces very little force at high shortening velocities.

One difficulty with a rubber-type contractile mechanism is the recent finding that the birefringence is largely form birefringence (Fig. 8) due to oriented structures in the spasmoneme. The birefringence from oriented polypeptide chains would be expected to be intrinsic birefringence, and thus the agreement between the optical measurements and classical rubber theory may be fortuitous. However, the fact that the spasmoneme is rubberlike cannot be overlooked and must reflect an interesting arrangement of the spasmoneme protein.

Another mechanism for spasmoneme contraction is suggested from the myonemes of the heterotrich ciliate *Stentor*. These myonemes are smaller than the spasmoneme but are ultrastructurally similar and undergo a birefringence change on contraction like that of the peritrich spasmoneme (Kristensen, Engdahl Nielsen and Rostgaard 1974). During contraction, *Stentor* can contract to 13% of its extended length, whereas the spasmoneme contracts to only 30%. The contracted myonemes in *Stentor* show a majority of 8–12-nm tubules in place of the 3–4-nm filaments in the extended state (Bannister and Tatchell 1972; Huang and Pitelka 1973; Kristensen, Engdahl Nielsen and Rostgaard 1974). This may be due to the tight helical coiling of the 3–4-nm filaments. No such structures have been observed in the spasmoneme of contracted peritrichs, though such a mechanism could well occur, thus generating less tightly coiled filaments which would be difficult to recognize. Such coiled filaments might have the long-range elasticity that has been observed (Weis-Fogh and Amos 1972). The elasticity of thin filaments from muscle has been measured recently (Oosawa et al. 1973). Thin filaments are composed of a double helix of actin monomers, with a troponin complex every seven monomers and a tropomyosin molecule in the groove between the actin helices (Huxley 1973). When calcium binds to troponin, the tropomyosin moves out of the groove and the thin filament becomes more flexible, the end-to-end distance decreasing to 85% of the precalcium length (Yanagida and Oosawa 1975). A similar change in the flexibility of spasmoneme filaments may occur, with a large decrease in the end-to-end length of the filaments, on the direct binding of calcium.

How could calcium act on spasmin to produce such a change in the filaments? When calcium is removed from the spasmoneme, the organelle can exert a pushing force. This could represent an active change in the angle of bonding between spasmin molecules, which in a helical system would produce a considerable change in overall length. Such a change in intersubunit bonding could easily be produced by a calcium-sensitive conformational change in spasmin. The exact mechanism by which calcium can produce contraction and extension in the spasmoneme may involve aspects of all three mechanisms so far discussed, but its elucidation will require further experimental studies on the spasmoneme proteins.

Acknowledgments

We thank Dr. S. E. Hitchcock for the sample of troponin C, Dr. J. Kendrick-Jones for rabbit skeletal muscle actin, and Dr. B. M. F. Pearse for a sample of purified pig brain tubulin. We also wish to thank the SRC for grants to Professor Weis-Fogh and to Drs. Echlin, Gupta, Hall and Morton of the Biological Microprobe Laboratory.

REFERENCES

Alema, S., P. Calissano, G. Rusca and A. Giuitta. 1973. Identification of a calcium-binding, brain specific protein in the axoplasm of squid giant axons. *J. Neurochem.* **20**:681.

Allen, R. D. 1973. Contractility and its control in peritrich ciliates. *J. Protozool.* **20**:25.

Amos, W. B. 1971. A reversible mechanochemical cycle in the contraction of *Vorticella. Nature* **229**:127.

———. 1972. Structure and coiling of the stalk in the peritrich ciliates *Vorticella* and *Carchesium. J. Cell Sci.* **10**:95.

———. 1975. Contraction and calcium binding in vorticellid ciliates. In *Molecules and cell movement* (ed. R. E. Stephens and S. Inoué), pp. 411–436. Raven Press, New York.

———. 1976. An apparatus for microelectrophoresis in polyacrylamide slab gels. *Anal. Biochem.* **70**:612.

Amos, W. B., L. M. Routledge and F. F. Yew. 1975. Calcium-binding proteins in a vorticellid contractile organelle. *J. Cell Sci.* **19**:203.

Amos, W. B., L. M. Routledge, T. Weis-Fogh and F. F. Yew. 1976. The spasmoneme and calcium dependent contraction in relation to specific calcium-binding proteins. In *Calcium in biological systems. 30th Symposium of the Society for Experimental Biology.* Cambridge University Press. (In press.)

Bannister, L. H. and E. C. Tatchell. 1968. Contractility and the fibre systems of *Stentor coeruleus. J. Cell Sci.* **3**:295.

———. 1972. Fine structure of the M fibres in *Stentor* before and after shortening. *Exp. Cell Res.* **73**:221.

Carasso, N. and P. Favard. 1966. Mise en évidence du calcium dans les myonèmes pédonculaires de ciliés péritriches. *J. Microscopie* **5**:759.

Close, R. 1965. The relation between intrinsic speed of shortening and duration of the active state in muscle. *J. Physiol.* **180**:542.

Dorrington, K. J., A. Hui, T. Hofmann, A. J. W. Hitchman and J. E. Harrison. 1974. Porcine intestinal calcium-binding protein: Molecular properties and the effect of binding calcium ions. *J. Biol. Chem.* **249**:199.

Engelmann, T. W. 1875. Contractilität und Doppelbrechung. *Pflügers Arch.* **11**:432.

Ettienne, E. M. 1970. Control of contractility in *Spirostomum* by dissociated calcium ions. *J. Gen. Physiol.* **56**:168.

Hall, T. A., H. C. Anderson and T. Appleton. 1973. The use of thin specimens for X-ray microanalysis in biology. *J. Microsc.* **99**:177.

Hall, T. A., H. O. E. Roeckert and R. L. Saunders. 1972. *X-ray microscopy in clinical and experimental medicine,* pp. 162, 170. Charles Thomas, Springfield, Illinois.

Hartshorne, D. J. and H. Y. Pyun. 1971. Calcium binding by the troponin complex and the purification and properties of troponin A. *Biochim. Biophys. Acta* **229**:698.

Hawkes, R. B. and D. V. Holberton. 1975. Myonemal contraction of *Spirostomum.* II. Some mechanical properties of the contractile apparatus. *J. Cell. Physiol.* **85**:595.

Head, J. F. and S. V. Perry. 1974. The interaction of the calcium-binding protein (troponin C) with bivalent cations and the inhibitory protein (troponin I). *Biochem. J.* **137**:145.

Hitchman, A. J. W., M. K. Kerr and J. E. Harrison. 1973. The purification of pig vitamin-D induced calcium binding protein. *Arch. Biochem. Biophys.* **155**:221.

Hoffmann-Berling, H. 1958. Der Mechanismus eines neuen, von der Muskelkontraktion verschiedenen Kontraktionzyklus. *Biochim. Biophys. Acta* **27**:247.

Huang, B. and D. R. Pitelka. 1973. The contractile process in the ciliate *Stentor coeruleus*. I. The role of microtubules and filaments. *J. Cell Biol.* **57**:704.

Huxley, H. E. 1973. Structural changes in actin- and myosin-containing filaments during contraction. *Cold Spring Harbor Symp. Quant. Biol.* **37**:361.

Jones, A. R., T. L. Jahn and J. R. Fonseca. 1970. Contraction of protoplasm. IV. Cinematographic analysis of the contraction of some peritrichs. *J. Cell. Physiol.* **75**:9.

Katchalsky, A. 1949. Rapid swelling and deswelling of reversible gels of polymeric acids by ionization. *Experientia* **5**:319.

Kretsinger, R. H. and C. E. Nockolds. 1973. Carp muscle calcium-binding protein. II. Structure determination and general description. *J. Biol. Chem.* **248**:3313.

Kristensen, B. I., L. Engdahl Nielsen and J. Rostgaard. 1974. Variations in myoneme birefringence in relation to length changes in *Stentor coeruleus*. *Exp. Cell Res.* **85**:127.

Laemmli, U. K. and M. Favre. 1973. Maturation of the head of bacteriophage T4. I. DNA packaging events. *J. Mol. Biol.* **80**:575.

Lake, J. A. and K. R. Leonard. 1974. Structure and protein distribution for the capsid of *Caulobacter crescentus* bacteriophage ϕCbK. *J. Mol. Biol.* **86**:499.

Murray, A. C. and C. M. Kay. 1972. Hydrodynamic and optical properties of troponin A. *Biochemistry* **11**:2623.

Oosawa, F., S. Fujime, S. Ishiwata and K. Mihashi. 1973. Dynamic property of F-actin and thin filament. *Cold Spring Harbor Symp. Quant. Biol.* **37**:277.

Ostwald, T. J. and D. H. MacLennan. 1974. Isolation of a high affinity calcium-binding protein from sarcoplasmic reticulum. *J. Biol. Chem.* **249**:974.

Rahat, M., Y. Pri-Paz and I. Parnas. 1973. Properties of stalk "muscle" contractions of *Carchesium sp. J. Exp. Biol.* **58**:463.

Roth, M. and A. Hampai. 1973. Column chromatography of amino acids with fluorescence detection. *J. Chromatog.* **83**:353.

Routledge, L. M., W. B. Amos, B. L. Gupta, T. A. Hall and T. Weis-Fogh. 1975. Microprobe measurements of calcium-binding in the contractile spasmoneme of a vorticellid. *J. Cell Sci.* **19**:195.

Schmidt, W. J. 1940. Die Doppelbrechung des stieles von *Carchesium,* inbesondere die optische-negative Schwankung seines Myonemes bei der Kontraktion. *Protoplasma* **35**:1.

Sillén, L. G. and A. E. Martell, eds. 1964. *Stability constants of metal-ion complexes.* Special Publication No. 17, pp. 697–698. The Chemical Society, London.

van Leeuwenhoek, A. 1676. Letter. *Phil. Trans. Roy. Soc. London B.* **12**:133.

Weis-Fogh, T. 1975. Principles of contraction in the spasmoneme of vorticellids. A new contraction system. In *Comparative physiology—Functional aspects of structural materials* (ed. L. Bolis, S. H. P. Maddrell and K. Schmidt-Nielsen), pp. 83–98. North-Holland, Amsterdam.

Weis-Fogh, T. and W. B. Amos. 1972. Evidence for a new mechanism of cell motility. *Nature* **236**:301.

Wolff, D. J. and F. L. Siegel. 1972. Purification of a calcium-binding phosphoprotein from pig brain. *J. Biol. Chem.* **247**:4180.

Yanagida, T. and F. Oosawa. 1975. Effect of myosin on conformational changes of F-actin in the thin filament *in vivo* induced by calcium ions. *Eur. J. Biochem.* **56**:547.

Section 2

MUSCLE CONTRACTION

Introductory Remarks:
The Relevance of Studies on
Muscle to Problems of Cell Motility

Hugh E. Huxley

MRC Laboratory of Molecular Biology
Cambridge CB2 2QH, England

In introducing this section on muscular contraction I have attempted, very briefly, first, to put the muscle field in historical perspective; second, to assess how work on muscle contraction may help our understanding of certain other forms of cell motility; and finally, to summarize some recent structural work on muscle that may be of relevance and interest at the present time.

It was proposed a little over 20 years ago that the proteins actin and myosin were located in separate filaments of defined length in striated muscle, and that these filaments were organized into separate but overlapping arrays which gave rise to the visible band pattern (Hanson and Huxley 1953). It was subsequently proposed that contraction was produced by a process in which the arrays of filaments, maintaining their individual lengths approximately constant, actively slid past each other (Huxley and Hanson 1954; Huxley and Niedergerke 1954) as a consequence of interactions between the actin and myosin in the region of overlap. It was suggested that the sliding force might be developed at a number of points acting in parallel in this region (Huxley and Niedergerke 1954), and that the entities responsible for producing the force might be cross bridges, which, projecting out sideways from the myosin filaments, might be able to attach to actin, pull the actin along a short distance, and detach again in a repetitive cycle (Huxley and Hanson 1955). These cross bridges were identified in the electron microscope in sections of muscle (Huxley 1953a, 1957; Huxley and Hanson 1956).

It took a considerable time for this "sliding-filament mechanism" to become generally accepted; in part because a number of the techniques used in the studies were new and unfamiliar, and hence the arguments and conclusions derived from them were not immediately convincing, although in fact they were very powerful ones. However by about the mid-1960's, many people were prepared to base their experiments on the probability that this mechanism was basically correct, and it became increasingly apparent that the sliding-filament model provided a very satisfactory conceptual framework in which to think about many different aspects of the muscle problem. With new experimental techniques making possible much more controlled and precise measurements on much better defined systems and with the influx of many more workers into the field, progress in recent years has been very rapid (see, for example, *Cold Spring Harbor Symp. Quant. Biol.* Vol. 37, 1973).

The main areas of particular interest at present are approximately as follows:

1. What are the precise details of the structural cycle that the myosin cross bridges undergo in order to develop force?
2. What are the details of the biochemical cycle of interaction of actin and myosin which produces ATP hydrolysis, and how are these biochemical steps related to the structural steps that generate force?
3. How are these processes switched on and off by calcium ions?
4. What are the biochemical and structural modulations and variations of this scheme that occur in different types of muscles?

These areas will be discussed by the authors of the various papers in this particular section.

Relationship to Cell Motility

The relationship between muscular contraction and certain other forms of cell motility only became convincing when it was demonstrated decisively that proteins very similar to actin and myosin in muscle could be identified in a variety of non-muscle cells. Most of the people associated with this work are contributors to this volume, and it would be superfluous for me to rehearse such recent history in detail. However, I think it is of some interest to recall that the first really persuasive evidence was structural in nature. I have in mind, in the first place, Hatano's electron micrographs of negatively stained filaments from slime mould that show, without any doubt, the familiar double-helical actin structure (Hatano, Totsuka and Oosawa 1967). This was followed by Ishikawa's identification (Ishikawa, Bischoff and Holtzer 1969), in a variety of whole glycerinated cells, using the heavy meromyosin (HMM) labeling technique, of filaments showing a characteristic arrowhead structure similar to that originally observed with negatively stained preparations of HMM-decorated rabbit actin (Huxley 1963). Subsequently, a very close identity was established between the decorated actin structures from hybrids of purified slime mould actin and rabbit myosin subfragment 1 (S_1) (Nachmias, Huxley and Kessler 1970), the purified amoeba actin and rabbit heavy meromyosin (Pollard et al. 1970), and the corresponding complexes formed from rabbit actin and myosin. This identity was all the more remarkable, and its implications all the more far-reaching, since it was established that the characteristic arrowhead appearance depended on a very specific structural relationship between the myosin heads and the actin monomers in the complex (Moore, Huxley and DeRosier 1970).

Once this central point had been established—that a protein very closely similar to the muscle protein actin in its structure and interactions existed in cells other than muscle—then a very rapid exploitation of the whole field took place, aided in particular by the newly developed technique of sodium dodecyl sulfate (SDS) gel electrophoresis for identifying particular protein components in the presence of a complex and unknown background of other protein species.

I think it is clear that there are basically two ways in which studies on muscle can be of interest to people working on other aspects of cell motility. The first one arises from the obvious fact that large amounts of protein are readily available from muscle and that methods have been worked out for obtaining the individual protein components in a state of high purity yet still retaining a large part, if not all, of their original biological activity. Thus there is now available a vast amount of experience, information and techniques that can be adapted to the study of analogous proteins from non-muscle cells, both in the preparation of such proteins and in the testing of them for specific properties related to their functional role. The pitfalls of such an approach are obvious, but the rapid rate of recent advances demonstrates its advantages. However, one should always bear in mind the possibility that proteins similar to familiar ones in muscle may, on occasion, be employed in a rather different role, and that there may be certain subtleties that we have not yet appreciated, to paraphrase Sir Francis Bacon.

The second way in which muscle studies may be of interest is also a very obvious one. Since actin and myosin assemble to form a sliding-filament mechanism in muscle and since that mechanism is driven by actin–myosin interaction at the cross bridges in muscle and, in addition, since closely similar proteins capable of virtually identical chemical and structural interactions are found in non-muscle motile systems, then there is a strong possibility that such systems operate by some analog of the sliding-filament mechanism. Indeed, it seems very plausible that the basic cyclic actin–myosin interaction originally developed in the course of evolution in simpler cellular systems, and that muscle represents an elaboration of a basic mechanism still utilized in a less extensively ordered form in non-muscle cells. The analogy between models of cytoplasmic streaming based on an active shearing mechanism and the structurally polarized sliding interaction between actin and myosin filaments in muscle is an attractive one, as I pointed out several years ago (Huxley 1963). The problem now is (1) to discover the exact structural arrangement of actin, and especially of myosin, in all the cells that contain these proteins and display various forms of motility; (2) to obtain hard experimental evidence as to whether the mechanism really does correspond to our expectations; and (3) to examine the proteins themselves to discover what special properties they may have which suit them for their particular role, whatever that may be. In particular, the interfacing between the motile, force-producing system within a cell and the external environment of that cell, whether it be a substrate and contacts with other cells (see Fig. 1 and Huxley 1973b) or molecules that become attached to the outside of the cell membrane, presents problems and possibilities of the very greatest interest.

Recent Structural Results on Striated Muscle

In this section I will review very briefly some aspects of our present understanding of the cross-bridge mechanism, which I hope may be of interest to those working on cell motility but who are not necessarily closely familiar with all the details of recent structural studies on muscle. I think a convenient

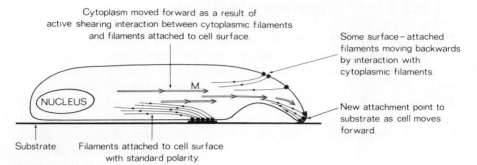

Figure 1
Diagrammatic representation of a mechanism by which active shearing forces developed between two sets of filaments could produce cytoplasmic streaming and cell movement.

point to begin would be with that particular type of sliding-filament mechanism known as the "swinging cross-bridge" model (Fig. 2) (Huxley 1969).

In the first place, this model incorporates all the essential features of the original sliding-filament mechanism and its subsequent refinements. Actin and myosin molecules are assembled into separate filaments of defined length (how the length is defined we do not as yet know). In the myosin filaments, the myosin molecules are arranged with their tails, which are oriented longitudinally and make up the backbone of the filaments, always pointing towards the midpoint of the filament. Thus there is a reversal of polarity at the center of the filaments and hence at the center of the A bands. If the myosin molecules can develop a longitudinal force by interacting with the actin filaments, the direction of that force must be related to the orientation of the myosin. Consequently, the reversal of structural polarity will ensure that the forces developed in either half of an A band are oppositely directed, as they would have to be to move the actin filaments toward each other. Similarly, if a stereospecific interaction between actin and myosin molecules is to take place, the actin monomers also have to be assembled into filaments with a strictly defined structural polarity. In this case, all the monomers in an I filament on one side of a Z line have to be oriented in the same direction, and that direction has to be reversed on the opposite side of the Z line (Fig. 3). There is ample experimental evidence that these arrangements are an essential feature of the actual construction of striated muscles (Huxley 1963). In the case of actin, the polarity is very easy to demonstrate by HMM or S_1 labeling. This provides a very useful tool for investigating the direction in which force might be developed in a particular group of actin filaments in various cellular situations, provided the direction of the arrowheads (pointing in the direction in which the actin will tend to move) can be clearly established. Only those myosin molecules having appropriate orientation will be able to interact with a given actin filament (or group of similarly oriented filaments).

In the muscle system, the heads of the myosin molecules (the S_1 subunits, two per molecule—again, we do not know if the double head represents an

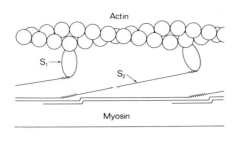

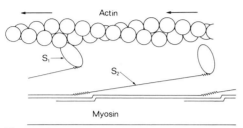

Figure 2

Active change in angle of attachment of cross bridges (S_1 subunits) to actin filaments could produce relative sliding movement between filaments maintained at constant lateral separation (for small changes in muscle length) by long-range force balance. Bridges can act asynchronously, since subunit and helical periodicities differ in the actin and myosin filaments.

 (*Top*) Left-hand bridge has just attached; other bridge is already partly tilted. (*Bottom*) Left-hand bridge has just come to end of its working stroke; other bridge has already detached and probably will not be able to attach to this actin filament again until further sliding brings helically arranged sites on actin into favorable orientation.

essential part of the mechanism or whether it is merely a consequence of the stable, two-chain, coiled-coil, α-helical structure of the myosin tails) project out sideways from the backbone of the myosin filaments and interact with actin in a cyclic manner. In the resting state it is supposed that the myosin heads, which are thought to be somewhat asymmetrical structures, are

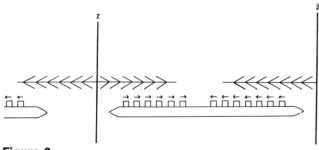

Figure 3

Diagrammatic illustration of structural polarity of cross bridges on myosin filaments, and of molecules of actin in the thin filaments, so that the interaction could produce sliding forces which would move the actin filaments toward each other in the center of the sarcomere.

oriented approximately perpendicular to the filaments. Upon activation, the heads begin to attach to neighboring actin monomers and when attached, they undergo some structural change that causes them to forcibly alter their angle of attachment, tilting over and pulling the actin filament along for a distance of the order 50–100 Å in the appropriate direction. At the end of the working stroke, the S_1 subunit detaches from that particular actin monomer and returns towards its original orientation, ready to attach to another actin monomer further out along the actin filament. Similar cycles of interaction are taking place at all the other cross bridges in the overlap area, so that a steady sliding motion is produced. From the work of Lymm and Taylor and others (see article by Taylor and Sleep, this volume) it is believed that the myosin heads split the ATP attached to them *before* they attach to actin, but that the split products (ADP and P_i) remain bound to them and are released during the force-generating stroke when actin and myosin are combined. The binding of a fresh molecule of ATP then dissociates the cross bridge at the end of the working stroke. This particular cross-bridge model differs from earlier ones in that it supposes that the myosin heads are attached relatively flexibly to the myosin filament backbone by the S_2 linkage, a section of the linear portion of the myosin molecule that is not strongly bonded into the assembly of the terminal light meromyosin (LMM) portions of the myosin molecules making up the shaft of the filaments. The function of the S_2 linkage is to transmit tension to the myosin filaments, yet allowing a considerable latitude in the exact lateral and radial position of the heads of the molecules relative to the myosin backbone, while still enabling them to make a precisely defined attachment to actin. In this way, the mechanism could operate over a range of side spacings between actin and myosin filaments (which in practice do occur at different sarcomere lengths); it would also allow some flexibility in the orientation of the actin helix relative to the myosin filament at which attachment could occur and even in the selection of which particular neighboring actin filament a cross bridge might attach to at a particular moment.

In this mechanism, the driving force that causes the bridge to tilt on its active stroke is *not* produced at the junctions between the myosin heads and the myosin filament backbone but at or near the interacting surface between the myosin head and actin. The return stroke in the unattached state is envisaged as one requiring only weak forces. Thus although the myosin head–actin filament interaction has to be the invariant feature of all such mechanisms— as indeed appears to be the case from the remarkable constancy of the appearance of HMM-decorated actin throughout nature—the model places much looser restrictions on the precise mode of assembly of the myosin filaments, provided their structural polarity is maintained. Considerable differences are in fact found in the diameters, lengths and exact packing arrangements of the myosin-containing filaments among different animal species and even within the same animal. What form the myosin assemblies take in non-muscle motile systems is still very uncertain, but it is noteworthy that in every case examined so far except one (*Acanthamoeba* myosin; Pollard and Korn 1973), the myosin molecules have the familiar "tadpole-like" structure (i.e., a globular region at one end of an approximately 1400-Å long tail) and in many cases will assemble to form bipolar filaments, often somewhat smaller than those from muscle myosin.

Control of the myosin head–actin interaction is affected in vertebrate striated muscle by the tropomyosin-troponin system, which forms part of the structure of the actin-containing filaments. The characteristics of this system are that in the presence of very low concentrations of free calcium ions (around 10^{-8} M or less), calcium is not bound by troponin, and attachment of myosin heads carrying the split products of ATP hydrolysis is prevented, so that the muscle remains in the relaxed, readily extensible state. Following stimulation, release of calcium previously sequestered by the sarcoplasmic reticulum raises the free calcium concentration in contact with the filaments towards (probably) about 10^{-5} M, calcium binding to troponin takes place, and a change occurs in the actin filaments which allows myosin head attachment to occur and contraction to proceed as already described. Removal of free calcium by the reticulum when stimulation ceases then causes the muscle to relax again.

In structural terms, there is a considerable body of evidence that these changes in the reactivity of the actin filaments are brought about by a movement of tropomyosin. Tropomyosin is a long, two-chain, α-helical molecule, and continuous strands, built up from the end-to-end aggregation of such molecules, are located along the whole length of the two, long-pitch, helical grooves in the actin filaments. In an "active" filament, the tropomyosin strands lie relatively nearer the center of the grooves. In the inhibited actin filament, they are found to lie further out from the center of the groove and nearer the probable position of the attached myosin heads. Thus it is possible that the system operates by a steric blocking mechanism. The evidence and arguments for this have been described quite extensively elsewhere (see, for example, Huxley 1971, 1973a; Spudich, Huxley and Finch 1972; Haselgrove 1973; Parry and Squire 1973; Wakabayashi et al. 1975), and the details lie outside the scope of the present discussion. What is important is that the system is a built-in part of the actin filament and presumably depends on a number of very specific interactions with actin for its operation. Although the identification of control mechanisms in non-muscle systems is still in a rather preliminary stage, it is noteworthy that in several cases, e.g., *Acanthamoeba* actin (Eisenberg and Weihing 1970), human platelet actin (Adelstein and Conti 1973) and physarum actin (Tanaka and Hatano 1972), it has been shown that the actin component can be controlled in its interactions with myosin by the troponin-tropomyosin system from muscle. This indicates a remarkable conservation of the actin structure and sequence, as indeed is already known from other studies, and it is difficult to believe that the capacity to be controlled in this way was an accidental property of primitive actins that was nevertheless conserved and eventually utilized in vertebrate striated muscle! It seems to me much more likely that analogous control mechanisms were present throughout the evolution of actin as we find it today and that such mechanisms will indeed be found to be still present in at least some of the non-muscle systems.

Current Work on the Cross-bridge Mechanism

I have said very little so far about the experimental evidence in support of the "swinging, tilting cross-bridge model." It is essentially of three kinds:

First, it has been shown, especially by the work of Lowey and collaborators (see, for example, Lowey et al. 1969), that the myosin molecule can be cleaved at two distinct regions along its length by appropriate proteolytic digestion. The three resultant subfragments have distinctly different properties. As already mentioned, the head subfragments (S_1 subunits) can interact with actin and split ATP. The two heads are located at one end of the molecule, and their polypeptide chains are continuous with the two chains that form the α-helical rod portion of myosin. The first portion of the rod (S_2), nearer to the heads and about 400–500 Å in length, has solubility properties quite different from the remaining 900 Å or so of the rod (LMM). The S_2 fragments are soluble at physiological ionic strength and do not interact with the LMM fragments (which form filamentous aggregates under these conditions). This indicates that the myosin heads are not directly attached to the backbone of the myosin filaments but are free to "hinge out" on the S_2 subunits. The susceptibility of the regions at either end of the S_2 rod portion to enzymatic attack suggests that these regions of the molecule have a different structure from the rest of the two-chain, coiled-coil α helix, which could be because here the chains are in a more open and flexible configuration. It is natural to suppose that these structural features, common to myosins from many different types of muscle, have a very definite functional significance; namely, the freedom of movement that they give to the myosin heads enabling them to attach to actin in a closely defined manner. It will be very interesting to see whether the tail region of non-muscle myosins also has two regions that differ in their solubility properties and are cleaved by proteolytic enzymes.

Second, there is good electron microscope evidence for a tilting bridge mechanism. Thus in insect indirect flight muscle, in which a much greater degree of synchrony of bridge action than in vertebrate muscle apparently obtains, the bridges *in rigor* can be clearly seen to be attached to the actin filaments in a characteristically tilted configuration (Reedy, Holmes and Tregear 1965), and the tilt is in such a direction as to move the end of the cross bridge attached to actin towards the center of the A band. The attachment of "free" myosin heads to actin can also be examined in the electron microscope by the negative staining technique. In studies of thin filaments "decorated" with S_1 in the absence of nucleotide (Huxley 1963; Moore, Huxley and DeRosier 1970), a tilted form of attachment is again apparent, with the direction of tilt (as seen on isolated Z segments) in the appropriate direction. It is very likely that these images correspond to the configuration adopted by the cross bridge at the end of its working stroke, when ADP and P_i have been released but before dissociation is produced by the arrival of a fresh molecule of ATP. The electron microscope (and X-ray) evidence indicates that the bridges in relaxed muscle are in a more perpendicular orientation relative to the filament axis, so the simplest supposition would be that they attached in this configuration at the beginning of the working stroke. The extent of movement this would imply, from obviously oversimplified geometrical considerations, would be about 70–80 Å, which accords well with the values obtained by Huxley and Simmons (1971, 1973) from observations on rapid mechanical transients in muscle.

One of the best, though by no means straightforward, methods of investigating the nature and behavior of the cross bridges is by low-angle X-ray diffraction, since a large part of the diagram comes from the cross bridges themselves and since it is possible to study a muscle by this technique under almost normal working conditions. The disadvantage of this technique—besides the inherent and well-known ambiguities of X-ray diagrams—is that the reflections from muscle are rather weak (about one-millionth as strong as the direct beam), and it has therefore been necessary to invest a considerable amount of time and effort in the technical innovations required to record the patterns with sufficient speed. This is not an appropriate occasion to discuss these developments; suffice it to say, we can now record changes in the strongest reflections with a time resolution of 10 msec and we can still make use of gains of X-ray intensity of several orders of magnitude in order to measure weaker reflections and to avoid having to use very long series of contractions. However, what of the results? In the diagram from a resting muscle there is a well-developed system of layer lines with a 429-Å axial repeat and a strong third-order meridional repeat at 143 Å (Huxley and Brown 1967; Elliott, Lowy and Millman 1967). This pattern arises from a regular helical arrangement of cross bridges on the thick filaments, with groups of cross bridges occurring at intervals of 143 Å along the length of the filaments with a helical repeat of 429 Å. The number of cross bridges in each group is not yet absolutely certain, but it is more likely to be three than two (Squire 1973). When a muscle goes into rigor, the axial X-ray diagram loses nearly all the features associated with the myosin filament helix and shows, instead, a pattern of reflections that can be indexed on the actin helix. This shows that the cross bridges, the S_1 heads of myosin, form a relatively flexible attachment to the myosin filament backbone and a relatively rigid attachment (in rigor anyway) to actin. The same conclusion can be drawn from a comparison of the very regular and highly ordered appearance in the electron microscope of negatively stained specimens of actin filaments decorated with S_1 and the very disordered appearance of the cross bridges on myosin filaments examined by the same technique (Huxley 1963, 1969). The characteristics of the two types of attachment suggest very strongly that the site at which force originates is at the interacting surface of the myosin head and the monomer on the actin filament.

Diagrams from contracting muscle may be recorded on film using a shutter so as to transmit the X-ray beam only when the muscle is being stimulated. A long series of tetani is necessary, with intervals in between them for recovery—usually 1-second tetani and 2-minute intervals. Such diagrams show that the whole pattern becomes very much weaker during contraction, indicating that the cross bridges are much less regularly arranged, as would be expected if they were undergoing asynchronous longitudinal or tilting movements (and possibly lateral ones too) during their tension-generating cycles of attachment to actin. If indeed this is the case, then the very rapid development of the active state of a muscle following stimulation should be accompanied by an equally rapid decrease in intensity of the layer-line pattern, and it is most important to find out if this is really the case. Recent technical developments have now made it possible to record the X-ray diagram with sufficient speed to investigate this question. Studying single twitches of frog sartorius muscle, Dr. John Haselgrove and I have found that the change in pattern (at 10°C) begins about 10–15 msec after stimulation and is half-complete by about 20 msec, a time when the externally measured tension (which in a normal isometric contraction is delayed behind the onset of the active state by the necessity to stretch the series' elastic elements before the internal activity can manifest itself as tension) has hardly begun to rise.

The equatorial part of the X-ray diagrams from muscles is also very informative. It is generated by the regular, side-by-side hexagonal lattice in which the filaments are arranged, and the relative intensity of the two principle reflections is strongly influenced by the lateral position of the cross bridges. Large changes occur between resting muscle and muscle in rigor (Huxley 1952, 1953, 1968), when a high proportion, if not all, of the cross bridges will be attached to actin. These have been interpreted as indicating that in a resting muscle, the cross bridges lie relatively closer to the backbone of the thick filaments, whereas when they attach to actin, they hinge further out and lie with their centers of mass nearer to the axes of the actin filaments at the trigonal positions of the hexagonal lattice.

Similar changes have been observed in contracting muscles (Haselgrove and Huxley 1973), though the extent of change is less and would correspond to about half the cross bridges being in the vicinity of the actin filaments at any one time. Again the observations are consistent with a model in which the cross bridges are undergoing during contraction a mechanical cycle of attachment to and detachment from the actin filaments. However, it should be appreciated that the parameter being measured is the average lateral position of the cross bridges, and this does not provide conclusive evidence about the proportion actually attached. Nevertheless, the changes do indicate that a substantial lateral movement of the cross bridges takes place during contraction, and as in the case of the layer-line changes, it is important to establish whether the movement occurs at a sufficiently rapid rate for it to arise from a force-generating attachment of cross bridges. Again this is a quantity that we can now measure, and Dr. Haselgrove and I have found that the expected changes in the equatorial X-ray diagram (decrease in intensity of [10] reflection, increase in intensity of [11] reflection) do indeed occur with great rapidity after stimulation. At 10°C, for example, the change is half-complete within about 20–30 msec, which accords well with the expected temporal characteristics of the active state.

These types of observations (and analogous ones of insect flight muscle— see, for example, Armitage, Tregear and Miller 1975) are thus beginning to provide moment-to-moment information about cross-bridge behavior which strongly supports the general features of the swinging cross-bridge model and which we are beginning to be able to compare with other kinetic measurements on muscle and muscle proteins. Eventually, this should make it possible to establish whether muscle behavior can be completely described in these terms. The X-ray technique is not, however, one that readily lends itself to the study in situ of the much smaller and less well oriented motile systems in non-muscle cells, even with very high power X-ray sources, and radiation damage is likely to be a severely limiting factor. The study of isolated components would appear to offer more promise, however, especially if more of the muscle and non-muscle proteins can be crystallized. Indeed, it may well turn out that the cellular motile systems will lead the way in this respect!

In general then, there are good prospects for further advances in the structural field in muscle, and the wide application this information may have throughout the whole of cell biology will make it all the more interesting to obtain.

REFERENCES

Adelstein, R. S. and M. A. Conti. 1973. The characterization of contractile proteins from platelets and fibroblasts. *Cold Spring Harbor Symp. Quant. Biol.* **37**:599.

Armitage, P. M., R. T. Tregear and A. Miller. 1975. Effect of activation by calcium on the X-ray diffraction pattern from insect flight muscle. *J. Mol. Biol.* **92**:39.

Eisenberg, E. and R. R. Weihing. 1970. Effect of skeletal muscle native tropomyosin on the interaction of amoeba actin with heavy meromyosin. *Nature* **228**: 1092.

Elliott, G. F., J. Lowy and B. M. Millman. 1967. Low-angle X-ray diffraction studies of living muscle during contraction. *J. Mol. Biol.* **25**:31.

Hanson, J. and H. E. Huxley. 1953. Structural basis of the cross-striations in muscle. *Nature* **172**:530.

Haselgrove, J. C. 1973. X-ray evidence for a conformational change in the actin-containing filaments of vertebrate striated muscle. *Cold Spring Harbor Symp. Quant. Biol.* **37**:341.

Haselgrove, J. C. and H. E. Huxley. 1973. X-ray evidence for radial cross-bridge movement and for the sliding filament model in actively contracting skeletal muscle. *J. Mol. Biol.* **77**:549.

Hatano, S., T. Totsuka and F. Oosawa. 1967. Polymerization of plasmodium actin. *Biochim. Biophys. Acta* **140**:109.

Huxley, A. F. and R. Niedergerke. 1954. Interference microscopy of living muscle fibres. *Nature* **173**:971.

Huxley, A. F. and R. M. Simmons. 1971. Proposed mechanism of force generation in striated muscle. *Nature* **233**:533.

———. 1973. Mechanical transients and the origin of muscular force. *Cold Spring Harbor Symp. Quant. Biol.* **37**:669.

Huxley, H. E. 1952. Investigations in biological structures by X-ray methods. The structure of muscle. Ph.D. thesis, University of Cambridge, Cambridge, England.

———. 1953a. Electron microscope studies of the organization of the filaments in striated muscle. *Biochim. Biophys. Acta* **12**.387.

———. 1953b. X-ray analysis and the problem of muscle. *Proc. Roy. Soc. B* **141**: 59.

———. 1957. The double array of filaments in cross-striated muscle. *J. Biophys. Biochem. Cytol.* **3**:631.

———. 1963. Electron microscope studies on the structure of natural and synthetic protein filaments from striated muscle. *J. Mol. Biol.* **7**:281.

———. 1968. Structural differences between resting and rigor muscle; evidence from intensity changes in the low-angle equatorial X-ray diagram. *J. Mol. Biol.* **37**:507.

———. 1969. The mechanism of muscle contraction. *Science* **164**:1356.

———. 1971. Structural changes during muscle contraction. *Biochem. J.* **125**:85.

———. 1973a. Structural changes in the actin- and myosin-containing filaments during contraction. *Cold Spring Harbor Symp. Quant. Biol.* **37**:361.

———. 1973b. Muscular contraction and cell motility. *Nature* **243**:445.

Huxley, H. E. and W. Brown. 1967. The low-angle X-ray diagram of vertebrate striated muscle and its behavior during contraction and rigor. *J. Mol. Biol.* **30**:383.

Huxley, H. E. and J. Hanson. 1954. Changes in the cross-striations of muscle during contraction and stretch and their structural interpretation. *Nature* **173**: 973.

————. 1955. The structural basis of contraction in striated muscle. *Symp. Soc. Exp. Biol.* **9**:228.

————. 1956. Preliminary observations on the structure of insect flight muscle. *Proceedings of the 1st European Regional Conference on Electron Microscopy,* Stockholm, p. 202. Almquist and Wiksell, Stockholm.

Ishikawa, H., R. Bischoff and H. Holtzer. 1969. Formation of arrowhead complexes with heavy meromyosin in a variety of cell types. *J. Cell Biol.* **43**:312.

Lowey, S., H. S. Slayter, A. G. Weeds and H. Baker. 1969. Substructure of the myosin molecule. I. Subfragments of myosin by enzymatic degradation. *J. Mol. Biol.* **42**:1.

Moore, P. B., H. E. Huxley and D. J. DeRosier. 1970. Three-dimensional reconstruction of F-actin, thin filaments and decorated thin filaments. *J. Mol. Biol.* **50**:279.

Nachmias, V. T., H. E. Huxley and D. Kessler. 1970. Electron microscope observations on actomyosin and actin preparations from *Physarum polycephalum* and on their interaction with heavy meromyosin subfragment I from muscle myosin. *J. Mol. Biol.* **50**:83.

Parry, D. A. D. and J. M. Squire. 1973. Structural role of tropomyosin in muscle regulation: Analysis of the X-ray diffraction patterns from relaxed and contracting muscles. *J. Mol. Biol.* **75**:33.

Pollard, T. D. and E. D. Korn. 1973. Acanthamoeba myosin. I. Isolation from *Acanthamoeba castellanii* of an enzyme similar to muscle myosin. *J. Biol. Chem.* **248**:4682.

Pollard, T. D., E. Shelton, R. R. Weihing and E. D. Korn. 1970. Ultrastructural characterization of F-actin isolated from *Acanthamoeba castellanii* and identification of cytoplasmic filaments as F-actin by reaction with rabbit heavy meromyosin. *J. Mol. Biol.* **50**:91.

Reedy, M. K., K. C. Holmes and R. T. Tregear. 1965. Induced changes in orientation of the cross-bridges of glycerinated insect flight muscle. *Nature* **207**:1276.

Spudich, J. A., H. E. Huxley and J. T. Finch. 1972. Regulation of skeletal muscle contraction. II. Structural studies of the interaction of the tropomyosin–troponin complex with actin. *J. Mol. Biol.* **72**:619.

Squire, J. M. 1973. General model of myosin filament structure. III. Molecular packing arrangements in myosin filaments. *J. Mol. Biol.* **77**:291.

Tanaka, H. and S. Hatano. 1972. Extraction of native tropomyosin-like substances from myxomycete plasmodium and the cross reaction between plasmodium F-actin and muscle native tropomyosin. *Biochim. Biophys. Acta* **257**:445.

Wakabayashi, T., H. E. Huxley, L. A. Amos and A. Klug. 1975. Three-dimensional image reconstruction of actin–tropomyosin complex and actin–tropomyosin–troponin T–troponin I complex. *J. Mol. Biol.* **93**:477.

Mechanism of Actomyosin ATPase

Edwin W. Taylor* and John A. Sleep

MRC Cell Biophysics Unit
London WC2, England

The isolation of actin and myosin from a variety of cell types (Pollard and Weihing 1974) and the demonstration of cytoplasmic streaming in extracts containing actomyosin (Oplatka and Tirosh 1973; Taylor et al. 1973) suggest that cell locomotion can be accomplished by a mechanism similar to muscle contraction. Much is known about the mechanical cycle in muscle and the biochemical cycle of muscle actomyosin, and it is likely that a complete understanding of these mechanisms will provide a basis for the explanation of streaming.

In the present context we wish to ask (1) what are the events in the cross-bridge cycle that are essential for force generation and movement, and (2) how could the cycle be utilized to produce streaming. Dr. Huxley (this volume) has described the presumed steps of attachment, motion and detachment of a cross bridge leading to sliding of the two filament lattices. Here a question arises: some asymmetry is necessary, which is supposed to be provided by a preferential attachment in a particular conformation, and this we will assume to be a perpendicular orientation of the cross bridge followed by a rotation to a second conformation in which a cross bridge makes a 45° angle to the thin filament. For simplicity, the system is described in terms of a single trajectory, but it should be remembered that this is the most probable one; in actual cycles, cross bridges may attach over a range of angles distributed about 90° and detach over a range around 45°. Is the asymmetry a property of the filament lattice or is it impressed on the reaction by kinetic considerations, namely, a higher rate of attachment at 90° than at 45° even though the free energy in the latter orientation must be lower? The primary purpose of kinetic studies is to determine the sequence of intermediate steps in the cycle of ATP hydrolysis and to relate these steps to the cross-bridge cycle inferred from structural studies. In addition, we wish to determine the

* Present address: Department of Biophysics and Theoretical Biology, The University of Chicago, Chicago, Illinois 60637.

rate constants of all steps, since this allows the free energy change to be determined at each step of the cycle.

Methods for Determination of Rate Processes

The reaction between an enzyme and substrate proceeds through a sequence of steps and involves an (initially) unknown number of intermediates. The intermediate states are detected by a change in a spectroscopic property of the complex (absorbance, fluorescence, light scattering), by ionization of an acidic or basic group (which is converted to an absorbance change of a pH indicator dye), or by direct chemical measurements of intermediates or products formed from the substrate. As a general rule, the number of detectable intermediates increases with the number of measurement techniques employed. The method of analysis of kinetic data to determine rate constants and intermediates is essentially the same for all types of measurements and can be illustrated by a simple example. The binding of ADP to myosin is a two-step process:

$$M + ADP \underset{}{\overset{k_1}{\rightleftharpoons}} M{\cdot}ADP \underset{}{\overset{k_2}{\rightleftharpoons}} M{\cdot}ADP^* + \varepsilon H^+.$$

For each step, there are rate constants k_i and k_{-i}, and the corresponding equilibrium constant $K_i = k_i/k_{-i}$. When myosin is mixed with ADP in a rapid mixing device, there is a change in fluorescence emission of tryptophan residues of the enzyme and a partial ionization of some side chain at pH 8. The asterisk refers to a state of enhance fluorescence, and ε is 0.3 to 0.4 moles per site at pH 8. If ADP is in excess, the differential equation describing the concentration of each enzyme state, M, M·ADP and M·ADP*, as a function of time after mixing is easily solved, and the solution is the sum of two exponential terms. Thus $[M{\cdot}ADP^*] = (M{\cdot}ADP^*)_f(1 - ae^{-\lambda_1 t} + be^{-\lambda_2 t})$, where the subscript "$f$" refers to the final concentration at equilibrium. The parameters λ_1 and λ_2 are constants for a given ADP concentration and have the dimensions of first-order rate constant ($\sec^{-1}$). The values are determined by fitting the experimental curve to the equation graphically or by means of a suitable computer program. The parameters are determined for a series of ADP concentrations, if possible extending to a concentration range such that λ_1 becomes independent of ADP concentration. The behavior as a function of ADP concentration can be understood by a simple argument. Ignoring the reverse reactions, the flow rate to the formation of the first intermediate M·ADP, expressed per mole of myosin, is k_1 (ADP): k_1 has the dimensions $M^{-1}\sec^{-1}$, and, consequently, k_1 (ADP) is a rate ($\sec^{-1}$). At a low ADP concentration we can arrange to have k_1 (ADP)$<< k_2$; thus M·ADP is converted to M·ADP* as fast as it is formed, and the rate of generation of the M·ADP* state is k_1 (ADP). The signal measured in the experiment, such as an increase in fluorescence, is proportional to the concentration of M·ADP*. Since this state is reached from another intermediate, there will be a time lag in forming it; which leads to the second exponential term (λ_2) in the rate equation. As the ADP concentration is increased, the apparent rate will increase linearly with ADP concentration so long as the inequality holds; but at high ADP concentrations we can have k_1 (ADP)$>> k_2$. Now, all M is converted to M·ADP in a very short time, and the lag disappears. The rate of formation of M·ADP* proceeds at the rate k_2, which is now independent of ADP concentration, and the time dependence takes the simple form $(1 - e^{-k_2 t})$. Therefore the rate constants k_1 and k_2 can be determined from the initial slope and final value of a plot of apparent rate versus ADP concentration. If the reverse reactions are included, the problem is only slightly more complicated. The maximum rate is $k_2 + k_{-2}$, and the initial slope is k_1 if $k_{-1} << k_2$ and $K_1 k_2$ if $k_{-1} >> k_2$. From studies on a number of enzymes it has been found that substrate-binding is a fast process, with rate constants approaching the diffusion-controlled limit $(k_1 \gtrsim 10^7 \text{ M}^{-1}\text{sec}^{-1})$, and the reverse reaction is also very fast. Therefore it is usually the case that $k_{-1} >> k_2$, and kinetic measurements give values of K_1 and k_2.

If the experimentally observable state corresponds to a change at the third step of a reaction, the actual rate of this step can be determined if $k_3 < k_2$. For example, the first three steps of the myosin reaction appear to be

$$\text{M} + \text{ATP} \xrightarrow{k_1} \text{M·ATP} \xrightarrow{k_2} \text{M·ATP**} \xrightarrow{k_3} \text{M·Pr**},$$

where the double asterisks indicate a state with increased fluorescence, and Pr refers to a state in which the bond in ATP has been hydrolized and both products are bound to the enzyme. Therefore measurements of fluorescence and inorganic phosphate formation can give K_1, k_2 and k_3 under favorable conditions. However since the steps are sequential if $k_3 > k_2$, the apparent rate of P_i formation at high ATP concentrations could not be larger than k_2. This appears to be the case at 20°C and illustrates a problem in kinetic analysis. The state is known to be present since the appearance of enzyme-bound P_i can be measured. However if a fast step occurs in the middle of the sequence and there is no method of detecting it, the corresponding intermediate will be missing in the kinetic scheme. One can be fairly confident that any scheme proposed will have less than the actual number of intermediates.

A second problem is the measurement of a branch in the pathway. A simple example is the dissociation of actomyosin by ATP versus hydrolysis of ATP. The simplest pathway is

$$\text{AM} + \text{ATP} \xrightarrow{1} \text{AM·ATP} \xrightarrow{2} \text{AM·Pr}$$

$$(-\text{A}) \Big\downarrow 3$$

$$\text{M·ATP} \xrightarrow{4} \text{M·Pr}.$$

The dissociation of AM can be measured by a change in turbidity, and the apparent rate of hydrolysis, by measurements of P_i. Measurements have to be made over a sufficient range of ATP concentrations to determine the actual rate constant of at least the slower step. Experiments have shown that the rate of dissociation (step 3) is many times faster than the rate of the hydrolysis step, which is essentially the same as for myosin itself. Consequently, the reaction follows the branch via steps 3 and 4.

A detailed scheme for myosin ATPase has been built up by the methods outlined here. For actomyosin, the evidence is less complete, but the essential features are relatively clear. However, there is one complication that has so far been ignored. With one exception, myosin isolated from all muscle and non-muscle sources has two heads and one ATP site per head. Most contraction models treat a cross bridge as a single entity, and it may turn out that interaction between heads is a necessary feature of the cycle. We will consider the mechanism for a single-headed molecule because it is considerably simpler to describe and much of the evidence has been obtained using subfragment 1.

Myosin Kinetic Scheme

The following scheme has been proposed by Trentham and collaborators (Bagshaw and Trentham 1974) and independently by Koretz and Taylor (1975):

$$M \underset{1}{\rightleftharpoons} M \cdot ATP \underset{2}{\rightleftharpoons} M \cdot ATP^{**} \underset{3}{\rightleftharpoons} M \cdot Pr^{**} \underset{4}{\rightleftharpoons} M \cdot ADP^* \underset{5}{\rightleftharpoons} M \cdot ADP \underset{6}{\rightleftharpoons} M.$$

$$\uparrow ATP \qquad \downarrow \varepsilon H^+ \qquad\qquad \downarrow H^+, P_i \qquad\qquad \uparrow \varepsilon H^+ \qquad \downarrow ADP$$

The values of the rate and equilibrium constants under roughly physiological conditions are shown in Table 1. The scheme has a number of interesting features. The first step is a rapid equilibrium, which is followed by a change in conformation affecting the environment of a tryptophan and one or more charged groups and leading to an extremely tight binding of substrate. The total binding constant for substrate is $K_1 K_2 = 10^{10} - 10^{12}$ M^{-1}, which is a larger change in standard free energy than for the hydrolysis of ATP itself. The actual hydrolysis step (3) for ATP on the enzyme has a small free energy change and may be associated with a further change in conformation, but

Table 1
Kinetic Constants for Myosin ATPase

Quantity	Value	Method	Reference
K_1	5×10^3 M^{-1}	H^+, fluorescence (calculated)	1,2,3
k_2	200–400 sec^{-1}	H^+, fluorescence, pH-dependent	1,2,3
K_2	$10^6 - 2.10^8$	reversal of reaction[a]	4,5
k_3	~ 200 sec^{-1}	P_i	2,6
K_3	5–10	stoichiometric reaction, burst size	7,8,9
k_4	.05 sec^{-1}	fluorescence, conductance, chromatography	3,10,11
K_4	~ 10	reversal of reaction	4
k_5	2 sec^{-1}	fluorescence, absorbance	1,3
k_{-5}	200–400 sec^{-1}	fluorescence for ADP binding, H^+	1,2,3
K_6	2×10^{-4} M	fluorescence, H^+ (calculated)	1,2,3

References: [1] Bagshaw et al. (1974); [2] Koretz and Taylor (1975); [3] J. Sleep and E. W. Taylor (in prep.); [4] Mannherz et al. (1974); [5] Wolcott and Boyer (1976); [6] Lymn and Taylor (1970); [7] Bagshaw and Trentham (1973); [8] E. W. Taylor (in prep.); [9] Taylor et al. (1970); [10] Bagshaw and Trentham (1974).

[a] The equilibrium constant for ATP hydrolysis (10^6 M^{-1}) and the relation $K_1 \cdot K_2 \ldots K_6 = K_{ATP}$ was used to obtain K_2. The conditions are 0.05–0.1 M KCl, pH 8, 20°C, excess Mg^{++}. K_2 and K_4, determined for subfragment 1 only, remaining values refer to heavy meromyosin and subfragment 1.

there is no direct evidence from the signals measured thus far. Both products are bound in the M·Pr** state, and the H^+, which should be released from ionization of the product $H_2PO_4^{-1}$, is not obtained until the product is released in step 4. This is a metastable state that decays at a very slow rate, i.e., the lifetime is 15 seconds, and the release of P_i from this state determines the steady-state rate of the enzyme. (Trentham distinguishes two steps here, M·Pr** $\rightleftharpoons$ M·ADP** + P_i and M·ADP** $\rightleftharpoons$ M·ADP*, the asterisks referring to the magnitude of fluorescence enhancement. But the second of these steps is much faster than the first and, consequently, cannot be observed in the forward reaction; we have therefore omitted it.) Step 4 also involves a conformation change, as detected by a difference in fluorescence enhancement. The last two steps appear to be the same as for the binding of ADP to myosin. The first binding step for ADP and ATP has essentially the same equilibrium constant, and the rate of the conformation change is also the same. The two processes differ in the very small value of k_{-2} compared to k_5.

In passing it should be noted that when a reaction mechanism is subjected to detailed analysis, the concept of a high energy phosphate bond disappears.

The important points about this series of maneuvers are the very tight binding of the substrate, which produces the double-asterisk state, and the rapid formation of M·Pr**, which is a metastable state. The large binding energy is presumably used for the dissociation of actomyosin, whereas the long-lived state is available for combination with actin. We would expect these properties to be retained by a non-muscle myosin.

Actomyosin Kinetic Scheme

ATP is both a substrate for actomyosin and a dissociator of the actomyosin complex. As was first recognized by Eisenberg and Moos (1968), the true maximum rate of actomyosin ATPase can only be determined by extrapolation to infinite actin concentration. The degree of activation is 200–500-fold. Kinetic analysis shows that ATP produces dissociation before it is hydrolyzed, and that a number of nonhydrolyzable polyphosphates, including pyrophosphate, ADP and AMPPNP, also give dissociation. Thus dissociation is not directly connected with hydrolysis. It was therefore suggested (Lymn and Taylor 1971) that the reaction cycle, in a simplified form, omitting conformation changes, is

$$
\begin{array}{ccc}
 & \xrightarrow{\quad 4 \quad} & \\
\text{AM} & & \text{AM·Pr} \\
 & -(\text{Pr}) & \\
\Big\uparrow 1 \!\!\!\Big\downarrow \begin{array}{c} +\text{ATP} \\ -\text{A} \end{array} & & \begin{array}{c} +\text{A} \end{array} \Big\uparrow 3 \!\!\!\Big\downarrow \\
\text{M·ATP} & \xrightarrow{\quad 2 \quad} & \text{M·Pr}
\end{array}
$$

This scheme, which is consistent with the biochemical evidence, is easily identified with the contraction cycle, since the dissociation and recombination of a cross bridge corresponds to dissociation of AM by ATP (step 1) and recombination of actin with the M·Pr state (step 3). Further studies have supported this scheme. A basic requirement of the model is that step 3 be rate-limiting at a finite actin concentration. Thus from a measurement of the apparent rate constant of step 3, the steady-state ATPase rate should be predicted quantitatively. An extensive study over a range of temperatures and ionic strengths (H. White and E. W. Taylor, in prep.) gave very good agreement between the measured and predicted values.

The simple Lymn-Taylor scheme has to be expanded to take into account the conformational states that have been distinguished in the myosin mechanism. The same branching pathway problem has to be reinvestigated, since we must now ask whether the change in conformation to the double-asterisk state precedes or follows dissociation. The system was studied by simultaneously measuring dissociation by turbidity and conformation change by fluorescence, using a stop-flow apparatus with two photomultipliers at right angles to the incident beam. Some signal processing is necessary since the turbidity change contributes to the light intensity of the fluorescence signal. The result is unequivocal at high ATP concentration, since dissociation is again much faster than the fluorescence change and there is no interference between signals. The experiments were done at 4°C, thereby increasing the ratio of the relative rates of the two processes.

In Figure 1 we have plotted the rate of dissociation (actomyosin), the rates of the fluorescence increases for actomyosin and myosin, and the rate of

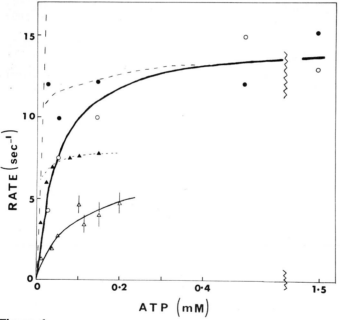

Figure 1

Rates of dissociation of actomyosin, fluorescence enhancement and phosphate formation by actomyosin and myosin. (– –•– –) Apparent rate of fluorescence enhancement for acto-S_1; (—○—) apparent rate of fluorescence enhancement for S_1; (– –▲– –) apparent rate of phosphate formation for acto-S_1; (—△—) apparent rate of phosphate formation for S_1; (– – –) apparent rate of acto-S_1 dissociation measured by turbidity change. All experiments: 4°C, pH 6.9, .05 M KCl, 10 mM $MgCl_2$.

P_i formation for actomyosin and myosin. The maximum rate for dissociation exceeds 700 sec^{-1}, and the rates for the fluorescence and phosphate steps are 13–15 sec^{-1} and about 7 sec^{-1}, respectively. Therefore the dissociation step must precede the conformation change. Furthermore, the myosin state released cannot be M·ATP. In this state, ATP is in rapid equilibrium (less than 1 msec) with free ATP in the medium; but dissociation of AM with [γ-^{32}P]-ATP, followed by quenching with excess unlabeled ATP at a time when dissociation is complete but only a small fraction of the fluorescence change has taken place, shows that the myosin-ATP state does not exchange with ATP in the medium. Thus the myosin is released in a state distinguishable from those occurring in the myosin pathway and is referred to as M·ATP†.

The complete cycle can be represented by

$$
\begin{array}{c}
\text{AM} \xleftarrow{\quad\quad 5 \quad\quad} \text{AMPr}^{**} \\
\end{array}
$$

AM ⟵————— 5 —————— AMPr**

$-P_i$

1 | −A   +A | 4

+ATP

M·ATP† ⟶ M·ATP** ⟶ M·Pr**.
     2   3

A shortcoming of this scheme is that the rate of release of products from actin has not been measured directly. Originally we assumed that this rate is the steady-state rate of actomyosin ATPase obtained by extrapolation to infinite actin. Studies by Eisenberg (Eisenberg and Kielley 1973) have shown that under some conditions, actomyosin is largely dissociated even at very high actin concentrations. This could be explained if there is a state that cannot combine with actin (the refractory state) missing between M·ATP** and M·Pr**.

This problem has not been settled satisfactorily but it does not greatly affect the general conclusions we wish to draw. The expanded cycle preserves the essential features of the Lymn-Taylor model and begins to take the conformational changes into account. The cycle has one new and interesting feature. We can equate steps 1 and 4 with the dissociation and recombination of a cross bridge. The bridge dissociates in the dagger state and rebinds in the double-asterisk state. The transition between these states takes place on the free cross bridge, the reverse transition occurs between attached states, although we cannot as yet assign a conformational label to AM. The kinetic scheme has the property we might expect to find in order for the cycle to be driven with spatial asymmetry; that is, attachment and detachment occur in different conformation states, and these states might correspond to the 90° and 45° conformations of the cross bridge.

This is no more than an appealing speculation yet it provides a basis for an asymmetric cycle, maintained by kinetic constraints, in which the asymmetry is in the myosin conformation rather than being a property of the lattice.

Two Heads

Much of the work on the cross-bridge mechanism has been done with the single-headed fragment, although studies in our laboratory have usually been

Troponin-Tropomyosin-dependent Regulation of Muscle Contraction by Calcium

John Gergely

Department of Muscle Research, Boston Biomedical Research Institute (02114)
Department of Neurology, Massachusetts General Hospital (02114)
and Department of Biological Chemistry, Harvard Medical School
Boston, Massachusetts 02115

The interactions between actin and myosin (see Huxley, this volume) that lead to tension development and contraction are regulated by the sarcoplasmic Ca^{++} concentration. The latter, in turn, is controlled by the sarcoplasmic reticulum (see MacLennan et al., this volume). This brief review will deal with the mechanism by which the increased Ca^{++} concentration in an active muscle relieves the inhibition of the actin–myosin interaction that exists in the relaxed state.

   Ca^{++} can exert its control on the actin–myosin interaction either by what is referred to as thin filament control, or by thick filament control, or both (Szent-Györgyi 1975; Lehman and Szent-Györgyi 1975). The term "thin filament control" refers to the fact that additional proteins required for the control mechanism—tropomyosin and troponin—are located in the thin filaments. In the case of thick filament control, which is prevalent in a large number of lower organisms, the effect of Ca^{++} is exerted directly on myosin (as discussed by Lehman, this volume). This paper is concerned with thin filament control.

   The possibility of calcium-induced contraction was clearly shown by Heilbrun and Wiercinski (1947). They found that microinjection of Ca^{++} into a muscle fiber produced contraction. In a sense, this observation was premature, since at that time, Ca^{++} control of the actin–myosin interaction did not fit into the accepted conceptual framework based on the known properties of myofibrillar proteins. About ten years later, the work of A. Weber and S. Ebashi and their colleagues firmly established the fact that the interaction of actin and myosin, manifested in various ways, namely, an increase in ATPase activity, superprecipitation, contraction of glycerinated fibers and in the presence of ATP, depended on a minute amount of Ca^{++} (for reviews, see Ebashi and Endo 1968; Ebashi, Endo and Ohtsuki 1969; Weber and Murray 1973; Perry 1973). The inhibitory effects of supraoptimal concentrations of ATP on ATPase activity and superprecipitation were recognized by Weber (1959) as being due to the chelation of small amounts of Ca^{++}

ions present in the system. The use of the chelating agent EGTA, whose affinity is about 10^6 times higher for Ca^{++} than Mg^{++}, made it possible to perform experiments under conditions where the free Ca^{++} concentration was precisely controlled (Ebashi 1960; Weber and Herz 1963).

Regulation by Tropomyosin Plus Troponin

In the early 1960's, S. Ebashi and colleagues, following an earlier brief report by Perry and Gray (1956) and some observations by Weber and Winicur (1961), discovered that highly purified actomyosin from rabbit skeletal muscle did not show dependence on Ca^{++} unless a specific component, at first called native tropomyosin (Ebashi 1963; Ebashi and Ebashi 1964), was present. In its presence, about 10^{-6} M Ca^{++} was required for superprecipitation and ATPase activity. Native tropomyosin in many ways resembled tropomyosin, a protein whose existence had been known about since 1946 (Bailey) but for which no role had been found in muscle, although since the early 1960's, beginning with the work of Hanson and Lowy (1963), it had been known that tropomyosin molecules are located in the thin filaments. Tropomyosin molecules consist of pairs of coiled-coil α helices; their length (40 nm) is such that as they run along the groove of the thin filament, each tropomyosin molecule is in contact with about seven actin monomers (Cohen et al. 1973; Stone et al. 1974; Squire 1975). It soon became clear that native tropomyosin contains Bailey's tropomyosin plus another component, which was given the name troponin (Ebashi and Kodama 1966). S. Ebashi and colleagues suggested that each tropomyosin is combined with a troponin (Ebashi, Endo and Ohtsuki 1969), a suggestion that has received ample confirmation. Early work indicated that control by calcium involves the combination of calcium with troponin (Ebashi and Endo 1968; Ebashi, Endo and Ohtsuki 1969), but progress toward a detailed understanding of the mechanism did not take place until more recently.

Understanding the mechanism depended on two things. One was the discovery by a number of investigators, utilizing X-ray and electron microscopic techniques, including three-dimensional reconstruction by means of optical diffraction of electron micrographs, that the position of tropomyosin in the thin filaments changes depending on the state of activation; tropomyosin in relaxed muscle is located farther away from the long-pitch groove of the actin filament and upon activation it moves closer to it (Huxley 1973; Haselgrove 1973; Vibert et al. 1972; Spudich, Huxley and Finch 1973; Parry and Squire 1973; Wakabayashi et al. 1975). The second was the finding that troponin is not a single entity (Hartshorne and Mueller 1968).

Subunits of Troponin

Initially it seemed that troponin could be fractionated into two components, but more detailed studies showed that troponin consists of three subunits, named TnI, TnC and TnT (ATPase-inhibitory, Ca^{++}-binding and tropomyosin-binding subunits, respectively), each having distinct characteristics (Greaser and Gergely 1971, 1973; Greaser et al. 1973; Ebashi, Ohtsuki and

Table 1

Subunits of Troponin

Name	MW[a]	Characteristic
TnC	18,000	Ca^{++}-binding
TnI	21,000	inhibition of actomyosin ATPase $\pm$ Ca^{++}
TnT	32,000	binding of tropomyosin

[a] Based on amino acid sequences and composition (see text).

Mihashi 1973). The amino acid sequences of TnC (Collins et al. 1973) and TnI (Wilkinson and Grand 1974) have been established and that of TnT (Pearlstone, Carpenter and Smillie 1975; Collins 1975) is well on its way to being established (Table 1). Thus it is possible to begin the analysis on the molecular and submolecular level of the mechanism by which Ca^{++} controls the contraction and relaxation of muscle. The three isolated subunits in troponin can be recombined and the original troponin activity restored (Greaser and Gergely 1971). This activity, as mentioned above, is manifested in rendering actomyosin calcium-sensitive in the presence of tropomyosin, or, to put it differently, actomyosin ATPase or superprecipitation is inhibited when Ca^{++} is chelated by EGTA.

One of the components, TnI, is in itself an inhibitor of ATPase activity of actomyosin; but this effect is independent of Ca^{++} (Wilkinson et al. 1974). The precise relation of this inhibition, from which the name TnI is derived, to the calcium-sensitive one found with natural Tn (unfractionated troponin) or with the reconstituted complex is not clear. TnT has strong affinity for tropomyosin and furnishes at least one anchoring point for the troponin complex on tropomyosin. Electron microscopic study of the staining pattern of tropomyosin paracrystals, with and without troponin, has led to the suggestion (Cohen et al. 1973; Ohtsuki 1974) that TnT binds to a region located about a third of the way in from the end of the tropomyosin molecule (Fig. 1). The amino acid sequence of tropomyosin and studies of methyl mercury-bind-

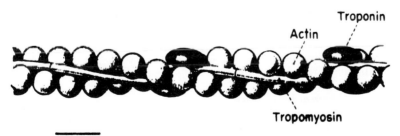

10 nm

Figure 1

Model for the thin filament structure proposed by Ebashi and colleagues. Note the relation of the troponin to each of the tropomyosin molecules. (Reprinted, with permission, from Ebashi et al. 1975.)

ing to paracrystals indicate that the TnT binding region contains the single cysteine residue of α-tropomyosin approximately 14 nm from the carboxyl terminus (Stone et al. 1974; Stewart 1975).

Ca-binding to TnC

TnC has been recognized as the Ca^{++} binding component that mediates the effect of Ca^{++} on the troponin complex, which results in the release of the actin-myosin inhibition. Hence studies of the interaction of Ca^{++} with TnC are important for achieving an understanding of the troponin–Ca^{++} interaction. Earlier studies of this kind suffered from various technical difficulties, including the lack of homogeneous preparations and appropriate methods for determining calcium-binding at very low free Ca^{++} concentrations. The availability of purified TnC and the use of calcium/EGTA buffers have made such measurements possible. It now appears that TnC binds 4 moles of Ca^{++} per molecule (Table 2) (Potter and Gergely 1975, see also for earlier references; for another view, see Honig and Reddy 1975). There is no indication of interaction among the calcium binding sites, which appear to divide into two classes, each containing a pair of sites. The two sites of higher affinity towards Ca^{++} also bind Mg^{++} and have been referred to as Ca^{++}-Mg^{++} sites (Potter and Gergely 1975). In contrast, the other two sites bind only Ca^{++}, hence the term Ca^{++}-specific sites. In the presence of higher concentrations of Mg^{++}, the apparent affinity of Ca^{++} to the Ca^{++}-Mg^{++} binding sites is reduced; and in the presence of 2 mM Mg^{++}, Ca^{++}-binding can be described in terms of four equivalent sites with the same binding constant. The Ca^{++} binding properties of TnC change when it is combined with the other components; for example, the combination of TnC with TnI has the same binding constants as the whole Tn complex. In addition to the four Ca^{++} binding sites, there are two more sites that bind Mg^{++} but not Ca^{++}, bringing the total of divalent cation binding sites in TnC to six. The role of the Mg^{++} binding sites that do not bind Ca^{++} is at the moment not known.

Subunit Interactions and the Mechanism of Regulation

TnC interacts not only with TnI, as shown by the solubilization of the latter at low ionic strength and by the neutralization of the inhibitory action of TnI,

Table 2
Divalent Cation Binding Sites of Troponin C

Type of site	Apparent binding constants		No. of sites
	Ca^{++} (M^{-1})	Mg^{++} (M^{-1})	
Ca^{++}-Mg^{++}	2×10^7	5×10^3	2
Ca^{++}	2×10^5	—	2
Mg^{++}	—	5×10^3	2

but also with TnT; TnT does not combine with TnI in the absence of TnC
(Greaser and Gergely 1973; Ebashi, Ohtsuki and Mihashi 1973; Drabikowski
and Dabrowska 1974). Thus it seems that TnC plays a central role in joining
the two other components in the troponin complex. The initial observations of
Hitchcock and Szent-Györgyi (1974) made it clear that the binding of the
TnC-TnI complex to the thin filaments was itself modulated by Ca^{++}: addition of Ca^{++} promoted dissociation, whereas binding was strengthened by its
removal. Further studies showed that the binding of the TnC-TnI complex
required both tropomyosin and actin, and the idea has been put forward
that in vivo troponin is attached in a two-pronged fashion to the thin filaments
(Margossian and Cohen 1973; Hitchcock, Huxley and Szent-Györgyi 1973;
Potter and Gergely 1974; Hitchcock 1975). According to this idea, one connection would be formed via the TnT component, which forms a strong link
with Ca^{++}-insensitive tropomyosin. The other attachment would be formed
by TnI binding to a site made up jointly of tropomyosin and actin (Potter and
Gergely 1974). According to this view, the binding of Ca^{++} to TnC would
lead to a release of TnI from one of the binding sites, leaving the other attachment, involving TnT, intact. Following this dissociation, a change in the position of tropomyosin would take place so that it would no longer block the
interaction of myosin heads with actin. The question of whether the release of
troponin from one of its binding sites would induce conformational changes in
tropomyosin or actin has not been settled (Tonomura, Watanabe and Morales
1969).

Conformational Changes

Extensive studies have been carried out on the changes that take place in
TnC upon combination with Ca^{++}. Combination with either Ca^{++} or Mg^{++}
produces large changes in structure, usually referred to as conformational, as
evidenced by changes in circular dichroism and optical rotation (Van Eerd and
Kawasaki 1972; Kawasaki and Van Eerd 1972; Murray and Kay 1972; Gergely
et al. 1974), in the mobility of attached spin labels (Potter et al. 1974; Ebashi
et al. 1974; Ohnishi, Maruyama and Ebashi 1975) and in the reactivity of the
Cys98 residue (Potter et al. 1975). Changes have also been reported in the optical properties of tyrosine (Van Eerd and Kawasaki 1972) and phenylalanine
(Head and Perry 1974) and in extrinsic chromophores (Potter et al. 1974;
Van Eerd and Kawasaki 1973), which are attributable to changes in their
environment resulting in turn from overall structural changes in the protein.
With the knowledge that binding sites of different affinities exist, the question
arises as to which of the binding sites are involved in the production of these
conformational changes. This question is of even greater interest now that the
primary sequence of troponin is known and homologies have been discovered
(Collins et al. 1973; Collins 1974) between troponin C and the calcium binding protein of carp muscle parvalbumin, whose three-dimensional structure
has been determined by X-ray diffraction (Kretsinger and Nockolds 1973).
Comparison of the amino acid sequence of TnC with that of parvalbumin
led to the conclusion that there are four regions in TnC that exhibit the same
features as the two calcium binding regions of parvalbumin, which is in agreement with the number of directly determined calcium binding sites of TnC.

In parvalbumin, these calcium binding regions consist of a so-called calcium binding loop flanked by two α-helical segments (Kretsinger and Nockolds 1973). On the basis of the chemical homology it has been suggested that the Ca^{++} binding regions in TnC have a similar structure (Collins et al. 1973; Collins 1974; Weeds and McLachlan 1974; Kretsinger and Barry 1975; Kretsinger 1975). One of these regions (III) contains the only sulfhydryl (cysteinyl) residue of troponin C, which makes it possible to recognize changes in that portion of the molecule. Thus it appears that when calcium-binding to the Ca^{++}-Mg^{++} sites takes place, the large conformational change is accompanied by changes in the environment of the sulfhydryl group (Potter et al. 1975), suggesting that at least one of the Ca^{++}-Mg^{++} binding sites is the site adjacent to the Cys98 of TnC. So far, very little, if any, evidence exists for conformational changes accompanying the binding of Ca^{++} to the Ca-specific sites.

Ca^{++} Dependence of Actin–Myosin Interaction—Cooperativity

The question naturally arises whether Ca-binding to all or only to some of the sites is involved in producing deinhibition of the actin–myosin interaction and whether the conformational changes that one observes in vitro are related to changes that would occur in vivo in conjunction with the repression of the actin–myosin interaction. Light can be thrown on this question by studying the Ca dependence of the ATPase activity of myofibrils, a system that contains actin, myosin, tropomyosin and troponin (Bremel and Weber 1972; Potter and Gergely 1975). By varying both the Ca^{++} and Mg^{++} concentrations and using the calcium binding constants determined for whole troponin, the ATPase activity can be calculated based on various assumptions regarding the requirement for calcium at different binding sites. According to Potter and Gergely (1975), Mg^{++} does not affect the range of Ca^{++} concentrations that activate the ATPase activity, and the experimental data can best be explained by assuming that Ca^{++} is required at the Ca^{++}-specific sites but either Ca^{++} or Mg^{++} may be present at the Ca^{++}-Mg^{++}-sites (Potter and Gergely 1975). However, tension development in muscle fiber preparations requires more Ca^{++} at higher Mg^{++} concentrations (Ebashi and Endo 1968; Donaldson and Kerrick 1975), suggesting that the Ca^{++}-Mg^{++} sites are involved. More work will be required to resolve this apparent contradiction.

Attention has been drawn to the cooperative nature of some of the interactions in the filament regulatory system. Calcium-binding itself does not show evidence of cooperative binding. The somewhat steeper dependence of myofibrillar ATPase activity on Ca^{++} concentration is satisfactorily described in terms of binding of Ca^{++} to multiple noninteracting sites, as indicated above. This apparent cooperativity in the effect of Ca^{++} can arise even though the binding process of Ca^{++} takes place without cooperativity; that is, the free energy of binding to the successive binding sites is additive. Weber and colleagues (Bremel, Murray and Weber 1973) have suggested cooperativity in the myosin–actin interaction to explain the observation that at low MgATP concentration, the combination of heavy meromyosin subfragment

1 and actin was independent of Ca^{++} even in the presence of tropomyosin-troponin. Accordingly, binding of a myosin head within the domain of the tropomyosin molecule facilitates the binding of another myosin head in the same domain, resulting in potentiation of the activating effect on the ATPase activity. This potentiation can be exerted either by rigor links, that is, myosin heads that do not bear ATP or the $ADP \cdot P_i$ intermediate complex (see Taylor and Sleep, this volume), or by myosin heads that do carry the intermediate complex. The physiological importance of such a mechanism is not fully clear. It would appear that there are nine cross bridges per 43 nm (Squire 1975) along the thick filaments, there being two thin filaments for each thin filament in cross section, with an actin monomer repeat of 27.5 nm. There could be, on the average, at most two cross bridges for eight actin monomers if one assumes that each cross bridge could always form a link with an actin monomer; but each cross bridge forms only one cross-link. The first of these assumptions neglects the restriction arising out of the geometry of the myosin projections on the thick filaments and the orientation of actin-myosin with the thin filaments. In fact, only about 40% to 50% of the actin cross bridges are attached in active muscle (Haselgrove and Huxley 1973), and hence the probability of there being several cross bridges attached within the domain of a single tropomyosin molecule (seven actin monomers) is small. However if the movement of one tropomyosin molecule influences that of others in the same thin filament, the effect of myosin-binding can extend beyond the domain of a single tropomyosin molecule, and cooperativity among cross bridges even at low average occupancy of actin sites would be possible.

Comparative Aspects

Some of the components of the tropomyosin-troponin system differ according to the type of muscle from which they have been obtained (Perry 1974). They share this property with other muscle proteins, notably myosin. In the case of myosin, a correlation exists between its ATPase activity and the velocity of muscle contraction (Bárány 1967; Close 1972). It is not as yet clear what the functional significance of the chemical and structural differences in tropomyosin and troponin is. Tropomyosin preparations contain two kinds of chains, α and β; the α chain is found in predominantly fast skeletal muscles, whereas the β chain is more abundant in slow skeletal muscles (Cummins and Perry 1973, 1974). While cardiac muscle myosin seems to resemble its slow skeletal muscle counterpart, cardiac tropomyosin contains only the α chain and is therefore similar to the fast skeletal muscle protein. Two of the troponin components, TnI and TnT, can be distinguished by their different electrophoretic mobility in cardiac muscle and in slow skeletal muscle (Tsukui and Ebashi 1973; Syska, Perry and Trayer 1974; Drabikowski, Dabrowska and Barylko 1975). The amino acid sequence of cardiac troponin C has been reported recently (Van Eerd and Takahashi 1975); on the basis of differences between it and that of skeletal muscle TnC in one of the regions presumed to be a Ca binding site it has been suggested that there would be only three calcium binding sites in the cardiac protein. Some preliminary binding data apparently support this conclusion. Again the physiological significance of the difference in chemical structure and calcium-binding remains to be seen. Lobster troponin C has recently been reported as having only one Ca^{++} binding site (Regenstein and Szent-Györgyi 1975). The type of regulation present in smooth muscle has not been settled. Several reports (Bremel 1974; Sobieszek and Bremel 1975) suggest that thick filament regulation involving a direct action of Ca^{++} on myosin exists, whereas others (Driska and Hartshorne 1975; Ebashi, Toyo-oka and Nonomura 1975; Ebashi et al. 1975) find evidence for a tropomyosin-troponin-dependent regulatory system having properties different from those found in thin-filament-regulated striated muscle.

Phosphorylation

Finally, the phosphorylation of components of the troponin system deserves a brief discussion. Phosphorylation of both TnI and TnT has been reported (for a review, see Perry et al. 1975). Both proteins can be phosphorylated by the calcium-activated phosphorylase kinase, and a 3'-5'-cyclic AMP-dependent protein kinase phosphorylates TnI at a second site. The phosphorylation of TnI is more pronounced in cardiac muscle. In TnI from fast skeletal muscle, both of the phosphorylation sites are blocked by TnC, whereas in cardiac TnI (Cole and Perry 1975), only the site phosphorylated by phosphorylase kinase is blocked. TnC also inhibits the phosphorylation of TnT, suggesting that in both TnI and TnT, there is a phosphorylation site in the region that interacts with TnC. The site phosphorylated by the cyclic AMP-dependent kinase appears to be in the region of the TnI molecule that is involved in the interaction with actin. It should be emphasized that notwithstanding the structural similarities between myosin light chains and TnC (Collins 1974; Potter et al. 1975), no counterpart of the phosphorylation of the so-called Nbs_2 (5, 5'-dithiobis (2-nitrobenzoic acid)) light chain of myosin (Perry et al. 1975) is found in TnC.

In view of the widespread involvement of phosphorylation in the regulation of various biological processes it is intriguing to speculate on the possible regulatory role of phosphorylation in the troponin-tropomyosin system and thus of the role of phosphorylation in the regulation of actin–myosin interaction (Cole and Perry 1975). Since the level of cyclic AMP is influenced by catecholamines, the possibility of catecholamine control of TnI phosphorylation arises. Epinephrine is known to increase the force of contraction in cardiac muscle, and according to a recent report, correlation exists between the inotropic effect of epinephrine and the phosphorylation of troponin I of rat heart (England 1975). While it would be premature to conclude a causal relation between these effects, additional work will undoubtedly be done on this important process.

Acknowledgments

Preparation of this paper was supported by grants from the National Institutes of Health (HL-5949), the National Science Foundation, and the Muscular Dystrophy Associations of America, Inc.

REFERENCES

Bailey, K. 1946. Tropomyosin: A new asymmetrical protein component of muscle. *Nature* **157**:368.

Bárány, M. 1967. ATPase activity of myosin correlated with the speed of muscle shortening. *J. Gen. Physiol.* **50**:197.

Bremel, R. D. 1974. Myosin linked calcium regulation in vertebrate smooth muscle. *Nature* **252**:405.

Bremel, R. D. and A. Weber. 1972. Cooperation within actin filaments in vertebrate skeletal muscle. *Nature New Biol.* **238**:97.

Bremel, R. D., J. M. Murray and A. Weber. 1973. Manifestations of cooperative behaviour in the regulated actin filament during actin-activated ATP hydrolysis in the presence of calcium. *Cold Spring Harbor Symp. Quant. Biol.* **37**:267.

Close, R. 1972. Dynamic properties of mammalian skeletal muscles. *Physiol. Rev.* **52**:129.

Cohen, C., D. L. D. Caspar, J. P. Johnson and K. Nauss. 1973. Tropomyosin-troponin assembly. *Cold Spring Harbor Symp. Quant. Biol.* **37**:287.

Cole, H. A. and S. V. Perry. 1975. The phosphorylation of troponin I from cardiac muscle. *Biochem. J.* **149**:525.

Collins, J. H. 1974. Homology of myosin light chains, troponin C and parvalbumin deduced from comparison of their amino acid sequences. *Biochem. Biophys. Res. Comm.* **58**:301.

———. 1975. Purification and analysis of the cyanogen bromide peptides of troponin T from rabbit skeletal muscle. *Biochem. Biophys. Res. Comm.* **65**:604.

Collins, J. H., J. Porter, M. Horn, G. Wilshire and N. Jackman. 1973. Structural studies on rabbit skeletal muscle troponin C: Evidence for gene replication and homology with calcium binding proteins from carp and hake muscles. *FEBS Letters* **36**:268.

Cummins, P. and S. V. Perry. 1973. The subunits and biological activity of polymorphic forms of tropomyosin. *Biochem. J.* **133**:765.

———. 1974. Chemical and immunochemical characteristics of tropomyosins from striated and smooth muscle. *Biochem. J.* **141**:43.

Donaldson, S. K. and J. Kerrick. 1975. Characterization of the effects of Mg^{2+} on Ca^{2+}-Sr^{2+} activated cation generation of skinned skeletal muscle fibers. *J. Gen. Physiol.* **66**:427.

Drabikowski, W. and R. Dabrowska. 1974. Interactions among the proteins of the thin filament. In *Proceedings of the 9th FEBS Meeting: Proteins of the contractile system* (ed. N. A. Biro), vol. 31, p. 85. Akadémiai Kiadó, Budapest.

Drabikowski, W., R. Dabrowska and B. Barylko. 1975. Comparison of cardiac muscle troponin. In *Basic functions of cations in myocardial activity, Recent advances in studies on cardiac structure and metabolism* (ed. A. Fleckenstein and N. S. Dhalla), vol. 5. p. 245. University Park Press, Baltimore.

Driska, S. and D. J. Hartshorne. 1975. The contractile proteins of smooth muscle. Properties and components of a Ca^{2+}-sensitive actomyosin from chicken gizzard. *Arch. Biochem. Biophys.* **167**:203.

Ebashi, S. 1960. Calcium binding and relaxation in the actomyosin system. *J. Biochem.* **48**:150.

———. 1963. Third component participating in the superprecipitation of natural actomyosin. *Nature* **200**:1010.

Ebashi, S. and F. Ebashi. 1964. A new protein component participating in the superprecipitation of myosin B. *J. Biochem.* **55**:604.

Ebashi, S. and M. Endo. 1968. Calcium ion and muscle contraction. *Prog. Biophys. Mol. Biol.* **18**:123.

Ebashi, S. and A. Kodama. 1966. Native tropomyosin-like action of troponin on trypsin-treated myosin B. *J. Biochem.* **60**:733.

Ebashi, S., M. Endo and I. Ohtsuki. 1969. Control of muscle contraction. *Q. Rev. Biophys.* **2**:351.

Ebashi, S., I. Ohtsuki and K. Mihashi. 1973. Regulatory proteins of muscle with special reference to troponin. *Cold Spring Harbor Symp. Quant. Biol.* **37**:215.

Ebashi, S., T. Toyo-oka and Y. Nomomura. 1975. Gizzard troponin. *J. Biochem.* **78**:859.

Ebashi, S., Y. Nonomura, T. Kitazawa and T. Toyo-oka. 1975. Troponin in tissues other than skeletal muscle. In *Calcium transport in contraction and secretion* (ed. E. Carafoli et al.), p. 405. North-Holland, Amsterdam.

Ebashi, S., T. Ohnishi, K. Maruyama and T. Fujii. 1974. Molecular mechanisms of the regulation of muscle contraction by the Ca-troponin system. In *Proceedings of the 9th FEBS Meeting: Proteins of the contractile system* (ed. N. A. Biro), vol. 31, p. 71. Akadémiai Kiado, Budapest.

England, P. J. 1975. Correlation between contraction and phosphorylation of the inhibitory subunit of troponin in perfused rat heart. *FEBS Letters* **50**:57.

Gergely, J., M. Greaser, B. Nagy and J. Potter. 1974. Troponin: The regulatory protein complex of muscle. In *Myocardial biology. Recent advances in studies on cardiac structure and metabolism* (ed. N. S. Dhalla), vol. 4, p. 281. University Park Press, Baltimore.

Greaser, M. and J. Gergely. 1971. Reconstitution of troponin activity from three protein components. *J. Biol. Chem.* **246**:4226.

———. 1973. Purification and properties of the components from troponin. *J. Biol. Chem.* **248**:2125.

Greaser, M., M. Yamaguchi, C. Brekke, J. Potter and J. Gergely. 1973. Troponin subunits and their interactions. *Cold Spring Harbor Symp. Quant. Biol.* **37**:235.

Hanson, J. and J. Lowy. 1963. The structure of F-actin and of actin filaments isolated from muscle. *J. Mol. Biol.* **6**:46.

Hartshorne, D. J. and H. Mueller. 1968. Fractionation of troponin into two distinct proteins. *Biochem. Biophys. Res. Comm.* **31**:647.

Haselgrove, J. C. 1973. X-ray evidence for a conformational change in the actin-containing filaments of vertebrate striated muscle. *Cold Spring Harbor Symp. Quant. Biol.* **37**:341.

Haselgrove, J. C. and H. E. Huxley. 1973. X-ray evidence for radial cross-bridge movement and for the sliding filament model in actively contracting skeletal muscle. *J. Mol. Biol.* **77**:549.

Head, J. F. and S. V. Perry. 1974. The interaction of the calcium-binding protein (troponin C) with bivalent cations and the inhibitory protein (troponin I). *Biochem. J.* **137**:145.

Heilbrun, L. V. and F. J. Wiercinski. 1947. The action of various cations on muscle protoplasm. *J. Cell. Comp. Physiol.* **29**:15.

Hitchcock, S. E. 1975. Regulation of muscle contraction: Binding of troponin and its components to actin and tropomyosin. *Eur. J. Biochem.* **52**:255.

Hitchcock, S. E. and A. G. Szent-Györgyi. 1974. Calcium dependent interaction of troponin with actin and tropomyosin. *Biophys. Soc. Abstr.*, p. 79a.

Hitchcock, S. E., H. E. Huxley and A. G. Szent-Györgyi. 1973. Calcium sensitive binding of troponin to actin-tropomyosin: A two site model for troponin action. *J. Mol. Biol.* **80**:825.

Honig, C. R. and Y. Reddy. 1975. Calcium, tropomyosin and actomyosin as a control of calcium binding by troponin. In *The cardiac sarcoplasm. Recent advances in studies on cardiac structure and metabolism* (ed. P. E. Roy and P. Harris), vol. 8, p. 233. University Park Press, Baltimore.

Huxley, H. E. 1973. Structural changes in the actin- and myosin-containing filaments during contraction. *Cold Spring Harbor Symp. Quant. Biol.* **37**:361.

Kawasaki, Y. and J.-P. Van Eerd. 1972. The effect of Mg++ on the conformation of the Ca++ binding component of troponin. *Biochem. Biophys. Res. Comm.* **49**:898.

Kretsinger, R. H. 1975. Hypothesis: Calcium modulated proteins contain EF hands. In *Calcium transport in contraction and secretion* (ed. E. Carafoli et al.), p. 469. North-Holland, Amsterdam.

Kretsinger, R. H. and C. D. Barry. 1975. The predicted structure of the calcium-binding component of troponin. *Biochim. Biophys. Acta* **405**:40.

Kretsinger, R. H. and C. E. Nockolds. 1973. Carp muscle calcium binding protein. II. Structure determination and general description. *J. Biol. Chem.* **248**: 3313.

Lehman, W. and A. G. Szent-Györgyi. 1975. Regulation of muscle contraction. Distribution of actin control and myosin control in the animal kingdom. *J. Gen. Physiol.* **66**:1.

Margossian, S. S. and C. Cohen. 1973. Troponin subunit interactions. *J. Mol. Biol.* **81**:409.

Murray, A. C. and C. M. Kay. 1972. Hydrodynamics and optical properties of troponin A. Demonstration of conformational change upon binding of calcium. *Biochemistry* **11**:2622.

Ohnishi, S., K. Maruyama and S. Ebashi. 1975. Calcium induced conformational changes and mutual interaction of troponin components as studied by spin labelling. *J. Biochem.* **78**:73.

Ohtsuki, I. 1974. Location of troponin in the thin filament and tropomyosin paracrystal. *J. Biochem.* **75**:753.

Parry, D. A. D. and J. M. Squire. 1973. Structural role of tropomyosin in muscle regulation: Analysis of the X-ray diffraction patterns from relaxed and contracting muscle. *J. Mol. Biol.* **75**:33.

Pearlstone, J. R., M. R. Carpenter and L. B. Smillie. 1975. The primary structure of troponin T: Cyanogen bromide fragments. *Fed. Proc.* **34**:539.

Perry, S. V. 1973. The control of muscle contraction. *Symp. Soc. Exp. Biol.* **37**: 531.

———. 1974. Variation in the contractile and regulatory proteins of the myofibril with muscle type. In *Exploratory concepts in muscular dystrophy II: Control mechanisms in development and function of muscle and their relationship to muscular dystrophy and related neuromuscular diseases* (ed. A. T. Milhorat), p. 319. Excerpta Medica, Amsterdam.

Perry, S. V. and T. C. Gray 1956. Ethylenediaminotetra-acetate and the adenosine-triphosphatase activity of the actomyosin system. *Biochem. J.* **64**:5p.

Perry, S. V., H. A. Cole, J. F. Head and F. J. Wilson. 1973. Localization and mode of action of the inhibitory protein component of the troponin complex. *Cold Spring Harbor Symp. Quant. Biol.* **37**:251.

Perry, S. V., H. A. Cole, M. Morgan, A. J. G. Moir and E. Pires. 1975. Phosphorylation of the proteins of the myofibril. In *Proceedings of the 9th FEBS Meeting: Proteins of the contractile system* (ed. N. A. Biro), vol. 31, p. 163. Akadémiai Kiadó, Budapest.

Potter, J. D. and J. Gergely. 1974. Troponin, tropomyosin and actin interactions in the Ca2+ regulation of muscle contraction. *Biochemistry* **13**:2697.

———. 1975. The calcium and magnesium binding sites on troponin and their role in the regulation of myofibrillar ATPase. *J. Biol. Chem.* **250**:4628.

Potter, J. D., P. Leavis, J. Seidel, S. Lehrer and J. Gergely. 1975. Interaction of divalent cations with troponin and myosin. In *Calcium transport in contraction and secretion* (ed. E. Carafoli et al.), p. 415. North-Holland, Amsterdam.

Potter, J. D., J. C. Seidel, P. Leavis, S. S. Lehrer and J. Gergely. 1974. Interaction of Ca++ with troponin. In *Calcium binding proteins* (ed. W. Drabikowski, H. Strzelecka-Golaszewska and E. Carafoli), p. 129. Elsevier, Amsterdam.

Regenstein, J. M. and A. G. Szent-Györgyi. 1975. Regulatory proteins of lobster striated muscle. *Biochemistry* **14**:917.

Sobieszek, A. and R. D. Bremel. 1975. Preparation and properties of vertebrate smooth muscle myofibrils and actomyosin. *Eur. J. Biochem.* **55**:49.

Spudich, J. A., H. E. Huxley and J. T. Finch. 1973. Regulation of skeletal muscle contraction. II. Structural studies of the interaction of the tropomyosin-troponin complex with actin. *J. Mol. Biol.* **72**:619.

Squire, J. M. 1975. Muscle filament structure and muscle contraction. *Annu. Rev. Biophys. Bioeng.* **4**:137.

Stewart, M. 1975. The location of the troponin binding site on tropomyosin. *Proc. Roy. Soc. London* **190**:257.

Stone, D., J. Sodek, P. Johnson and L. B. Smillie. 1974. Tropomyosin: Correlation of amino acid sequence and structure. In *Proceedings of the 9th FEBS Meeting: Proteins of the contractile system* (ed. N. A. Biro), vol. 31, p. 125. Akadémiai Kiadó, Budapest.

Syska, H., S. V. Perry and L. P. Trayer. 1974. A new method of preparation of troponin I (inhibitory protein) using affinity chromatography. Evidence for three different forms of troponin I in striated muscle. *FEBS Letters* **40**:253.

Szent-Györgyi, A. G. 1975. Ca^{2+} regulation of muscle contraction. *Biophys. J.* **15**:707.

Tonomura, Y., S. Watanabe and M. Morales. 1969. Conformational changes in the molecular control of muscle contraction. *Biochemistry* **8**:2171.

Tsukui, R. and S. Ebashi. 1973. Cardiac troponin. *J. Biochem.* **73**:1119.

Van Eerd, J.-P., and Y. Kawasaki. 1972. Ca^{2+} induced conformational changes in the $Ca+^2$ binding component of troponin. *Biochem. Biophys. Res. Comm.* **47**:859.

———. 1973. Effect of calcium (II) on the interaction between the subunits of troponin and tropomyosin. *Biochemistry* **12**:4972.

Van Eerd, J.-P. and K. Takahashi. 1975. The amino acid sequence of bovine cardiac troponin C. Comparison with rabbit skeletal troponin C. *Biochem. Biophys. Res. Comm.* **64**:122.

Vibert, P. J., J. C. Haselgrove, J. Lowy and F. R. Paulsen. 1972. Structural changes in actin-containing filaments of muscle. *J. Mol. Biol.* **71**:757.

Wakabayashi, T., H. E. Huxley, L. A. Amos and A. Klug. 1975. Three-dimensional image reconstruction of actin-tropomyosin complex and actin-tropomyosin-troponin T-troponin I complex. *J. Mol. Biol.* **93**:477.

Weber, A. 1959. On the role of calcium in the activity of adenosine 5'-triphosphate hydrolysis by actomyosin. *J. Biol. Chem.* **234**:2764.

Weber, A. and R. Herz. 1963. The binding of calcium to actomyosin systems in relation to their biological activity. *J. Biol. Chem.* **238**:599.

Weber, A. and J. M. Murray. 1973. Molecular control mechanisms in muscle contraction. *Physiol. Rev.* **53**:612.

Weber, A. and S. Winicur. 1961. The role of calcium in the superprecipitation of actomyosin. *J. Biol. Chem.* **236**:3198.

Weeds, A. G. and A. D. McLachlan. 1974. Structural homology of myosin alkali light chains. *Nature* **252**:646.

Wilkinson, J. M. and R. J. A. Grand. 1974. The amino acid sequence of troponin I from rabbit white skeletal muscle. In *Proceedings of the 9th FEBS Meeting:*

Proteins of the contractile system (ed. N. A. Biro), vol. 31, p. 137. Akadémiai Kiadó, Budapest.

Wilkinson, J. M., S. V. Perry, H. A. Cole and I. P. Trayer. 1972. The regulatory proteins of the myofibril. Separation and biological activity of the components of the inhibitory factor preparations. *Biochem. J.* **127**:215.

Calcium Regulation in Invertebrate Muscles

William Lehman

Department of Physiology, Boston University School of Medicine
Boston, Massachusetts 02118

Two distinctly different calcium regulatory systems exist in the animal kingdom. In one, sites on actin are blocked in the absence of calcium, such that myosin cannot interact with the thin filaments, and in the other, myosin is directly affected by calcium and in its absence cannot interact with actin. In chordate striated muscles, the thin-filament-linked system occurs, and there is no evidence, based on ATPase studies, for a myosin control. This form of control is rare among invertebrates, and in most cases a dual control by both systems exists in a particular muscle. In some phyla, such as molluscs and echinoderms, the myosin control occurs alone, and troponin is either absent or present in very small amounts (see Lehman and Szent-Györgyi 1975, for a detailed categorization).

Regulated invertebrate thin filaments contain, in addition to actin and tropomyosin, three troponin subunits with the same general characteristics as in vertebrate troponin. The stoichiometry of actin:tropomyosin:TnC is 7:1:1, but invertebrate troponin binds only one calcium (versus four for rabbits; Potter and Gergely 1975), and the amino acid composition of its TnC is substantially less acidic than that of vertebrate TnC. (Regenstein and Szent-Györgyi 1975; W. Lehman, J. M. Regenstein and A. L. Ransom, in prep.). In spite of these differences, functional interphyletic hybrids can be made using appropriate mixtures of arthropod and vertebrate troponin subunits, indicating a common overall mechanism and probably a common evolutionary ancestry (Lehman 1975).

Myosin regulation, as characterized using scallop myosin, probably involves cooperativity between subfragment 1 heads of individual myosins or among the myosins along the thick filament. Evidence supporting this contention is as follows: the ATPase dependence of scallop actomyosin on calcium is very sharp (Lehman and Szent-Györgyi 1975); the scallop acto-HMM ATPase is calcium-sensitive, whereas the acto-S_1 ATPase is not (Szent-Györgyi, Szentkiralyi and Kendrick-Jones 1973); removal of one of two EDTA "regulatory" light chains from scallop myosin completely desensitizes the scallop actomyosin ATPase to calcium (J. Kendrick-Jones, E. M. Szentkiralyi and A. G. Szent-György, in prep.). Readdition of the scallop regulatory light chain, the rabbit DTNB light chain or light chains from several calcium-insensitive myosins reconfers a calcium dependency on desensitized scallop myosin (Szent-Györgyi, Szentkiralyi and Kendrick-Jones 1973; Kendrick-Jones 1974; J. Kendrick-Jones, E. M. Szentkiralyi and A. G. Szent-Györgyi, in prep.). Since vertebrate myosins do not appear to be calcium regulated, it is probably their heavy chains that have lost this function, whereas the light chains retain their apparent regulatory potential.

REFERENCES

Kendrick-Jones, J. 1974. Role of myosin light chains in calcium regulation. *Nature* **249**:631.

Lehman, W. 1975. Hybrid troponin reconstituted from vertebrate and arthropod subunits. *Nature* **255**:424.

Lehman, W. and A. G. Szent-Györgyi. 1975. Regulation of muscular contraction. Distribution of actin control and myosin control in the animal kingdom. *J. Gen. Physiol.* **66**:1.

Potter, J. and J. Gergely. 1975. The calcium and magnesium binding sites on troponin and their role in the regulation of the myofibrillar adenosine triphosphatase. *J. Biol. Chem.* **250**:4628.

Regenstein, J. M. and A. G. Szent-Györgyi. 1975. Regulatory proteins of lobster striated muscle. *Biochemistry* **14**:917.

Szent-Györgyi, A. G., E. M. Szentkiralyi and J. Kendrick-Jones. 1973. The light chains of scallop myosin as regulatory subunits. *J. Mol. Biol.* **74**:179.

Composition, Structure and Biosynthesis of Sarcoplasmic Reticulum

David H. MacLennan, Peter S. Stewart,
Elżbieta Zubrzycka and Paul C. Holland*

Banting and Best Department of Medical Research
Charles H. Best Institute, University of Toronto
Toronto, Ontario M5G 1L6 Canada

In muscle tissues, regulation of the concentration of free Ca^{++} in the cytoplasm regulates contraction (Ebashi and Endo 1968; Ebashi, Endo and Ohtsuki 1969). In mammalian skeletal muscle, Ca^{++} is bound to troponin with a constant of $1.3 \times 10^6 \ \text{M}^{-1}$ (Ebashi, Kodama and Ebashi 1968). Binding of Ca^{++} to troponin releases steric restraints imposed by tropomyosin on the interaction between actin and myosin heads; dissociation of the Ca^{++}-troponin complex permits reimposition of these restraints (Huxley 1973). In molluscan muscle (Lehman, Kendrick-Jones and Szent-Györgyi 1973; Kendrick-Jones 1974) and in smooth muscle (Bremel 1974), Ca^{++} binds directly to myosin to stimulate muscle contraction. Structural aspects of the interaction of Ca^{++} with these forms of myosin are unknown, although changes have been observed upon interaction of skeletal muscle myosin with Ca^{++} (Morimoto and Harrington 1974).

Regulation of Intracellular Calcium

Systems have evolved which regulate intracellular Ca^{++} with great speed and efficiency. Of these, the sarcoplasmic reticulum of skeletal muscle is probably the most specialized. In skeletal muscle (Bennett and Porter 1953), each myofibril is encased in a fenestrated sheath of sarcoplasmic reticulum membranes that can make up as much as 18% of the muscle cell volume (Birks and Davey 1969). The sheath is differentiated into repeating structures approximately the same length as the sarcomere, consisting of expanded cisternal elements separated by narrow longitudinal elements. Invaginations of the muscle cell plasma membrane, the transverse tubular system, adjoin the sarcoplasmic reticulum in the cisternal regions (Porter and Palade 1957). Three separate compartments of different ionic compositions are thereby

* Present address: Department of Biochemistry, University of Saskatchewan, Saskatoon, Saskatchewan S7N OWO, Canada.

established in the muscle cell: the cytoplasm, the interior of the sarcoplasmic reticulum and the interior of the transverse tubules, which is continuous with the extracellular space. In response to a nerve stimulus, a wave of depolarization proceeds through the transverse tubule (Huxley 1971), eliciting release of Ca^{++} from the sarcoplasmic reticulum (Jobsis and O'Connor 1966). There does not appear to be electrical activity in the sarcoplasmic reticulum. Release of Ca^{++} raises intracellular Ca^{++} concentrations to about 50 μM (Ebashi and Endo 1968; Ebashi, Endo and Ohtsuki 1969), saturating troponin, and leading to the initiation of contraction. During relaxation, the sarcoplasmic reticulum sequesters Ca^{++} through the action of a Ca^{++}-transporting ATPase enzyme (Ebashi and Lipmann 1962; Hasselbach 1964). Since the threshold for Ca^{++} activation of this enzyme is about 0.3 μM, intracellular Ca^{++} will be maintained at this concentration when muscle is at rest. The sarcoplasmic reticulum clearly has the advantage in the intracellular struggle for free Ca^{++} since its affinity for Ca^{++} is an order of magnitude greater than that for troponin.

There are other Ca^{++}-regulating systems in cells. In vertebrates, the plasma membrane is a barrier between an extracellular Ca^{++} concentration of 2–3 mM and an intracellular Ca^{++} concentration of less than 1 μM (Brinley 1973). Clearly this membrane plays a role in maintaining the differential. Mitochondria also possess the capacity for Ca^{++} transport (Lehninger 1970), and their role may be major not only in smooth (Batra 1973) and heart muscle (Carafoli 1972), but in all cells. Ca^{++} injected into non-muscle cells is rapidly sequestered in areas rich in mitochondria but not in areas poor in mitochondria (Rose and Loewenstein 1975).

Characterization of Skeletal Muscle Sarcoplasmic Reticulum

Isolated sarcoplasmic reticulum retains the ability to transport and sequester Ca^{++}, although the induction of Ca^{++} release has not been simulated in vitro (Hasselbach 1964; Ebashi and Endo 1968; Ebashi, Endo and Ohtsuki 1969). Two moles of Ca^{++} are transported for every mole of ATP hydrolyzed by the $Ca^{++} + Mg^{++}$-dependent ATPase of this membrane (Hasselbach 1964). We have isolated this enzyme (MacLennan 1970) and found that the active form consists of an ATPase protein of molecular weight 102,000 (MacLennan et al. 1971), a proteolipid of molecular weight 12,000, which has two moles of fatty acid covalently bonded per mole of enzyme (MacLennan et al. 1973), and a complex mixture of phospholipid and neutral lipid (MacLennan et al. 1971). This unit transports Ca^{++} when incorporated into excess phospholipid vesicles (Racker 1972).

The ATPase contains active sites. A site of ATP hydrolysis is phosphorylated in a Ca^{++}-dependent reaction and dephosphorylated in a Mg^{++}-dependent reaction (Yamamoto 1972; MacLennan and Holland 1975). This site is found in the 102,000 MW protein (Martonosi and Halpin 1971). The same protein has Ca^{++} carrier activity acting as a Ca^{++}-selective ionophore in a lipid bilayer placed in an electrical field (Shamoo and MacLennan 1974). These two sites have been separated in tryptic fragments of the enzyme (Stewart, MacLennan and Shamoo 1976; Shamoo et al. 1976). Clearly an important aspect in the understanding of Ca^{++} transport will involve an understanding of the interaction between these two sites.

ATP hydrolysis in sarcoplasmic reticulum requires phospholipid (Kielley and Meyerhof 1948; Ebashi and Lipmann 1962; Martonosi 1963). This lipid requirement can be met by 30 moles of dioleyl lecithin per mole of enzyme (Warren et al. 1975). The phospholipid appears to form an annulus around the enzyme, screening it from neutral lipid.

The proteolipid has a hydrophilic amino acid composition, but, by virtue of bound fatty acid, it has the solubility characteristics of a lipid (MacLennan et al. 1973.) The molecule, which is rich in glutamate residues, might act as a mobile Ca^{++} carrier in the lipid component of the sarcoplasmic reticulum membrane. However, it did not act as an ionophore under conditions where ionophoretic activity was readily measurable with the larger ATPase protein (Shamoo and MacLennan 1974). The proteolipid does appear to affect the efficiency of Ca^{++} transport since, in reconstituted systems, its addition to depleted ATPase preparations raised the Ca^{++}:ATP ratio from less than 1 to as high as 2 (Racker and Eytan 1975). It is possible, however, that the role of the proteolipid in transport is indirect. We have suggested that the proteolipid might play a structural role, its amphipathic structure providing a nucleus around which the larger intrinsic ATPase protein and phospholipid might associate (MacLennan et al. 1973).

The sarcoplasmic reticulum contains two other major proteins, calsequestrin (MacLennan and Wong 1971) and the high-affinity Ca^{++}-binding protein (Ostwald and MacLennan 1974). They are both water-soluble, extrinsic proteins, being released upon destruction of membrane integrity either with detergent (MacLennan and Wong 1971) or through removal of divalent metals with chelators (Duggan and Martonosi 1970). Calsequestrin binds 43 moles of Ca^{++} per mole with medium affinity (MacLennan and Wong 1971). The high-affinity Ca^{++}-binding protein binds 25 moles of Ca^{++} per mole with low affinity and 1 mole of Ca^{++} per mole with an affinity equal to that of troponin (Ostwald and MacLennan 1974). Calsequestrin undoubtedly acts as a Ca^{++} acceptor in the interior of the sarcoplasmic reticulum, lowering the free Ca^{++} concentration from an estimated 20 mM (if it were free) (Sandow 1970) to perhaps 500 μM. The role of the high-affinity Ca^{++}-binding protein is unknown.

The sarcoplasmic reticulum membrane has several structural features. In thin section, the vesicles appear as single, trilaminar boundary membranes enclosing a space containing, in most cases, a small amount of filamentous material (Ebashi and Lipmann 1962). Negative staining of the vesicles reveals particles, 40 Å in diameter, extending about 60 Å from the surface (Ikemoto et al. 1968), whereas freeze-fracture reveals 80–90-Å globules buried in the hydrophobic interior of the membrane (Deamer and Baskin 1969). These globules are asymmetrically arranged, most of them being associated with the exterior leaflet.

The purified ATPase forms vesicular structures (MacLennan et al. 1971). They retain surface particles (Stewart and MacLennan 1974) and intramembrane particles (MacLennan et al. 1971). Since only the ATPase of 102,000 MW can account for these two features, we have suggested that it consists of an external portion, making up the surface particles, and an internal portion, making up the intramembrane particles (Stewart and MacLennan 1974). The filamentous material removed on purification of the ATPase is probably calsequestrin, which precipitates as fibers in the presence of Ca^{++} (Stewart and MacLennan 1974; Meissner 1975). It may also include the high-affinity Ca^{++}-binding protein. No special structural feature can be assigned to the proteolipid, although its association with the lipid bilayer is virtually a certainty.

Biosynthesis of the Sarcoplasmic Reticulum

The development of sarcoplasmic reticulum during myogenesis offers unique opportunities for the study of membrane biogenesis. Muscle-cell precursors, referred to as myoblasts, are relatively undifferentiated and can be grown in primary culture (Holtzer et al. 1972) or continuous lines (Yaffe 1968). After a finite period in culture, the cells fuse to form elongated multinucleated myotubes, the equivalent in vitro of muscle fibers in vivo. Coincident with fusion, the cells begin to synthesize muscle-specific proteins, such as myosin, creatine phosphokinase, phosphorylase, adenylate kinase and myoglobin (Shainberg, Yagil and Yaffe 1971; Paterson and Strohman 1972; Yaffe and Dym 1973; Kagen and Freedman 1974). Fusion and initiation of synthesis of these proteins occurs over a period as short as 10 hours, particularly in primary cultures.

Ezerman and Ishikawa (1967) examined the morphology of muscle cells in culture and found rough endoplasmic reticulum in myoblasts. After fusion, the rough endoplasmic reticulum began to form vesicular structures at various growing points. The structures ultimately became recognizable, in myotubes, as extensive and well-differentiated sarcoplasmic reticulum. These observations suggest that the sarcoplasmic reticulum is formed de novo in a differentiation step that begins about the time of myoblast fusion. The exploitation of a similar differentiation step proved exceedingly useful in earlier studies of mitochondrial biosynthesis (Tzagoloff, Rubin and Sierra 1973). Moreover, as with mitochondria, the proteins of sarcoplasmic reticulum have been isolated and are well characterized structurally and functionally (MacLennan and Holland 1975). Antibodies have been raised against these proteins, and the use of these antibodies makes it possible to analyze several biosynthetic aspects of the individual proteins on a microscale.

Among the most important questions remaining regarding membrane assembly are the following: Where are insoluble intrinsic proteins such as the ATPase and the proteolipid synthesized? How are they transported and incorporated into the proper membrane in the proper proportion and orientation? How do soluble proteins such as calsequestrin and the high-affinity Ca^{++}-binding protein find their way to their sites of localization in the interior of the membrane? And finally, what controls the differentiation of membranes within cells?

Although the answers to these questions are long-term goals, we have initiated the study by examining the temporal pattern of synthesis of the ATPase in rat skeletal muscle cells in primary culture (Holland and MacLennan 1976). We plated the cells in 150-mm dishes in Dulbecco's modified

Eagle's medium containing 10% horse serum and 0.5% chick embryo extract (Yaffe 1973). At various time periods after plating, we pulse-labeled the cells for 2 hours with 4,5-[³H]leucine in leucine-free medium. Cells were then extracted in 1 ml of 150 mM NaCl, 10 mM sodium phosphate, pH 7.0, and 0.5% Triton. The extract was incubated for 48 to 72 hours with 10 μg of purified rat sarcoplasmic reticulum ATPase, to act as a carrier, and a twofold excess of rabbit anti-ATPase serum in a solution of 0.5% Triton, 0.5% deoxycholate and 0.2% sodium dodecyl sulfate. Immunoprecipitates were washed and subjected to disc gel electrophoresis (Weber and Osborn 1969). The ATPase protein band was clearly visible in the gel after staining. Slices from the top of the gel were counted, and the radioactivity in the ATPase was obtained by subtraction of background counts. Figure 1 shows an example of a gel obtained from an immunoprecipitate from pulse-labeled cells after incubation with anti-ATPase serum.

From such gel patterns we have been able to calculate relative rates of synthesis of the ATPase at different stages of differentiation. Figure 2 shows that there was very little ATPase synthesis prior to a burst of fusion occurring in standard medium at about 50 hours after plating. Following fusion, the rate of synthesis was elevated severalfold and continued at a high rate for

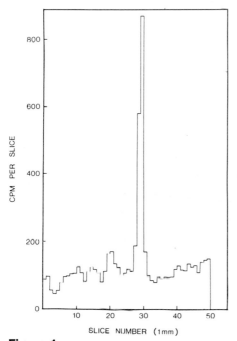

Figure 1

Radioactivity profile of an SDS-polyacrylamide gel of an antibody precipitate of the sarcoplasmic reticulum ATPase obtained from cell cultures labeled with 4,5-[³H]leucine for 2 hr starting after 97 hr in culture. The immunoprecipitate was obtained from 1 ml of a 0.5% Triton extract of cells from a 150-mm dish. The solution also contained 150 mM NaCl, 10 mM sodium phosphate, pH 7.0, 0.2% SDS, 0.5% deoxycholate, 10 μg of carrier ATPase and 0.3 ml of anti-ATPase serum.

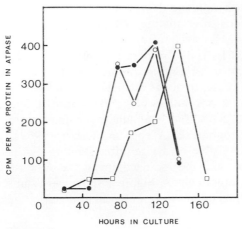

Figure 2

Synthesis of the ATPase during growth in different media. Cultures were pulsed for 2 hr with 4,5-[³H]leucine starting at the times after plating indicated. Triton extracts of the cells were analyzed for incorporation of radioactive leucine into the ATPase as described in Fig. 1. Fusion began at 50 hr with standard medium, at 80 hr with FE medium, and did not occur with low Ca^{++} medium: ($\bullet - \bullet$) standard medium; ($\circ - \circ$) low Ca^{++} medium; ($\square - \square$) FE medium.

several days, falling off as the cultures deteriorated. Synthesis of the ATPase is thus an early event in myogenesis, comparable, temporally, to synthesis of muscle structural and glycolytic proteins (Shainberg, Yagil and Yaffe 1971; Yaffe and Dym 1973). A common control probably exists for the synthesis of all these proteins.

We have examined some factors known to affect fusion and synthesis of other muscle proteins. A switch to medium containing 20% fetal calf serum and 8% chick embryo extract (FE medium) at any time during the first 40 hours in culture has been found to promote proliferation and to delay, but not prevent, fusion (Yaffe 1971). If cells are switched back to standard medium, a lag period equivalent to about one generation time of 18 hours is required before fusion begins. If cultures are exposed to FE medium subsequent to a 40-hour preincubation, fusion is not delayed. Thus the capacity for fusion appears to be initiated after the first 40 hours in culture, and, once initiated, it is not reversible by the enriched medium. When the synthesis of the ATPase in FE medium was examined, we noted that rapid rates of ATPase synthesis were also postponed to about 80 hours after plating (Fig. 2). Fusion and formation of the ATPase were therefore synchronous.

Depletion of Ca^{++} in the growth medium prevents fusion (Yaffe 1971). Low Ca^{++}, however, has a less subtle effect than does FE medium. If cells are changed to low Ca^{++} prior to the onset of fusion, they will not fuse, and if fusion is in progress, it can be halted by a switch to low Ca^{++}. Restoration of normal Ca^{++} permits fusion within 1 to 2 hours. These observations have led Yaffe (1971) to suggest that Ca^{++} depletion interferes directly with membrane processes leading to fusion, so that even though cells have developed the full potential for fusion, they cannot do so.

We also examined the effects of low Ca^{++} on fusion and on synthesis of the ATPase. We found that low Ca^{++} prevented fusion at all time periods, but that synthesis of the ATPase was unaffected by the lack of fusion. Figure 2 shows that the temporal pattern of ATPase synthesis was the same in normal and in low Ca^{++} media. This clear separation between fusion and ATPase synthesis shows that they are independent events in the course of muscle cell differentiation. Multinucleation, therefore, cannot be responsible for initiation of membrane biogenesis, and control systems must be looked for elsewhere.

Degradation rates during all three culture conditions were constant over the period studied (Holland and MacLennan 1976). The loss of radioactivity from the ATPase occurred with a half-life of 20 hours. The half-life of total protein, on the other hand, was 40 hours. Since the half-life of the ATPase protein was ten times longer than the labeling period, incorporation of radioactivity into the enzyme reflected synthesis and not degradation.

We have also investigated the temporal sequence of formation of the extrinsic protein calsequestrin. Figure 3 is an example of an antibody precipitate of this protein from pulse-labeled cells. Since calsequestrin comprises only 10–15% of the total sarcoplasmic reticulum protein, whereas the ATPase comprises nearly two-thirds of the total protein, its synthesis was correspondingly lower than that of the ATPase. Figure 4 shows that the rate of calsequestrin synthesis was enhanced eight- to ninefold following fusion of myoblasts in standard medium. It appears, therefore, that both intrinsic and extrinsic membrane proteins are synthesized in the same temporal sequence.

An interesting side issue of our investigations is the question of whether the synthesis of the ATPase can be detected in non-muscle cells. We have looked at two cells that are not differentiated muscle cells but which are closely related to differentiated muscle cells: myoblasts and muscle fibroblasts (Abbott et al. 1974). Figure 2 shows that a low level of ATPase synthesis occurred in myoblasts before gross fusion and gross morphological differentiation were evident. However, some multinucleation occurred in myoblast cul-

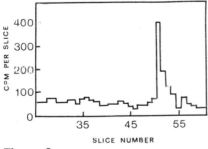

Figure 3
Radioactivity profile of an SDS-polyacrylamide gel of an antibody precipitate of calsequestrin obtained from cell cultures labeled with 4,5-[³H]leucine for 2 hr starting after 116 hr in culture. The immunoprecipitate was obtained from 2 ml of a 0.5% Triton extract of cells from two 150-mm dishes. The solution also contained 150 mM NaCl, 10 mM sodium phosphate, pH 7.0, 0.2% SDS, 0.5% deoxycholate, 16 μg of carrier calsequestrin and 0.5 ml of anticalsequestrin serum.

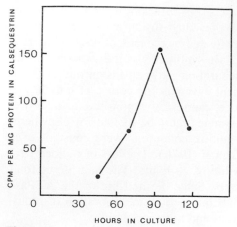

Figure 4

Synthesis of calsequestrin during growth in standard medium. Cultures were pulsed for 2 hr with 4,5-[³H]leucine starting at the times after plating indicated. Triton extracts of the cells were analyzed for incorporation of radioactive leucine into calsequestrin as described in Fig. 1. Fusion began at about 50 hr.

tures prior to the period of rampant fusion, which may account for some of the ATPase synthesis in these early cultures. It is possible, nevertheless, that some sarcoplasmic reticulum proteins as well as some muscle structural proteins, such as myosin, actin and tropomyosin, are formed in these muscle fiber precursors. On the other hand, we were unable to detect any synthesis of ATPase in pure primary cultures of muscle fibroblasts (Holland and MacLennan 1976). Therefore if there is a synthesis of sarcoplasmic reticulum in these cells, it will have to be detected by more sensitive techniques. It would be interesting to scan other cells for the possible synthesis of sarcoplasmic reticulum proteins. However, non-muscle cells may well have made use of other mechanisms for Ca^{++} regulation and may not possess a Ca^{++}-dependent ATPase.

Acknowledgments

The expert technical assistance of Elizabeth Zachwieja is gratefully acknowledged. Original research described in this paper was supported by the Medical Research Council of Canada and by the Muscular Dystrophy Association of Canada. P. C. Holland was a Charles H. Best Foundation postdoctoral fellow. E. Zubrzycka is a fellow of the Muscular Dystrophy Association of Canada.

REFERENCES

Abbott, J., J. Schiltz, S. Dienstman and H. Holtzer. 1974. The phenotyptic complexity of myogenic clones. *Proc. Nat. Acad. Sci.* **71**:1506.

Batra, S. 1973. The role of mitochondrial calcium uptake in contraction and relaxation of the human myometrium. *Biochim. Biophys. Acta* **305**:428.

Bennett, H. and K. R. Porter. 1953. An electron microscope study of sectioned breast muscle of the domestic fowl. *Amer. J. Anat.* **93**:61.

Birks, R. I. and D. F. Davey. 1969. Osmotic responses demonstrating the extracellular character of the sarcoplasmic reticulum. *J. Physiol.* **202**:171.

Bremel, R. D. 1974. Myosin linked calcium regulation in vertebrate smooth muscle. *Nature* **252**:405.

Brinley, F. J. 1973. Calcium and magnesium transport in single cells. *Fed. Proc.* **32**:1735.

Carafoli, E. 1972. Mitochondria in the contraction and relaxation of heart. In *Myocardial biology. Recent advances in studies on cardiac structure and metabolism* (ed. N. S. Dhalla), vol. 4, p. 393. University Park Press, Baltimore.

Deamer, D. W. and R. J. Baskin. 1969. Ultrastructure of sarcoplasmic reticulum preparations. *J. Cell Biol.* **42**:296.

Duggan, P. F. and A. Martonosi. 1970. Sarcoplasmic reticulum. IX. The permeability of sarcoplasmic reticulum membranes. *J. Gen. Physiol.* **56**:147.

Ebashi, S. and M. Endo. 1968. Calcium ion and muscle contraction. *Prog. Biophys. Mol. Biol.* **18**:125.

Ebashi, S. and F. Lipmann. 1962. Adenosine triphosphate-linked concentration of calcium ions in a particulate fraction of rabbit muscle. *J. Cell Biol.* **14**:389.

Ebashi, S., M. Endo and I. Ohtsuki. 1969. Control of muscle contraction. *Q. Rev. Biophys.* **2**:351.

Ebashi, S., A. Kodama and F. Ebashi. 1968. Troponin. I. Preparation and physiological function. *J. Biochem.* **64**:465.

Ezerman, E. B. and H. Ishikawa. 1967. Differentiation of the sarcoplasmic reticulum and T system in developing chick skeletal muscle *in vitro*. *J. Cell Biol.* **35**:405.

Hasselbach, W. 1964. Relaxing factor and the relaxation of muscle. *Prog. Biophys. Mol. Biol.* **14**:167.

Holland, P. C. and D. H. MacLennan. 1976. Assembly of the sarcoplasmic reticulum. I. Biosynthesis of the adenosine triphosphatase in rat skeletal muscle cell culture. *J. Biol. Chem.* **251**:2030.

Holtzer, H., H. Weintraub, R. Mayne and B. Mochan. 1972. The cell cycle, cell lineages, and cell differentiation. *Curr. Topics Devel. Biol.* **7**:229.

Huxley, A. F. 1971. The activation of striated muscle and its mechanical response. *Proc. Roy. Soc. London* **B 178**:1.

Huxley, H. E. 1973. Structural changes in the actin- and myosin-containing filaments during contraction. *Cold Spring Harbor Symp. Quant. Biol.* **37**:361.

Ikemoto, N., F. A. Sreter, A. Nakamura and J. Gergely. 1968. Tryptic digestion and localization of calcium uptake and ATPase activity in fragments of sarcoplasmic reticulum. *J. Ultrastruc. Res.* **23**:216.

Jobsis, F. F. and M. J. O'Connor. 1966. Calcium release and reabsorption in the sartorius muscle of the toad. *Biochem. Biophys. Res. Comm.* **25**:246.

Kagen, L. J. and A. Freedman. 1974. Studies on the effects of acetylcholine, epinephrine, dibutyryl cyclic adenosine monophosphate, theophylline and calcium on the synthesis of myoglobin in muscle cell cultures estimated by radioimmunoassay. *Exp. Cell Res.* **88**:135.

Kendrick-Jones, J. 1974. Role of myosin light chains in calcium regulation. *Nature* **249**:631.

Kielley, W. W. and O. Meyerhof. 1948. Studies on adenosinetriphosphatase of muscle. II. A new magnesium-activated adenosinetriphosphatase. *J. Biol. Chem.* **176**:591.

Lehman, W., J. Kendrick-Jones and A. G. Szent-Györgyi. 1973. Myosin-linked regulatory systems: Comparative studies. *Cold Spring Harbor Symp. Quant. Biol.* **37**:319.

Lehninger, A. L. 1970. Mitochondria and calcium ion transport. The fifth jubilee lecture. *Biochem. J.* **119**:129.

MacLennan, D. H. 1970. Purification and properties of an adenosine triphosphatase from sarcoplasmic reticulum. *J. Biol. Chem.* **245**:4508.

MacLennan, D. H. and P. C. Holland. 1975. Calcium transport in sarcoplasmic reticulum. *Annu. Rev. Biophys. Bioeng.* **4**:377.

MacLennan, D. H. and P. T. S. Wong. 1971. Isolation of a calcium-sequestering protein from sarcoplasmic reticulum. *Proc. Nat. Acad. Sci.* **68**:1231.

MacLennan, D. H., P. Seeman, G. H. Iles and C. C. Yip. 1971. Membrane formation by the adenosine triphosphatase of sarcoplasmic reticulum. *J. Biol. Chem.* **246**:2702.

MacLennan, D. H., C. C. Yip, G. H. Iles and P. Seeman. 1973. Isolation of sarcoplasmic reticulum proteins. *Cold Spring Harbor Symp. Quant. Biol.* **37**:460.

Martonosi, A. 1963. The activating effect of phospholipids on the ATPase activity and Ca^{++} transport of fragmented sarcoplasmic reticulum. *Biochem. Biophys. Res. Comm.* **13**:273.

Martonosi, A. and R. A. Halpin. 1971. Sarcoplasmic reticulum. X. The protein composition of sarcoplasmic reticulum membranes. *Arch. Biochem. Biophys.* **144**:66.

Meissner, G. 1975. Isolation and characterization of two types of sarcoplasmic reticulum vesicles. *Biochim. Biophys. Acta* **389**:51.

Morimoto, K. and W. F. Harrington. 1974. Evidence for structural changes in vertebrate thick filaments induced by calcium. *J. Mol. Biol.* **88**:693.

Ostwald, T. J. and D. H. MacLennan. 1974. Isolation of a high affinity calcium-binding protein from sarcoplasmic reticulum. *J. Biol. Chem.* **249**:974.

Paterson, B. and R. C. Strohman. 1972. Myosin synthesis in cultures of differentiating chicken embryo skeletal muscle. *Devel. Biol.* **29**:113.

Porter, K. R. and G. E. Palade. 1957. Studies on the endoplasmic reticulum. III. Its form and distribution in striated muscle cells. *J. Biophys. Biochem. Cytol.* **3**:269.

Racker, E. 1972. Reconstitution of a calcium pump with phospholipids and a purified Ca^{++}-adenosine triphosphatase from sarcoplasmic reticulum. *J. Biol. Chem.* **247**:8198.

Racker, E. and E. Eytan. 1975. A coupling factor from sarcoplasmic reticulum required for the translocation of Ca^{2+} ions in a reconstituted Ca^{2+} ATPase pump. *J. Biol. Chem.* **250**:7533.

Rose, B. and W. R. Loewenstein. 1975. Permeability of cell junction depends on local cytoplasmic calcium activity. *Nature* **254**:250.

Sandow, A. 1970. Skeletal muscle. *Annu. Rev. Physiol.* **32**:87.

Shainberg, A., G. Yagil and D. Yaffe. 1971. Alterations of enzymatic activities during muscle differentiation *in vitro*. *Devel. Biol.* **25**:1.

Shamoo, A. E. and D. H. MacLennan. 1974. A Ca^{++}-dependent and -selective ionophore as part of the $Ca^{++} + Mg^{++}$-dependent adenosinetriphosphatase of sarcoplasmic reticulum. *Proc. Nat. Acad. Sci.* **71**:3522.

Shamoo, A. E., T. E. Ryan, P. S. Stewart and D. H. MacLennan. 1976. Localization of ionophore activity in a 20,000 molecular weight fragment of the adenosine triphosphatase of sarcoplasmic reticulum. *J. Biol. Chem.* (in press).

Stewart, P. S. and D. H. MacLennan. 1974. Surface particles of sarcoplasmic reticulum membranes; structural features of the adenosine triphosphatase. *J. Biol. Chem.* **249**:985.

Stewart, P. S., D. H. MacLennan and A. E. Shamoo. 1976. Isolation and characterization of tryptic fragments of the adenosine triphosphatase of sarcoplasmic reticulum. *J. Biol. Chem.* **251**:712.

Tzagoloff, A., M. S. Rubin and M. F. Sierra. 1973. Biosynthesis of mitochondrial enzymes. *Biochim. Biophys. Acta* **301**:71.

Warren, G. B., M. D. Houslay, J. C. Metcalfe and N. J. M. Birdsall. 1975. Cholesterol is excluded from the phospholipid annulus surrounding an active calcium transport protein. *Nature* **255**:684.

Weber, K. and M. Osborn. 1969. The reliability of molecular weight determinations by dodecyl sulfate-polyacrylamide gel electrophoresis. *J. Biol. Chem.* **244**: 4406.

Yaffe, D. 1968. Retention of differentiation potentialities during prolonged cultivation of myogenic cells. *Proc. Nat. Acad. Sci.* **61**:477.

―――. 1971. Developmental changes preceding cell fusion during muscle differentiation *in vitro. Exp. Cell Res.* **66**:33.

―――. 1973. Rat skeletal muscle cells. In *Tissue culture: Methods and applications* (ed. P. Kruse and M. K. Patterson), p. 106. Academic Press, New York.

Yaffe, D. and H. Dym. 1973. Gene expression during differentiation of contractile muscle fibers. *Cold Spring Harbor Symp. Quant. Biol.* **37**:543.

Yamamoto, T. 1972. The Ca^{2+}-Mg^{2+}-dependent ATPase and the uptake of Ca^{2+} by the fragmented sarcoplasmic reticulum. In *Muscle proteins, muscle contraction and cation transport* (ed. Y. Tonomura), p. 305. University of Tokyo Press, Tokyo.

Vertebrate Smooth Muscle: Ultrastructure and Function

Andrew P. Somlyo, Avril V. Somlyo,
Francis T. Ashton and Julien Vallières

Pennsylvania Muscle Institute of the University of Pennsylvania;
Department of Pathology, Presbyterian-University of Pennsylvania Medical Center;
and Departments of Physiology and Pathology, School of Medicine
University of Pennsylvania, Philadelphia, Pennsylvania 19174*

Functional diversity is one of the most important features of vertebrate smooth muscles. Activation of smooth muscles by a rise in cytoplasmic calcium may be triggered by action potentials, by graded depolarization or through pharmacomechanical coupling independent of surface membrane potential (A. V. Somlyo and Somlyo 1968; for review, see Bohr 1973; A. P. Somlyo 1975). The source of activator calcium may be through the influx of extracellular cation or the release of intracellular stores (A. P. Somlyo and Somlyo 1971; A. P. Somlyo et al. 1971; Devine, Somlyo and Somlyo 1972), and the relative contribution of these two sources may vary in different smooth muscles and under different experimental conditions. Similarly, whereas both the uptake of calcium into intracellular organelles and its extrusion into the extracellular space may contribute to relaxation, the relative capacities and contributions of the various sinks to the removal of the activator calcium have not been quantitated.

The regulatory site of the calcium-sensitive actomyosin ATPase of smooth muscle (Sparrow et al. 1970; Rüegg 1971) has not been established unequivocally. There is evidence favoring the existence of a tropomyosin-troponin-like regulatory system (Ebashi et al. 1966; Ebashi, Toyo-oka and Nonomura 1976) and, alternatively, of a myosin regulation similar to that found in molluscs (Bremel 1974; Hartshorne et al. 1976).

The ultrastructural studies of smooth muscle to be summarized here, based on the calcium-activated interaction of actin with myosin, bear on two problems common to motile systems: the identification of the intracellular calcium storage sites and the cellular organization of the contractile proteins. Since some of these questions remain to be answered in non-muscle systems of cell motility, it may be appropriate to note that the recognition of a functional (Ca-accumulating) sarcoplasmic reticulum and the consistent demonstration

* Mailing address: 51 N. 39th St., Philadelphia, Pennsylvania 19104.

of organized myosin in vertebrate smooth muscle are relatively recent developments (e.g., A. P. Somlyo and Somlyo 1968; cf. Devine, Somlyo and Somlyo 1972; A. P. Somlyo et al. 1973).

MATERIALS AND METHODS

Unless otherwise indicated, smooth muscle for electron microscopy was incubated in a modified Krebs' solution for 30 minutes at 37°C prior to fixation with 2% glutaraldehyde and 4.5% sucrose–0.075 M cacodylate buffer (Devine, Somlyo and Somlyo 1972). In some experiments, tannic acid was also added to the glutaraldehyde fixative (Ashton, Somlyo and Somlyo 1975). Postfixation in 2% osmium tetroxide in 0.05 M cacodylate buffer, serial dehydration, and embedding in Spurr's resin (Spurr 1969) have been described elsewhere (Devine, Somlyo and Somlyo 1972; Ashton, Somlyo and Somlyo 1975). Electron microscopy at conventional (75 kV or 80 kV) accelerating voltages was performed on a Hitachi HU 11E or Philips EM301 electron microscope. Intermediate high voltage stereo electron microscopy was done on a JEM200 electron microscope equipped with a side-entry goniometer stage permitting rotation and a tilt of ± 60°. Details of the serial sectioning procedure, including the determination of section thickness, and the staining procedure used for intermediate high voltage stereo electron microscopy have been described in detail elsewhere (Ashton, Somlyo and Somlyo 1975).

Mitochondria were isolated from bovine mesenteric vein and from main pulmonary artery and their rates of calcium uptake studied with dual-wavelength spectroscopy employing arsenazo 3 (for 1–25 μM calcium) or murexide (for Ca above 25 μM) as the indicators. Details of the methods of isolation, the experimental procedures used and the functional parameters of the mitochondria have been published elsewhere (Vallières, Scarpa and Somlyo 1975).

Prior to cryoultramicrotomy, rabbit portal-anterior mesenteric vein strips were depolarized for 30 minutes in a high K solution (Devine, Somlyo and Somlyo 1972) containing calcium to increase calcium influx. The strips were placed on silver pins and quenched in liquid nitrogen slush or in freon 22 chilled with liquid nitrogen. Sections were cut with the specimen at −100°C to −140°C using an LKB cryoultramicrotome, modified in our laboratory, with knife temperature at −100°C. Some of the preparations were stained with osmium vapor by placing osmium crystals in vacuo, and others were examined unstained.

Electron probe X-ray microanalysis was done on a Philips EM301 electron microscope with a modified side-entry goniometer stage containing a 30-mm^2 Kevex Si (Li) X-ray detector (160 eV resolution), interfaced with a Kevex 5100 multichannel analyzer, and a Tracor Northern NS 880 system used for computer analysis of spectra stored on magnetic tape by a TN 1000 tape recorder. The sensitivity of the system (at 100 sec counting time, approx 1000 cps calibrated with 100-nm thick sections) is 10 mM/kg K, or a minimal detectable mass of 10^{-18} g (Shuman and Somlyo 1975).

RESULTS

Sarcoplasmic Reticulum and Mitochondria

The sarcoplasmic reticulum forms an interlacing system of continuous central and peripheral tubules occupying approximately 2–7.5% of cytoplasmic volume in different smooth muscles (A. P. Somlyo et al. 1971; Devine, Somlyo and Somlyo 1972). In turtle smooth muscle, the tubules communicate with occasional saclike expansions of the sarcoplasmic reticulum (A. P. Somlyo et al. 1971). Extracellular markers, such as ferritin, colloidal lanthanum or horseradish peroxidase, enter the surface vesicles (caveolae) but not the sarcoplasmic reticulum. In some regions, i.e., the surface couplings, the peripheral sarcoplasmic reticulum approaches the surface membrane: the 10–12-nm space between the two membrane systems is traversed by somewhat periodic, electron-opaque processes (Fig. 1).

The accumulation of divalent cations by the sarcoplasmic reticulum can be demonstrated by incubating smooth muscle, prior to fixation, in a solution in which strontium replaces calcium (A. V. Somlyo and Somlyo 1971); under these conditions, electron-opaque deposits of strontium are found in the lumen of the sarcoplasmic reticulum (Fig. 2a) and in mitochondria (Fig. 2a,2b). Mitochondria also accumulate barium in association with phosphorus (3Ba/4P) during barium contractures of smooth muscle; Ba uptake is inhibited by oligomycin and anoxia (A. P. Somlyo et al. 1974).

The kinetics of respiratory substrate-supported calcium accumulation by mitochondria isolated from smooth muscle are shown in Figure 3. The functional parameters of these mitochondria (respiratory control ratio of up to about 9, ADP/O ratios from 2 to 3, and state-3 oxygen consumption of up to 117 nmoles O_2/min/mg protein) were comparable to those of well-

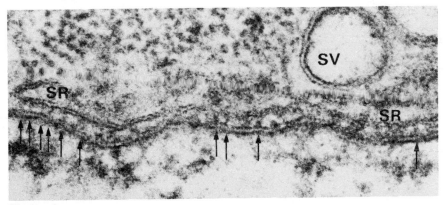

Figure 1

Sarcoplasmic reticulum (SR)–surface membrane coupling. This high-magnification view of a transverse section of rabbit portal-anterior mesenteric vein shows the close relationship between the two membrane systems and the periodic densities (arrows) connecting the gap between the plasma membrane and outer lamina of the SR membrane. SV, surface vesicle. 250,000×. (Reprinted, with permission, from A. P. Somlyo and Somlyo 1975.)

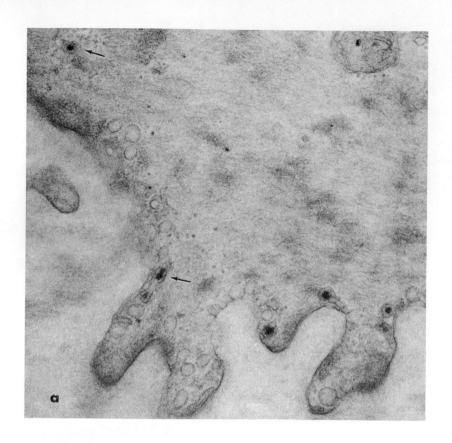

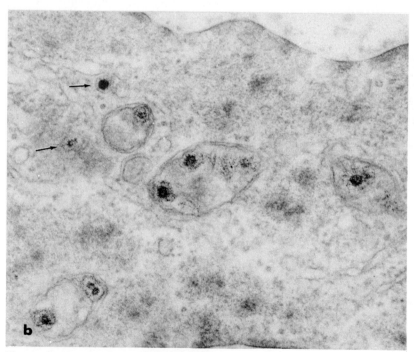

Figure 2 (*See facing page for legend*)

168

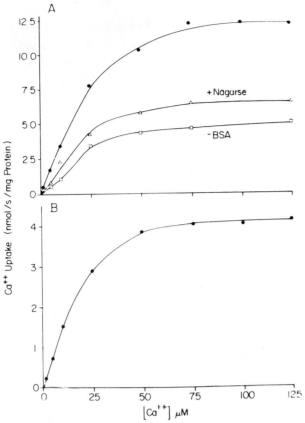

Figure 3
Rates of Ca++ uptake by bovine main pulmonary artery and mesenteric vein mitochondria at various Ca++ concentrations. Mitochondria (0.2–0.7 mg/ml) were incubated at 26°C in 100 mM sucrose, 50 mM KCl, 2 mM MgCl$_2$, 5 mM sodium succinate, 3 μM rotenone, 5 mM sodium acetate, 20 mM MOPS (pH 7.2) and 100 μM murexide or 40 μM arsenazo. Rates were measured with arsenazo for Ca++ concentrations ranging from 1–25 μM and with murexide for higher concentrations. (*A*) Mitochondria isolated from pulmonary artery in the presence of 0.2% bovine serum albumin (•), in the absence of albumin (□) and after exposure to proteinase (△). (*B*) Mesenteric vein mitochondria. (Reprinted, with permission, from Vallières, Scarpa and Somlyo 1975.)

Figure 2
(*a*) Oblique section of smooth muscle cell in guinea pig portal-anterior mesenteric vein incubated in Krebs' solution containing 10 mM Sr++ for 1 hr prior to fixation. Electron-opaque deposits (arrows) of strontium are present in both the peripheral and central portions of the SR. 190,500×. (*b*) Portion of a smooth muscle cell in a rabbit main pulmonary artery that had been depolarized with a high K (136 mM) solution containing 10 mM Sr++ for 50 min and then placed in Krebs' solution containing 10 mM Sr++ for an additional 10 min before fixation. Deposits of Sr++ are present in the mitochondria and in the SR (arrows). 60,000×. (Reprinted, with permission, from A. V. Somlyo and Somlyo 1971. © 1971 by the American Association for the Advancement of Science.)

preserved mitochondria isolated from other tissues. The mitochondrial content of bovine mesenteric vein smooth muscle, as estimated from cytochrome c oxidase activity (Vallières, Scarpa and Somlyo 1975), was approximately 8% of cytoplasmic volume. The apparent K_m of calcium accumulation was about 17 μM (Fig. 3), and mitochondria could accumulate up to 200 nmoles calcium/mg mitochondrial protein without uncoupling.

Cryoultramicrotomy of portal-anterior mesenteric vein permits electron probe X-ray microanalysis of unfixed, adult smooth muscle. In sections that are approximately 100 nm thick, the nucleus and the mitochondria can be readily recognized (Fig. 4). Since in these preliminary studies we only rarely obtained unequivocal images of the rather narrow sarcoplasmic reticulum tubules, we also analyzed the perinuclear space as it is continuous with the sarcoplasmic reticulum. Electron probe X-ray microanalysis of such sections of rabbit portal-anterior mesenteric vein, depolarized prior to fixation (see

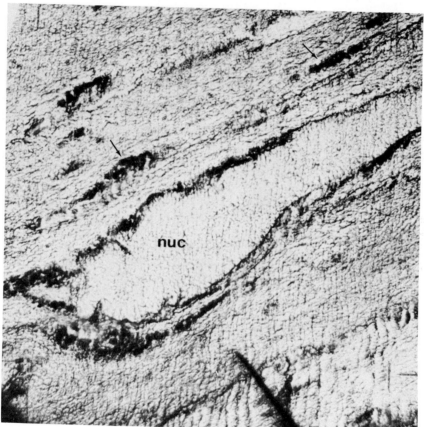

Figure 4

Low-magnification (21,000×) electron micrograph of frozen thin section of portal-anterior mesenteric vein smooth muscle showing nucleus (nuc) and mitochondria (arrow). There is some evidence of intracellular ice crystal formation and slight compression during sectioning. Unfixed tissue sectioned at −100°C and removed without floatation. Vapor-stained with osmium after drying.

Methods), showed the presence of significant Ca peaks over some of the mitochondria and the perinuclear space (Fig. 5). It should be emphasized that the examples shown here are those of relatively large calcium signals, indicative of high concentrations. In other mitochondria, the calcium peaks were lower or absent.

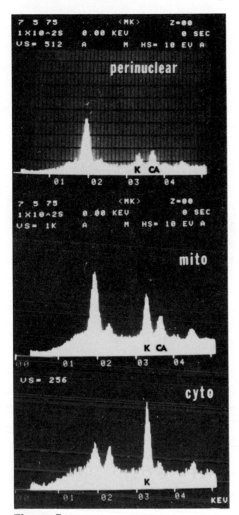

Figure 5

Representative X-ray spectra obtained from frozen thin sections of rabbit portal-anterior mesenteric vein placed in depolarizing (high K) solution containing 2.5 mM calcium for 30 min prior to freezing. The spectra shown are examples of those obtained over the region corresponding to (a) the perinuclear space and nuclear membranes of Fig. 4, (b) mitochondrion and (c) cytoplasm. There are highly significant calcium peaks over both the perinuclear space (continuous with the SR) and the mitochondrion. Note the peak at 3.59 keV is the potassium K_β and not the calcium K_a, which is at 3.69 keV. Section obtained from the adventitial surface of rabbit portal-anterior mesenteric vein were vapor-stained with osmium in vacuo after drying.

Contractile Apparatus and Intermediate Filaments

The major elements of the contractile apparatus are shown in a 0.22-μm thick transverse section taken at 200 kV (Fig. 6). There are thick (approx 15 nm diam) myosin filaments, surrounded in the best-organized examples by an orbit of thin (5–8 nm) actin filaments, with a thin to thick ratio of approx-

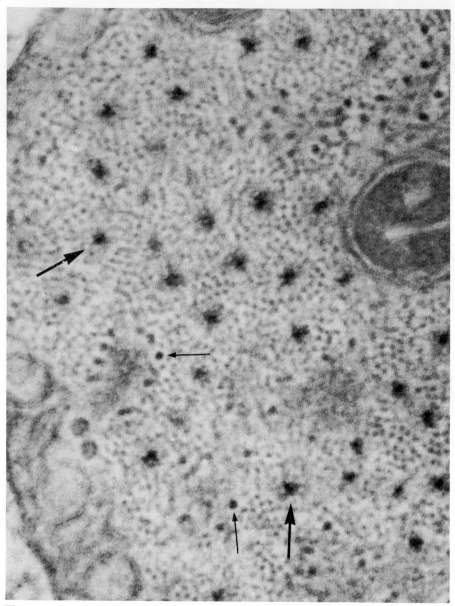

Figure 6

Transverse 220-nm thick section of rabbit portal-anterior mesenteric vein. Thick myosin filaments (large arrows) are surrounded by thin actin filaments. Intermediate 10-nm filaments (small arrows) are associated with dense bodies or are in bundles (upper right). 200 kV; 200,000×.

imately 15:1. Dense bodies are approximately 100 nm in diameter and are commonly surrounded by intermediate (approx 10 nm diam) filaments. The length of the myosin filaments has been measured in stereoscopic views of semi-thick (140–160 nm thick) longitudinal sections with intermediate high voltage electron microscopy (Fig. 7; Ashton, Somlyo and Somlyo 1975).

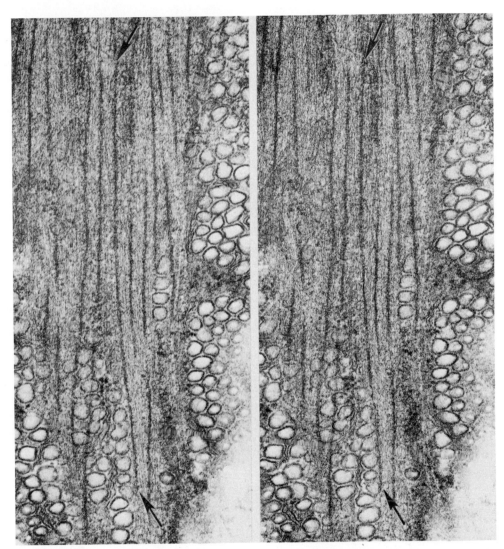

Figure 7
Stereo electron micrograph of 160-nm thick longitudinal section cut near the surface of a vascular smooth muscle fiber. A 2.3-μm long myosin filament (ends marked by arrows) is completely included within the section. Note also, below the upper arrow, the dense body with associated 10-nm filaments. A microtubule runs along the right side of the cell adjacent to the surface vesicles; another enters the picture from the left near the top of the micrograph. Eight percent tannic acid in the fixative; lead citrate-stained section. Tilt ± 10°. 50,000×. (Reprinted, with permission, from Ashton, Somlyo and Somlyo 1975.)

The combination of higher than conventional accelerating voltage and stereo-scopic examination of tilt pairs provides a three-dimensional view of entire thick filaments in relatively thick sections. With this method it is possible to

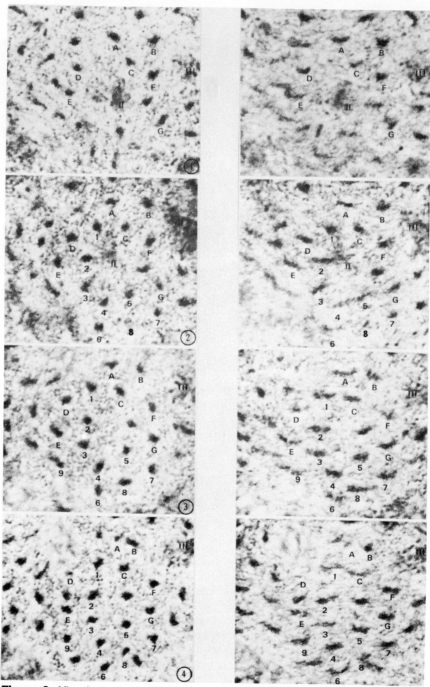

Figure 8 (*See facing page for legend*)

avoid both the artifactual shortening of the filaments, which would otherwise be produced by even minor (3–5°) misalignment of the thick filaments relative to the plane of thin (50–80 nm thick) sections, and the misinterpretation of two filaments overlapped end-to-end within the focal plane of a relatively thick section. In portal-anterior mesenteric vein smooth muscle, the length of the myosin filaments was 2.2 μm $\pm$ 0.14 s.d. The length of the thick filaments was independently verified through reconstruction of serial transverse sections (Fig. 8). Reconstruction of eight consecutive serial sections (0.47 μm thick) showed complete filaments present throughout five or six of the nonterminal sections (2 through 7). Considering the possible stagger of the thick filaments, a 2.2-μm long thick filament would be expected to be included in either five or six consecutive (approx 0.5 μm thick) sections; therefore, the results of serial reconstruction are compatible with the filament length measured in longitudinal sections. In the serial transverse sections we also unequivocally verified the taper of the thick filaments: the diameters of the thick filaments were significantly less ($P < 0.01$) at the ends than at the shank (14.6 nm $\pm$ 0.25 s.e., $n = 20$) of the same filaments. Examination of the stereo pairs also showed that without the use of a tilt stage, short, ribbonlike structures representing oblique views of the filaments could be clearly misinterpreted. Other variations in shape, such as triangular profiles, may be related to the molecular assembly of myosin into filaments (Pepe 1975) or to variability of preservation during fixation. It will be necessary to examine in greater detail the cross-sectional profiles in serial transverse sections to determine whether the triangular shape is consistently observed at any given region of the thick filaments.

To date, the preservation of the cross bridges has been imperfect. However, in our most well preserved material it is evident that the cross bridges are not only on two opposite faces of a mini-ribbonlike structure (Fig. 9), but their appearance is at least compatible with a helical arrangement (A. V. Somlyo et al. 1976). High-resolution views of the substructure of thick filaments, as well as examination of the serial sections, rule out the presence of the intermediate (10 nm diam) filaments as a core of the myosin elements (see Discussion).

The relatively regular arrangement, approximately 60–80-nm lattice, of

Figure 8

Stereo electron micrographs of portions of four of eight 0.47-μm thick serial transverse sections. Thick filaments present in the first section are labeled A–G. A group of thick filaments, labeled 1–9, starts in sections 2 and 3. These filaments are complete in this set of eight sections, ending in sections 6 and 7. A dense body (II) is found in sections 1 and 2. Actin filaments are seen in the subsequent sections in the regions continuous with dense body II. Another dense body (III) is continuous throughout the eight sections. Note particularly in section 4 that due to the oblique orientation, the right side view gives the appearance of short ribbons. Note that the profile of filament D becomes very small in section 4 and is absent in the subsequent section (not shown), showing the taper of the myosin filaments. Not stained with lead. Left-eye view, tilt + 15°; right-eye view, tilt + 5°. 80,000×. (Reprinted, with permission, from Ashton, Somlyo and Somlyo 1975.)

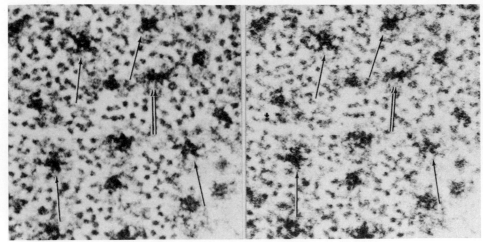

Figure 9

Stereo electron micrograph of 70-nm thick transverse section of vascular smooth muscle showing four thick filaments with cross bridges. Although a fifth filament (double arrows) appears to be a "mini-ribbon," the stereo image shows that it is an obliquely oriented filament. Section stained with saturated aqueous uranyl acetate and lead citrate; stereo angle ± 5°. 250,000×. (Reprinted, with permission, from Ashton, Somlyo and Somlyo 1975.)

the thick filaments in the transverse plane of rabbit portal-anterior mesenteric vein smooth muscle (Rice et al. 1971) was also evident in this material. In serial transverse sections we have also found some regularity of the arrangement of the thick filaments along the longitudinal axis. Thus the thick filaments started and ended not randomly, but in groups of three to five. Planar projections of five consecutive transverse sections also showed that the thick filaments were placed in parallel (see Fig. 1 in Ashton, Somlyo and Somlyo 1975).

The major question concerning dense bodies has been whether they serve as attachment sites for the thin (actin) filaments. The insertion of thin filaments on plasma-membrane-bound dense bodies is readily demonstrated (Pease and Molinari 1960; Gabella 1973; A. P. Somlyo et al. 1973). Preliminary studies reported elsewhere (A. V. Somlyo et al. 1976) show that S_1 subfragments attached to thin filaments in this region form arrowheads pointing away from plasma membrane, as would be expected at the sites of thin filament insertion (Huxley 1963).

The relationship of the thin filaments to the cytoplasmic dense bodies has been somewhat more difficult to define, partly due to the frequently overlapping image of the intermediate filaments. Examination of stereo views of longitudinal sections of cytoplasmic dense bodies (Fig. 10), combined with parallax measurement, established the insertion of actin filaments on at least some of these structures. Profiles comparable in size and spacing to those of actin filaments have also been found in dense bodies in serial transverse sections, and the termination of a dense body in serial transverse sections is frequently replaced in the next section by thin filaments (Ashton, Somlyo and Somlyo 1975). However, there are also regions of dense bodies that are bypassed by the thin filaments and in which the entry of actin filaments cannot be demonstrated. The length of the dense bodies appears to be variable, and the structures show some branching and confluence.

DISCUSSION

Sarcoplasmic Reticulum, Mitochondria and the Golgi Apparatus

The properties of the sarcoplasmic reticulum (SR) of smooth muscle, an intracellular system that can accumulate divalent cations and form couplings with the surface membrane, are compatible with a role in excitation-contraction coupling qualitatively similar to that of the SR in striated muscle. The association of the SR of smooth muscle with glycogen (A. P. Somlyo et al. 1971) and the presence of a rough sarcoplasmic reticulum, continuous with the smooth-surfaced system (A. P. Somlyo and Somlyo 1975), are suggestive of additional, metabolic and morphogenetic (Ross and Klebanoff 1971) functions. Significant variations in the volume of the SR occur between different smooth muscles, and there is some correlation between the ability of smooth muscle to contract in the absence of extracellular calcium and a relatively large SR content (A. P. Somlyo et al. 1971; Devine, Somlyo and Somlyo 1972). However, the larger amounts of SR found in large elastic arteries usually include rough membranes, and it is not known whether the cation storing capacities of the rough and smooth SR are identical.

The surface couplings are the most likely sites for excitation of the surface membrane to be communicated to the SR, thereby resulting in calcium release. Some of the contents of the SR may also be transferred to the extracellular space through these couplings or through other regions of the SR that are fenestrated around the surface vesicles (Devine, Somlyo and Somlyo 1972); the evidence for these suggestions is purely morphological. It is probable that localized associations with the surface membrane are a common characteristic of the endoplasmic reticulum in all cell systems in which this organelle plays a role in regulating cellular calcium levels.

The ability of the SR of smooth muscle to accumulate divalent cations is shown by the uptake of strontium during incubation with this cation (A. V. Somlyo and Somlyo 1971; A. P. Somlyo et al. 1974) and by the localization of calcium to the perinuclear space and SR. Considering that in portal-anterior mesenteric vein and taenia coli smooth muscles, the SR amounts to, at most, 3% of the cytoplasmic volume (Devine, Somlyo and Somlyo 1972; Popescu et al. 1974) and *assuming* that its total calcium content is similar (about 25 mM) to that of the SR of striated muscle (Ebashi and Endo 1968), the calcium capacity of the SR in these smooth muscles is about 0.8 mM/kg cell. Cellular calcium exceeding this amount would have to be stored in other organelles. As noted above, however, the volume of SR in other smooth muscles may be greater.

Mitochondria isolated from smooth muscle and in situ can accumulate divalent cations (A. V. Somlyo and Somlyo 1971; A. P. Somlyo et al. 1974;

Vallières, Scarpa and Somlyo 1975), suggesting that they also may contribute to the regulation of cytoplasmic calcium levels, at least as reserve sinks during sustained calcium influx. The close (4–5 nm) contacts formed between peripheral mitochondria and the surface vesicles have been suggested as pos-

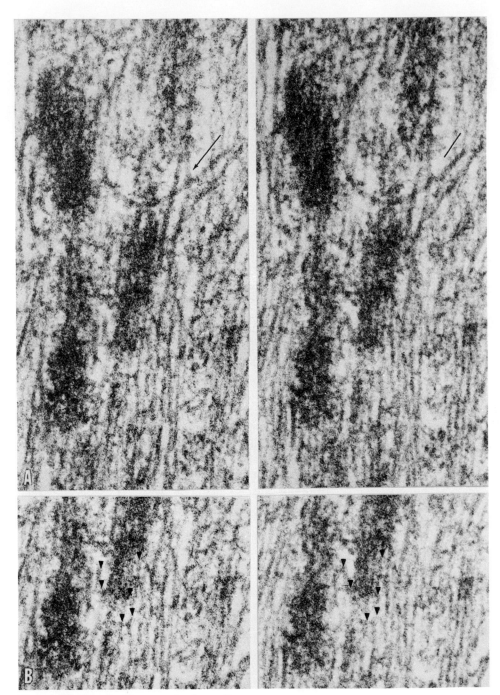

Figure 10 (*See facing page for legend*)

sible sites of extrusion of mitochondrial Ca into the extracellular space (A. P. Somlyo et al. 1974; Wootton and Goodford 1975). The rates of calcium uptake by mitochondria isolated from different smooth muscles appear to vary (Vallières, Scarpa and Somlyo 1975), but it is not known whether these differences also exist in situ. Mitochondria constitute between 5–10% of the cell volume in vertebrate smooth muscle, and isolated mitochondria accumulate up to 600 nM Ca/mg mitochondrial protein supported by respiratory substrate and up to 1800 nM Ca/mg mitochondrial protein in the presence of ATP (Vallières, Scarpa and Somlyo 1975). Therefore it can be estimated that mitochondria can store from 3 to 18 mM Ca/kg cell, values clearly in excess of normal cellular calcium content. Rather high mitochondrial calcium concentrations are suggested by the results of electron probe X-ray analysis of cultured, freeze-dried vascular smooth muscle cells (Garfield and Somlyo 1975; A. P. Somlyo et al. 1976), but these findings may represent special properties of immature cells or result from somewhat less than ideal culture conditions. Electron probe analysis of freeze-dried sections of adult vascular smooth muscle can provide the most direct evidence regarding mitochondria acting as intracellular calcium stores, at least during the sustained influx produced by prolonged depolarization, but our preliminary results are not yet sufficient to give a definitive answer to this question. However, in view of the relatively low rates of calcium accumulation by mitochondria at low calcium concentrations (Fig. 3) it is likely that the SR, rather than mitochondria, is responsible for the regulation of the spontaneous twitch contractions in spontaneously active smooth muscles.

The Golgi apparatus of smooth muscle is composed of the characteristic stacked cisternae and vesicles, located at the nuclear pole in adult tissues and more dispersed in cultured cells (A. P. Somlyo et al. 1975). There is no persistent structural communication between the Golgi apparatus and the SR: marked swelling of the Golgi by the ionophorus antibiotic X537A can occur without any swelling of the rough or smooth SR (A. P. Somlyo et al. 1975). Of course, this finding does not preclude intermittent functional communication between the two systems (Palade 1975). The X537A-induced swelling is associated with inhibition of [^{35}S]sulfate and [^{3}H]glucosamine incorporation into glycosaminoglycans, suggesting that the Golgi in smooth muscle, as in other cell systems (for review, see A. P. Somlyo et al. 1975), participates in glycosaminoglycan synthesis.

Figure 10

(*A*) Stereo electron micrograph of 50-nm thick longitudinal section of a vascular smooth muscle cell showing dense bodies. Thin filaments attach to the two lower portions of the dense bodies, and there are several prominent intermediate filaments (arrows) around the upper portions. (*B*) The lower portion of *A*. Arrowheads indicate the points where parallax measurements were made: the arrowheads to the left indicate the thickness of the section, those to the upper right, the top and bottom of the dense body; the lower arrowheads show that the thin filaments measured run into, rather than above or below, the dense body. Two percent tannic acid in glutaraldehyde. Section stained with 5% uranyl acetate in absolute ethanol and lead citrate. Stereo angle ± 20°. 160,000×. (Reprinted, with permission, from Ashton, Somlyo and Somlyo 1975.)

Filaments and Dense Bodies

The preservation of the myofilaments in smooth muscle is significantly improved through the avoidance of swelling of the tissues to be fixed for electron microscopy. In swollen smooth muscle fibers, the thick, and, to a lesser extent, the thin, filaments are destroyed during osmium fixation (A. P. Somlyo, Devine and Somlyo 1971; Jones, Somlyo and Somlyo 1973), although the fibers show normal contractility in the swollen state. The destruction of the filaments by the fixative can be avoided by osmotic measures that prevent or reverse swelling (Jones, Somlyo and Somlyo 1973). We have found it useful also to add sucrose (usually 4.5% in 0.075 M cacodylate) to the fixative buffer to maintain isotonicity, since the osmotic responses of cells are not abolished by aldehyde fixatives (Jones, Somlyo and Somlyo 1973; A. P. Somlyo and Somlyo 1975).

Thin filaments represent the organized form of actin in vertebrate smooth muscle, and, at least in normal adult smooth muscle, we find no amorphous regions suggestive of large amounts of actin in nonfilamentous form. The ratio of thin (actin) to thick (myosin) filaments is approximately 15:1 (A. P. Somlyo et al. 1973); this ratio is in agreement with that obtained from SDS gels of smooth muscle (Tregear and Squire 1973; Murphy, Herlihy and Megerman 1974).

The plasma-membrane-bound dense bodies are clearly attachment sites of the actin filaments and hence analogous to the Z lines of striated muscle. Thin filaments also insert on at least some of the cytoplasmic dense bodies, although on others the insertion of thin filaments cannot be unequivocally demonstrated. Such regions may represent extensions of dense bodies on which thin filaments insert outside the plane of a given section. Alternatively, the dense bodies, relatively nonspecifically stained and somewhat amorphous structures, may not represent a single functional and morphological entity in vertebrate smooth muscle (for discussion, see A. P. Somlyo and Somlyo 1975). The staining of at least some dense bodies with anti-α-actinin antibodies (Schollmeyer et al. 1973) is further evidence of the similarity between these structures and the Z lines.

Myosin is organized in 2.2-μm long filaments (in rabbit portal-anterior mesenteric vein) that have tapered ends and appear in groups of three to five. The parallel arrangement of the thick filaments and their relatively great length probably contribute to the ability of smooth muscle to develop tension at least equivalent to that developed by striated muscle, in spite of the lower concentration of myosin in smooth muscle (Murphy, Herlihy and Megerman 1974; Ashton, Somlyo and Somlyo 1975). The precise arrangement of the cross bridges, suggestive of a helix in our best material, cannot be specified until better preservation is obtained. The possibility of myosin being organized into "mini-ribbons" bearing cross bridges of opposing polarity, only on two opposite faces, has been ruled out. The proposals put forth by Small and Squire (1972), that ribbons are the normal, organized form of myosin in vertebrate smooth muscle, that the intermediate filaments are the core elements of organized myosin, and that the dense bodies are degraded myosin, have now been shown to be in error.

The intermediate filaments are clearly distinct from the thin and thick myofilaments. They resemble similar filaments found in a variety of muscle and non-muscle cells under normal and abnormal conditions (Ishikawa, Bischoff and Holtzer 1968; Cooke and Chase 1971; A. P. Somlyo et al. 1971, 1973; Shelanski et al. 1971; Wuerker and Kirkpatrick 1972; Goldman and Knipe 1973; Rice and Brady 1973; Holtzer et al. 1974). Although their resemblance to the cytoskeleton of some invertebrate muscles has been pointed out earlier (Cooke and Chase 1971; A. P. Somlyo et al. 1971), further functional evidence to support this view is required.

Acknowledgments

This work was supported by National Institutes of Health Grant HL 15835 to the Pennsylvania Muscle Institute. We want to thank Mr. John Silcox for excellent technical assistance.

REFERENCES

Ashton, F. T., A. V. Somlyo and A. P. Somlyo. 1975. The contractile apparatus of vascular smooth muscle: Intermediate high voltage stereo electron microscopy. *J. Mol. Biol.* **98**:17.

Bohr, D. F. 1973. Vascular smooth muscle updated. *Circ. Res.* **32**:665.

Bremel, R. D. 1974. Myosin linked calcium regulation in vertebrate smooth muscle. *Nature* **252**:405.

Cooke, P. H. and R. H. Chase. 1971. Potassium chloride-insoluble myofilaments in vertebrate smooth muscle cells. *Exp. Cell Res.* **66**:417.

Devine, C. E., A. V. Somlyo and A. P. Somlyo. 1972. Sarcoplasmic reticulum and excitation-contraction coupling in mammalian smooth muscle. *J. Cell Biol.* **52**:690.

Ebashi, S. and M. Endo. 1968. Calcium ion and muscle contraction. *Prog. Biophys. Mol. Biol.* **18**:123.

Ebashi, S., H. Iwakura, H. Nakajuna, R. Nakamura and Y. Ooi. 1966. New structural proteins from dog heart and chicken gizzard. *Biochem. Z.* **345**:201.

Ebashi, S., T. Toyo-oka and Y. Nonomura. 1976. Gizzard troponin. In *Biochemistry of smooth muscle* (ed. N. L. Stephens). University Park Press, Baltimore, Maryland. (In press.)

Gabella, G. 1973. Fine structure of smooth muscle. *Phil. Trans. Roy. Soc. B.* **265**:7.

Garfield, R. E. and A. P. Somlyo. 1975. Electron probe analysis and ultrastructure of cultured, freeze-dried vascular smooth muscle. In *Proceedings 33rd Annual Meeting EMSA*, p. 558. Claitor's Publishing Division, Los Angeles.

Goldman, R. D. and D. M. Knipe. 1973. Functions of cytoplasmic fibers in non-muscle cell motility. *Cold Spring Harbor Symp. Quant. Biol.* **37**:523.

Hartshorne, D., L. Abrams, M. Aksoy, R. Dabrowska, S. Driska and E. Sharkey. 1976. Molecular basis for the regulation of smooth muscle actomyosin. In *Biochemistry of smooth muscle* (ed. N. L. Stephens). University Park Press, Baltimore, Maryland. (In press.)

Holtzer, H., J. Croop, M. Gershon, and A. P. Somlyo. 1974. Effects of cytochalasin-B and colcimide on cells in muscle cultures. *Amer. J. Anat.* **141**:291.

Huxley, H. E. 1963. Electron microscope studies on the structure of natural and synthetic protein filaments from striated muscle. *J. Mol. Biol.* **7**:281.

Ishikawa, H., R. Bischoff and H. Holtzer. 1968. Mitosis and intermediate-sized filaments in developing skeletal muscle. *J. Cell Biol.* **38**:538.

Jones, A. W., A. P. Somlyo and A. V. Somlyo. 1973. Potassium accumulation in smooth muscle and associated ultrastructural changes. *J. Physiol.* **232**:247.

Murphy, R. A., J. T. Herlihy and J. Megerman. 1974. Force-generating capacity and contractile content of arterial smooth muscle. *J. Gen. Physiol.* **64**:691.

Palade, G. 1975. Intracellular aspects of the process of protein synthesis. *Science* **189**:347.

Pease, D. C. and S. Molinari. 1960. Electron microscopy of muscular arteries; pial vessels of the cat and monkey. *J. Ultrastruc. Res.* **3**:447.

Pepe, F. A. 1975. Structure of muscle filaments from immunohistochemical and ultrastructural studies. *J. Histochem. Cytochem.* **23**:543.

Popescu, L. M., I. Diculescu, U. Zelck and I. Ionescu. 1974. Ultrastructural distribution of calcium in smooth muscle cells of guinea-pig taenia coli. *Cell Tissue Res.* **154**:357.

Rice, R. V. and A. C. Brady. 1973. Biochemical and ultrastructural studies on vertebrate smooth muscle. *Cold Spring Harbor Symp. Quant. Biol.* **37**:429.

Rice, R. V., G. M. McManus, C. E. Devine and A. P. Somlyo. 1971. A regular organization of thick filaments in mammalian smooth muscle. *Nature New Biol.* **231**:242.

Ross, R. and S. J. Klebanoff. 1971. The smooth muscle cell. I. In vivo synthesis of connective tissue. *J. Cell Biol.* **50**:159.

Rüegg, J. D. 1971. Smooth muscle tone. *Physiol. Rev.* **51**:201.

Schollmeyer, J. S., D. E. Goll, R. M. Robson and M. H. Stormer. 1973. Localization of a-actinin and tropomyosin in different muscles. *J. Cell Biol.* **59**:306a.

Shelanski, M. L., S. Albert, G. H. DeVries and W. T. Norton. 1971. Isolation of filaments from brain. *Science* **174**:1242.

Shuman, H. and A. P. Somlyo. 1975. Quantitative EDS of ultra-thin biological sections. In *Proceedings 10th Annual Conference Microbeam Analysis Society,* Las Vegas, Nevada. Abstr. #41.

Small, J. V. and J. M. Squire. 1972. Structural basis of contraction in vertebrate smooth muscle. *J. Mol. Biol.* **67**:117.

Somlyo, A. P. 1975. Vascular smooth muscle. In *Cellular pharmacology of excitable tissues* (ed. T. Narahashi), pp. 360–407. Charles C. Thomas, Springfield, Illinois.

Somlyo, A. P. and A. V. Somlyo. 1968. Vascular smooth muscle. I. Normal structure, pathology, biochemistry and biophysics. *Pharmacol. Rev.* **20**:197.

―――. 1971. Electrophysiological correlates of the inequality of maximal vascular smooth muscle contraction elicited by drugs. In *Vascular neuroeffector systems* (ed. J. A. Bevan et al.), p. 216. S. Karger, Basel.

―――. 1975. Ultrastructure of smooth muscle. In *Methods in pharmacology* (ed. E. E. Daniel and D. M. Paton), vol. 3, p. 3. Plenum Press, New York.

Somlyo, A. P., C. E. Devine and A. V. Somlyo. 1971. Thick filaments in unstretched mammalian smooth muscle. *Nature New Biol.* **233**:218.

Somlyo, A. P., C. E. Devine, A. V. Somlyo and S. R. North. 1971. Sarcoplasmic reticulum and the temperature-dependent contraction of smooth muscle in calcium-free solutions. *J. Cell Biol.* **51**:722.

Somlyo, A. P., C. E. Devine, A. V. Somlyo and R. V. Rice. 1973. Filament organization in vertebrate smooth muscle. *Phil. Trans. Roy. Soc. B.* **265**:223.

Somlyo, A. P., R. E. Garfield, S. Chacko and A. V. Somlyo. 1975. Golgi organelle response to the antibiotic X537A. *J. Cell Biol.* **66**:425.

Somlyo, A. P., A. V. Somlyo, C. E. Devine and R. V. Rice. 1971. Aggregation of thick filaments into ribbons in mammalian smooth muscle. *Nature New Biol.* **231**:243.

Somlyo, A. P., A. V. Somlyo, C. E. Devine, P. D. Peters and T. A. Hall. 1974. Electron microscopy and electron probe analysis of mitochondrial cation accumulation in smooth muscle. *J. Cell Biol.* **61**:723.

Somlyo, A. P., J. Vallières, R. E. Garfield, H. Shuman, A. Scarpa and A. V. Somlyo. 1976. Calcium compartmentalization in vascular smooth muscle: Electron probe analysis and studies on isolated mitochondria. In *Biochemistry of smooth muscle* (ed. N. L. Stephens). University Park Press, Baltimore, Maryland. (In press.)

Somlyo, A. V. and A. P. Somlyo. 1968. Electromechanical and pharmacomechanical coupling in vascular smooth muscle. *J. Pharmacol. Exp. Ther.* **159**:129.

———. 1971. Strontium accumulation by sarcoplasmic reticulum and mitochondria in vascular smooth muscle. *Science* **174**:955.

Somlyo, A. V., F. T. Ashton, L. Lemanski, J. Vallières and A. P. Somlyo. 1976. Filament organization and dense bodies in vertebrate smooth muscle. In *Biochemistry of smooth muscle* (ed. N. L. Stephens). University Park Press, Baltimore, Maryland. (In press.)

Sparrow, M. P., L. C. Maxwell, J. C. Rüegg and D. F. Bohr. 1970. Preparation and properties of a calcium ion-sensitive actomyosin from arteries. *Amer. J. Physiol.* **219**:1366.

Spurr, A. R. 1969. A low viscosity epoxy resin embedding medium for electron microscopy. *J. Ultrastruc. Res.* **26**:31.

Tregear, R. T. and J. M. Squire. 1973. Myosin content and filament structure in smooth and striated muscle. *J. Mol. Biol.* **77**:279.

Vallières, J., A. Scarpa and A. P. Somlyo. 1975. Subcellular fraction of smooth muscle. I. Isolation, substrate utilization and Ca^{++} transport by main pulmonary artery and mesenteric vein mitochondria. *Arch. Biochem. Biophys.* **170**:659.

Wootton, G. S. and P. J. Goodford. 1975. An association between mitochondria and vesicles in smooth muscle. *Cell Tissue Res.* **161**:119.

Wuerker, R. B. and J. B. Kirkpatrick. 1972. Neuronal microtubules, neurofilaments, and microfilaments. *Int. Rev. Cytol.* **33**:45.

Structural and Functional Features of Isolated Smooth Muscle Cells

Fredric S. Fay

Department of Physiology, University of Massachusetts Medical Center
Worcester, Massachusetts 01605

The cellular and subcellular events underlying the contraction of smooth muscle are poorly understood, especially when compared to the insights we now have into the mechanism of force generation in skeletal muscle. The difference in the state of knowledge arises, at least in part, from the uncertainty that attends the analysis of studies on multicellular preparations, on which almost all of our information about smooth muscle is based. Smooth muscle tissues are characterized by complex electrical, mechanical and metabolic interactions between muscle and non-muscle elements, and it is these complexities that have often clouded the interpretation of studies on whole tissues. Hence in the last few years, several investigators (Bagby et al. 1971; Bagby and Fisher 1973; Fay and Delise 1973; Purves et al. 1973, 1974; Singer and Fay 1974; Small 1974; Bagby 1974; Fay, Cooke and Canaday 1976; Fay 1976) have attempted to investigate the physiology of smooth muscle by use of single isolated smooth muscle cells. The studies described here represent an attempt to characterize the contractile process in isolated smooth muscle cells and to derive an understanding of the mechanisms underlying the development of force in smooth muscle. These studies were performed on single smooth muscle cells obtained from the stomach muscularis of *Bufo marinus* by a modification (Fay and Delise 1973) of the technique originally described by Bagby et al. (1971). The single smooth muscle cells obtained by this technique appear to be basically unaltered by the isolation procedure, as judged by the similarity of many of their physiological (Bagby et al. 1971; Bagby and Fisher 1973; Bagby 1974; Fay, Cooke and Canaday 1976; Murray, Reed and Fay 1975), pharmacological (Fay and Singer 1974; Honeyman and Fay 1975), metabolic (Fay, Cooke and Canaday 1976; Honeyman and Fay 1975) and ultrastructural (Fay and Delise 1973) properties when compared to those found in smooth muscle within intact tissue.

185

STRUCTURE OF SMOOTH MUSCLE CELLS

Relaxed and Contracted Smooth Muscle Cells

In order to obtain some insight into the mechanism underlying force generation, the structure of isolated smooth muscle cells was examined as a function of contractile state. Individual smooth muscle cells on a glass slide at room temperature were fixed with 2.5% glutaraldehyde in amphibian physiological saline in the relaxed state or at various times after brief extracellular electrical stimulation. All cells on which ultrastructural studies were performed were observed continuously with differential interference contrast (Nomarski) optics both before and during the fixation procedure. Thus the contractile state of cells whose structures were studied was precisely defined.

A comparison of the surface features of two smooth muscle cells, one fixed in the relaxed state (Fig. 1) and the other fixed at the point of maximum contraction (Fig. 2), reveals that contraction is associated with a marked evagination of portions of the cell membrane. The development of these evaginations may be recorded on movie film in *unfixed* material using Nomarksi optics. Figure 3 shows portions of such a movie sequence, revealing that evaginations appear early during active shortening but disappear again during the earliest phases of relaxation following cessation of electrical stimulation. Analysis of many similar movie records indicates that on the average, cells were completely covered by evaginations upon shortening to 57 ± 2% of their original length (mean ± S.E.; 19 cells). In contrast, the evaginations were completely absent when the cells had reextended to 44 ± 2% of their original length (mean ± S.E.; 14 cells). On the average, cells are fully contracted when they have shortened to 35 ± 2% of their original length (mean ± S.E.; 39 cells). The impression obtained from movie records,

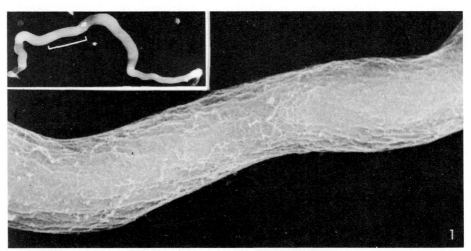

Figure 1
Scanning electron micrograph of an isolated smooth muscle cell fixed at its resting, unstimulated length. Note the generally smooth surface of this cell. 5700×. Inset at upper left shows entire cell. 570×. (Reprinted, with permission, from Fay, Cooke and Canaday 1976.)

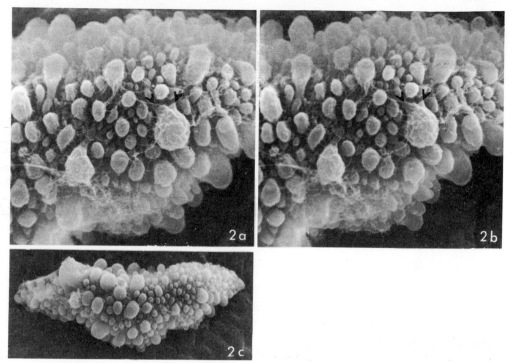

Figure 2

Scanning electron micrograph of an isolated smooth muscle cell fixed after maximal shortening in response to brief electrical stimulation. Cell had contracted to 35% of its initial length (L_i) at the time of fixation. (c) Entire cell: micrographs a and b are a stereo pair of the left end of the cell. Stereo pair may be viewed with binocular viewer (available from E. F. Fullam, Inc., Schenectady, N.Y.) to obtain a three-dimensional image of the cell. Note that most evaginations are connected to the main body of the cell by a slender neck; occasionally evaginations appear connected by multiple stalks (darts). (a,b) 6000×; (c) 1700×. (Reprinted, with permission, from Fay, Cooke and Canaday 1976.)

that evaginations are seen primarily during the active process of contraction and are lost again early during relaxation and reextension of the cell, is also supported by scanning electron microscopic observations of over 60 cells fixed at various defined stages during contraction and subsequent relaxation following electrical stimulation (Fay, Cooke and Canaday 1976).

In order to understand the basis for the striking changes in the surface of the isolated cells during the contractile cycle, the internal features of the smooth muscle cells were examined using the transmission electron microscope. A comparison of the ultrastructural features of single smooth muscle cells fixed in the relaxed or fully contracted stage is presented in Figures 4 and 5. The plasma membrane in the relaxed cell (Fig. 4a,b) consists of two areas alternating over the entire length of the cell—one contains amorphous material subtending the membrane proper (plasma membrane dense bodies), and the other, numerous micropinocytotic vesicles. In appropriately oriented sections, thin filaments appear to merge with these plasma membrane densities

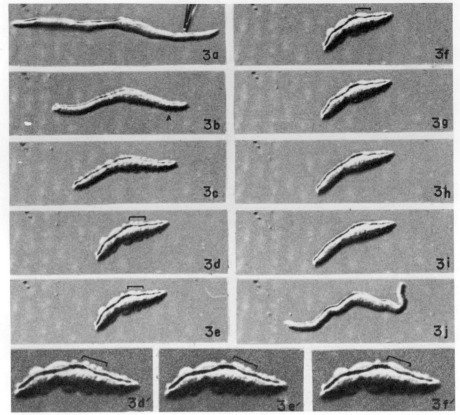

Figure 3

Ten frames from a film record showing the contractile response of an isolated smooth muscle cell to brief electrical stimulation. Time intervals: (a) just prior to stimulation; (b) 1 sec after initiation of contraction, cell at 66% of its initial length (L_i); (c) 2 sec after initiation of contraction, 51% L_i; (d, d') 9 sec after initiation of contraction, 38% L_i; (e, e') 10 sec after initiation of contraction, 38% L_i; (f, f') 11 sec after initiation of contraction, 38% L_i—cell maximally contracted; (g) 20 sec after initiation of contraction, 40% L_i; (h) 26 sec after initiation of contraction, 44% L_i; (i) 31 sec after initiation of contraction, 45% L_i; (j) 10 min after initiation of contraction, 74% L_i. Note the appearance of evaginations after shortening to 66% L_i (b, darts), whereas the loss of such evaginations during relaxation is almost complete at 45% L_i. In frames d, d', e, e', f and f', three small evaginations (brackets) are seen to fuse in a stepwise manner. (a–j) 202×; (d', e', f') 299×. (Reprinted, with permission, from Fay, Cooke and Canaday 1976.)

(Fay, Cooke and Canaday 1976). The surface of the fully contracted cell is characterized by numerous evaginations, whose plasma membrane and cytoplasm differ markedly from the rest of the cell (Fig. 4c,d). Plasma membrane dense bodies are not associated with the membrane surrounding the evaginations but are almost always found in the furrows between the evaginations. The cytoplasm contained within the evaginations is finely granular and generally devoid of myofilaments. The myofilaments contained within the nonevagi-

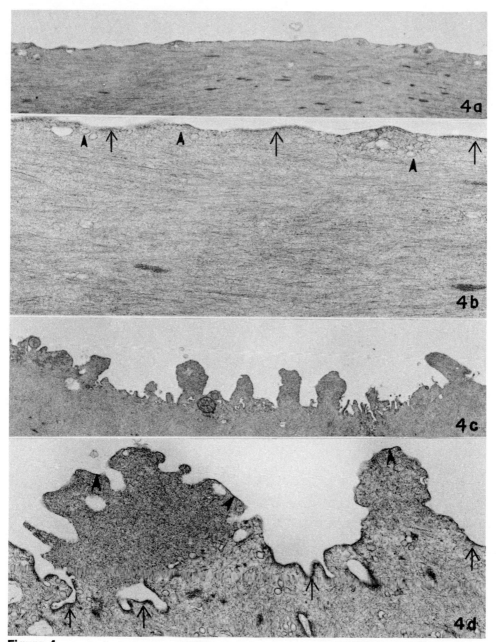

Figure 4

(*a*) Electron micrograph showing the smooth surface contour of a relaxed cell. (*b*) The plasma membrane of a relaxed cell consists of areas that are either subtended by amorphous dense material (arrows) or differentiated into micropinocytotic vesicles (darts). (*c*) Electron micrograph showing the rough surface contour of a contracted cell. (*d*) The plasma membrane of a contracted cell is subtended by amorphous dense material along the bases and "neck regions" of the surface evaginations (arrows). The membrane covering the evaginations proper (darts) is not subtended by these dense areas and, in general, it does not have the micropinocytotic vesicles which characterize the plasma membrane of relaxed cells. (*a,c*) 8370×; (*b,d*) 23,400×. (Reprinted, with permission, from Fay, Cooke and Canaday 1976.)

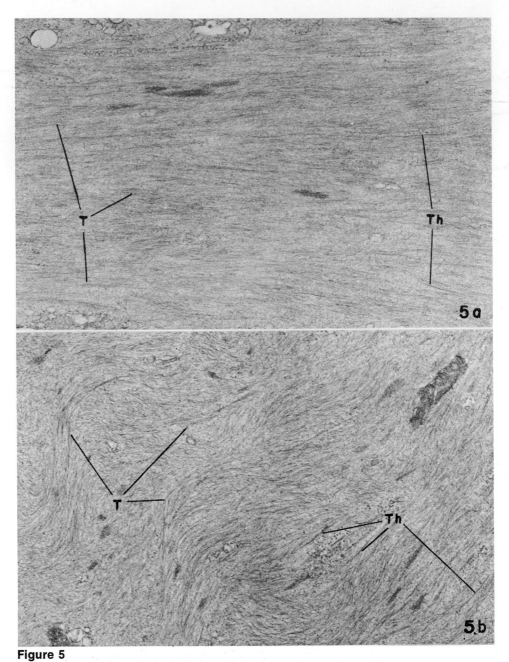

Figure 5
(a) Cytoplasm of relaxed cell showing the uniformly parallel orientation of the thick (Th) and thin (T) myofilaments. (b) Cytoplasm of a contracted cell showing the disorientation of the thick and thin myofilaments. Both a and b are longitudinal sections; the long axis of each cell would be parallel to the horizontal axis of each electron micrograph. 25,200×. (Reprinted, with permission, from Fay, Cooke and Canaday 1976.)

nated portions of the fully contracted cell are highly disoriented (Fig. 5b), whereas in relaxed cells, the filaments run generally parallel to the cell's long axis (Fig. 5a).

Model for the Arrangement of Contractile Units within the Smooth Muscle Cell

The observed changes in the surface and ultrastructure of isolated smooth muscle cells during the contractile cycle are readily explained by the model for the arrangement of contractile elements shown in Figure 6. This scheme represents an amplification of the general suggestion of Pease and Molinari (1960) that plasma membrane dense bodies act as anchoring sites for contractile units within the smooth muscle cell. Two additional features pointed out by this scheme are that contractile units run for relatively short lengths within the cell, and that the contractile units become more obliquely oriented relative to the long axis of the cell as it shortens. According to this model, the formation of evaginations is the consequence of two opposing tendencies: (a) the need to increase cellular diameter in order to accommodate the volume displaced by cell shortening and (b) the vectorial component of the force of the contractile units pulling in on the cell membrane at the points of attachment to the cell membrane, i.e., the plasma membrane dense bodies. Thus only those portions of the membrane not acted upon by the contractile elements are free to move outward. The evaginations do not appear to represent a means of accommodating excess surface membrane in going from a cylindrical (relaxed cell) to a more spherical (contracted cell) geometry, as suggested by others (Lane 1965; Kelly and Rice 1969). If this were true, then the appearance of evaginations ought to be correlated solely with cell

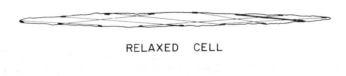

RELAXED   CELL

FULL   CONTRACTION

Figure 6
Schematic representation of a smooth muscle cell showing how contractile units are attached to the cell surface. The densities along the cell membrane represent plasma membrane dense bodies, and the lines connecting them represent the contractile units. One of these lines has been widened to facilitate identification in the two different contractile states. Note that contractile units run for only relatively short lengths within the cell, and that the angle between contractile units and the long axis of the cell increases during contraction.

length and be unaffected by the contractile state, which is not the case. Rather, the evaginations form early during contraction but disappear again at short lengths during the earliest stages of relaxation (Fig. 3).

The proposed scheme is supported by a variety of other observations: (1) Fisher and Bagby (1974) have observed a decrease in birefringence associated with the long axis of the cell when the isolated smooth muscle cells shorten following stimulation. The birefringence change is consistent with filament reorientation of the sort proposed. (2) Fay and Delise (1973) have noted the ability of the cells to shorten in a highly localized manner in response to low-level, localized stimulation—an expected property if contractile units connect relatively closely spaced points on the cell membrane. (3) A 20% decrease in volume has been measured during maximal contraction of the isolated cells (Fay and Delise 1973). This finding also supports the model predicting that an increase in intracellular pressure and the consequent loss of volume ought to be associated with contraction. An increase in intracellular pressure is a consequence of the proposed arrangement of contractile units, since there is an inwardly directed component force of the contractile elements applied over the entire extent of the cell membrane during contraction.

FUNCTIONAL MECHANICS IN SINGLE CELLS

Isometric Contractile Reponses of Single Isolated Smooth Muscle Cells

In order to test and refine this scheme further we have recently initiated studies of the mechanical behavior of single isolated smooth muscle cells. Eventually such studies should provide a crucial test of the proposed model, since one would expect marked changes in force-generating capacity to be associated with changes in cell length. Specifically, as the cell shortens and contractile units become very oblique relative to the long axis of the cell, the force measured along that axis ought to drop due to a decrease in the vector component of force along the axis of measurement. In a more general sense, studies of the mechanics of single cells provide a vital test of the validity of any scheme for the arrangement of contractile units, since in order for a scheme to be valid, it must be able to account for the force-generating properties of the cell it hopes to describe.

Two major problems had to be overcome to perform these studies. First, because the maximum force to be expected from a single smooth muscle cell[1] is so low (1–2 mg), force transducers currently used for studying the mechanics of relatively more forceful (10–100 times) single skeletal muscle fibers (Gordon, Huxley and Julian 1966b) are unsuitable for these studies. Thus a photooptical transducer was developed (Canaday and Fay 1976) which is very stable (drift = ± 6.2 μg/hr) yet has sufficient sensitivity to detect forces as low as a few micrograms. A second problem was posed by the ab-

---

[1] In calculating the expected maximum force of a single smooth muscle cell we have assumed a cell diameter of 10 μ and a maximum force of 2 kg/cm[2], similar to that of an intact smooth muscle strip (Lundholm and Mohme-Lundholm 1966).

sence of a tendon in the enzymatically isolated single smooth muscle cells, which made attachment of a force-measuring device quite difficult. This problem has been overcome by the development of a technique (Fay 1976) whereby a cell is tied in a knotlike manner around two probes (Fig. 7a). One probe is attached to a micromanipulator and stepping hydraulic microdrive (David Kopf Instruments), which allows for controlled changes in cell length; the other probe is attached to the force transducer.

Time Course and Magnitude of Contraction of Single Smooth Muscle Cells

A typical contractile response to brief electrical stimulation of a single smooth muscle cell, mounted in the manner described above, is shown in Figure 7b. Characteristically, peak force was reached in a few seconds, whereas relaxation had a $t_{1/2}$ of 3 seconds. The maximum force recorded to date from a single cell is 2.6 kg/cm^2. This is somewhat greater than the maximum force recorded from any intact tissue (2.0 kg/cm^2; Herlihy and Murphy 1973) and quite comparable to the maximum force produced by a single skeletal muscle fiber (Gordon, Huxley and Julian 1966b). In light of the much lower myosin content in smooth muscle (20% of that in rabbit skeletal muscle; Murphy, Herlihy and Megerman 1974) it is rather remarkable that similar forces can be generated. The proposed oblique orientation of filaments in smooth muscle might provide at least a partial explanation for the discrepancy. With this arrangement, there are more force-generating elements effectively in parallel than if the same elements ran end-to-end, in series, as they do in skeletal muscle (Rosenblueth 1965).

The latency between a single electrical shock of supramaximal intensity and the onset of force development in the isolated cells at 22°C is much longer (416 ± 35 msec; mean ± s.e.; $n = 35$ cells) than found in skeletal muscle under similar conditions (2 msec in frog skeletal muscle at room temperature; Sandow and Preiser 1964). A typical latency of 620 msec intervenes between electrical activation and the onset of force generation (Fig. 7c). In the 35 isolated single smooth muscle cells studied, the latencies ranged from 130–1000 msec. In an earlier report of these studies (Fay 1975), even longer latencies were noted, but subsequent reexamination of the force recordings containing these super-long latencies (> 2000 msec) revealed that in these experiments the cells were not maximally activated, as judged by very low values of peak dP/dt during the contraction. The rate-limiting step responsible for the long latencies is as yet unknown. Two potential links between excitation and contraction can be eliminated as causes for the long latency. The time required for propagation of an action potential along a single smooth muscle fiber is no more than a few milliseconds (Tomita 1967) and thus it cannot be rate-limiting. Second, the time required for diffusion of Ca^{++} through the sarcoplasm from its site of release to the myofibrils also could not be responsible for the long latencies. In considering the latter possibility, a "worst case" has been assumed, i.e., where Ca^{++} is released at the inner surface of the plasma membrane and has to diffuse to the center of the fiber (5 μ). Diffusional equilibration would be 90% complete within 10 msec (Hill 1945), a much shorter time than the observed mean latency

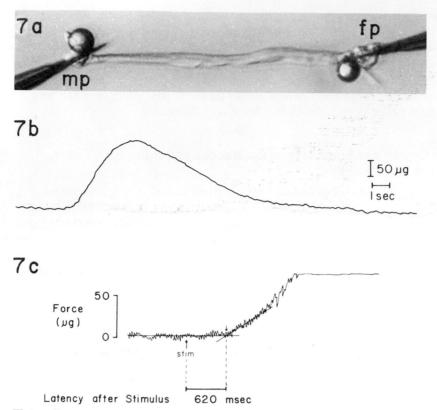

Figure 7
(*a*) Photomicrograph of an isolated smooth muscle cell prepared for isometric recording of force. The cell is tied in a knotlike manner around a probe at each end. One probe (fp) is attached to the force transducer, and the other probe (mp) is attached to a micromanipulator which allows for defined changes in cell length. For details of the attachment protocol, see Fay (1976). A cell was stimulated electrically either from a point source (3 M NaCl-filled extracellular microelectrode) or by Ag-AgCl paddles positioned parallel to the long axis of the cell. Phase-contrast photomicrograph. 575×.
(*b*) Isometric recording of the contractile response of a single isolated smooth muscle cell to electrical stimulation. The cell was stimulated for 1 second with 1-msec pulses (10^{-5} amp, cathodal; 10 pps) applied extracellularly via a micropipette filled with 3 M NaCl and set very close to the cell surface. The peak force achieved during contraction was 235 μg. (*c*) Latency of the onset of active force development in a single isolated smooth muscle cell following electrical stimulation. The cell was stimulated with a single, 1-msec electrical pulse (10^{-2} amp) between two Ag-AgCl paddles placed on either side of the cell. The time at which force began to increase in response to electrical activation (downward arrow) was estimated by the intersection of two straight lines, one of which was drawn through the force record prior to stimulation, the other being drawn through the initial portion of the record where force began to increase. The delay between electrical activation and onset of active force development in this cell was 620 msec. The flat, right-hand portion of the force record results from saturation of the amplifier at the high gain used. (Reprinted, with permission, from Fay 1976).

194

(416 msec). The long latency may reflect limits imposed by a Ca^{++} gate on the rate of movement of Ca^{++} into the sarcoplasm. It may also reflect the time required for Ca^{++} to interact with and/or activate the contractile apparatus.

Dependence of Active and Passive Force on Cell Length

Studies of the length dependence of resting and active force in single smooth muscle cells have also been initiated. As discussed previously, such studies eventually ought to provide an important test of the proposed model. In order to facilitate these experiments, provisions were made to (a) accurately determine changes in cell length during each experiment and (b) obtain cells that in the resting state produce little, if any, active force. In order to accurately define cell length, surface membrane markers, visible in the light microscope, were employed (Fay 1976). Several markers were usually placed on a cell and a pair of markers chosen which bounded a portion of the fiber that appeared to behave in a uniform manner. For all the cells described here, changes in contractile function will be related to such a specifically defined portion of the fiber and not to overall cell length. For the purpose of investigating the relationship of passive force and cell length, fully relaxed cells were obtained by incubation in the presence of 10 μM isoproterenol. Honeyman and Fay (1975) have previously shown that isoproterenol at this concentration has a potent inhibitory effect on contractile activity of the isolated smooth muscle cells, which can, however, be overcome by electrical or other stimuli.

The relationship between resting force and cell length of a typical cell is shown in Figure 8. This cell was stretched in a series of steps so that at the greatest degree of extension, the portion of the fiber between the markers was stretched over four times its original length. The passive force of the fiber at all lengths studied was extremely low and showed no detectable trend with increases in length. The scatter of points probably results from the fact that the force levels were near the resolution limits of the force transducer. In Figure 9, the passive mechanical behavior of 17 cells is compared to that of a typical visceral smooth muscle. In order to compare the isolated cells with an intact tissue, the force per cross-sectional area, rather than absolute force, was plotted as a function of cell length. Passive force in the isolated cells increased somewhat as the cells were stretched beyond their initial length but remained relatively constant as length was further increased. This behavior is strikingly different from that seen in whole tissue, where force characteristically increases abruptly as the tissue is stretched beyond 1.3 times its initial length. The relationship of length and resting force in single smooth muscle cells is also considerably flatter than similar curves obtained on single skeletal muscle fibers (Gordon, Huxley and Julian 1966a).

The differences in mechanical behavior of single smooth muscle cells and either single skeletal muscle fibers or intact smooth muscle tissue may reflect differences in the number of extracellular fibrous elements associated with the three different preparations. Fibrous elements are found in abundance in intact smooth muscle tissues and also remain associated with single skeletal

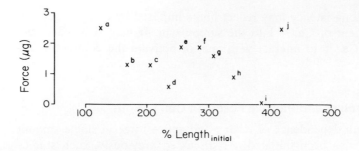

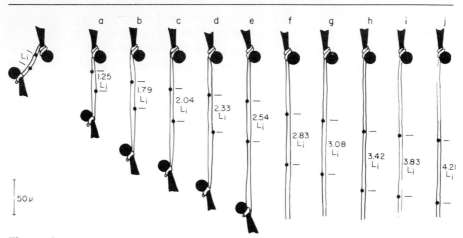

Figure 8

Length vs. passive tension in an isolated single smooth muscle cell. The lower portion shows the outline of the cell and the position of two marker beads at 11 different cell lengths; these images were obtained from video recordings made through a phase-contrast microscope during the experiment. The interbead distance before stretching the cell is denoted as L_i; all other interbead distances are expressed relative to it. The force existing in the fiber as a function of interbead distance is plotted in the upper portion of the figure. Note that the force is extremely low at all lengths. The scatter of points probably results from the fact that the force levels are near the resolution limits of the transducer. (Reprinted, with permission, from Fay 1976.)

muscle fibers following their mechanical isolation from whole tissue (Rapoport 1972). On the other hand following enzymatic isolation of single smooth muscle cells, few, if any, extracellular fibrous elements remain (Fig. 4). Extracellular fibrous elements are believed responsible for the steep increase in resting forces at large degrees of stretch in single skeletal muscle fibers (Rapoport 1972). Thus the difference in passive mechanical behavior of single skeletal and smooth muscle fibers probably results from differences in the extent of associated extracellular fibrous elements. Similarly, the isolated smooth muscle cells exhibit far less resistance to stretch than does intact smooth muscle tissue, probably because of differences in the extent of associated extracellular fibrous elements. Thus the passive mechanical properties

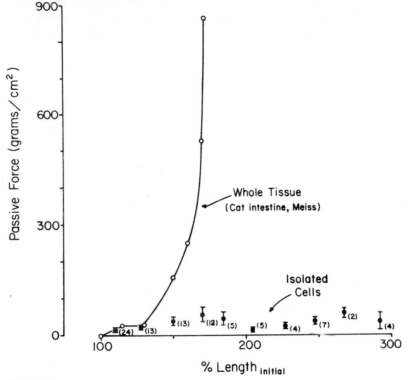

Figure 9

The relationship between length and passive tension in 17 isolated single smooth muscle cells and in intact visceral smooth muscle. The data for intact tissue are those Meiss (1971) obtained with cat intestine, and the force per cross-sectional area has been calculated assuming that the intact preparation retains constant volume as its length is changed. The data for the isolated cells have been grouped into length intervals of 20%. The points show the mean; the vertical bars indicate the standard error. The numbers in parentheses indicate the number of observations. (Reprinted, with permission, from Fay 1976.)

of intact smooth muscle tissues appear not to originate from the smooth muscle cells themselves. One other possible explanation for the difference in the passive mechanical properties of isolated smooth muscle cells and whole tissue is that the starting length of isolated cells may be considerably less than the initial length of cells within the intact tissue. However, this appears to be unlikely since the difference in passive mechanical properties is still observed when the data for both isolated cells and intact tissue are normalized against the length for optimal active force development rather than the initial length.

The experiments relating cell length to both active and passive mechanical behavior are in a most preliminary state. To date, a length versus active tension analysis has been successfully performed on only three cells. The major problem impeding more rapid progress is our inability to obtain a reproducible response for a standard stimulus. Until this problem has been solved

we have resorted to doing the length versus active tension analysis by changing the length of the cells that are subjected to continuous, low-frequency
electrical stimulation $(0.2–0.5 \ sec^{-1})$ for approximately 1 minute. Specifically, once the contractile force developed by a smooth muscle cell in
response to such activation has reached a plateau, the cell is allowed to
shorten by a fixed amount, and one then waits until a new stable force is
attained. This procedure is repeated several times in under 1 minute. Single
smooth muscle cells can produce reasonably stable forces in response to continuous activation for this period of time.

The length versus active force curve of three cells subjected to this procedure is compared in Figure 10 to that of intact tissue. Also included is
data for the passive mechanical properties of these three cells. The most
striking difference between the mechanical behavior of intact tissue and isolated cells is the relative absence of resting force in the isolated cells over
the range of lengths where whole tissue exhibits large resting forces. Detailed
comparison of the active curve of the isolated cells with that of whole tissue
is premature. Both show an optimum, and there is a suggestion that force
declines with increases in length less abruptly in the isolated cells than in the
whole tissue. The extremes of length over which active force can be produced
in the isolated cells are not yet fully defined. As a minimum estimate we know
that highly stretched cells can shorten so that the distance between markers
on the cell surface in the fully shortened cells is 10% of that seen prior to
initiation of contraction. Thus force can apparently be produced over a tenfold range of length.

SUMMARY

Ultrastructural studies of isolated single smooth muscle cells during shortening in response to electrical stimulation reveal that contraction is associated
with the reversible formation of surface membrane evaginations and a reorientation of myofibrils. In the relaxed cell, myofibrils run generally parallel to
the long axis of the cell; upon active shortening, the filaments become more
obliquely oriented. The evaginations in the contracted cells are characterized
by membrane free of dense bodies and cytoplasm devoid of myofilaments.
These changes in structure with active shortening are explained by a model
proposed for the deployment of contractile units within the smooth muscle
cell. The essential feature of the model is that contractile units connect relatively closely spaced dense bodies along the plasma membrane. According to
this scheme, the orientation of contractile units relative to the long axis of the
cell varies with cell length; in a fully shortened cell, such units are more
obliquely oriented than in an extended, relaxed, cell.

Techniques have been developed for recording the isometric contractile
force of a single isolated smooth muscle cell. Studies of the isometric contractile responses reveal several additional features that must be accommo-
dated in subsequent development of this or any other scheme. First, the
contractile response to electrical activation has a rather long latency—several
hundred milliseconds. Second, the cell exhibits a very low resting force,

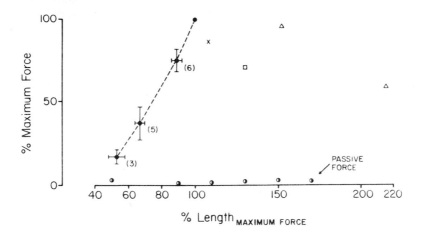

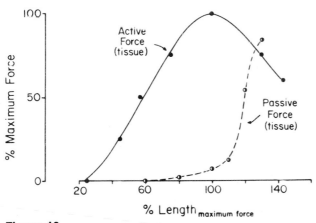

Figure 10

Relationships between length and active and passive force in three isolated single smooth muscle cells and in intact smooth muscle tissue. The upper portion shows the data from single isolated cells. The length-force data for each cell have been normalized, denoting the maximum active force as 100% and the length at which this force was developed as 100%. The data between 40% and 100% of the length for optimal force development have been grouped into length intervals of 20%. The points show the mean ± s.e.; the numbers in parentheses indicate the number of observations. Since less information is available beyond the peak of the curve, individual points have been plotted, each symbol representing a different cell. The passive force in each of the three cells was also determined as a function of length and has been included, grouping the data into length intervals of 20%. The points indicate the mean resting force as a percent of the maximum developed force of each cell. The lower portion of the figure reflects a consensus, gleaned from the literature (Lundholm and Mohme-Lundholm 1966; Meiss 1971; Herlihy and Murphy 1973), of the mechanical behavior of intact smooth tissues. (Reprinted, with permission, from Fay 1976.)

which shows little dependence on cell length. Third, a single smooth muscle cell is capable of generating active forces of almost 3 kg/cm^2. Fourth, the magnitude of active force generation is dependent on cell length, exhibiting a length for maximum force development; at lengths greater or less than this, force declines. Finally, cells are capable of generating at least some active force over a tenfold range of length.

Acknowledgments

I wish to acknowledge the collaboration of Dr. Peter H. Cooke in the transmission electron microscopy reported in this paper. This work was supported, in part, by a grant from the National Institutes of Health (HL 14523). F. S. Fay is the recipient of a Research Career Development Award from the National Institutes of Health (K04 HL 00048).

REFERENCES

Bagby, R. 1974. Time course of isotonic contraction in single cells and muscle strips from *Bufo marinus* stomach. *Amer. J. Physiol.* **227**:789.

Bagby, R. M. and B. A. Fisher. 1973. Graded contractions in muscle strips and single cells from *Bufo marinus* stomach. *Amer. J. Physiol.* **225**:105.

Bagby, R. M., A. M. Young, R. S. Dotson, B. A. Fisher and K. McKinnon. 1971. Contraction of single smooth muscle cell from *Bufo marinus. Nature* **234**:351.

Canaday, P. G. and F. S. Fay. 1976. An ultrasensitive isometric force transducer for single smooth muscle cell mechanics. *J. Appl. Physiol.* (in press).

Fay, F. S. 1976. Mechanical properties of single isolated smooth muscle cells. *INSERM Symp.* (1975) **50**: (in press).

Fay, F. S. and C. M. Delise. 1973. Contraction of isolated smooth muscle cells—Structural changes. *Proc. Nat. Acad. Sci.* **70**:641.

Fay, F. S. and J. J. Singer. 1974. Isolated smooth muscle cells in suspension. II. Response to pharmacological agents. *Fed. Proc.* **33**:435.

Fay, F. S., P. H. Cooke and P. G. Canaday. 1976. Contractile properties of isolated smooth muscle cells. In *Physiology of smooth muscle* (ed. E. Bulbring and M. F. Shuba), pp. 249–264. Raven Press, New York.

Fisher, B. A. and R. M. Bagby. 1974. Alteration of the birefringence of single smooth muscle cells during contraction. *Fed. Proc.* **33**:435.

Gordon, A. M., A. F. Huxley and F. J. Julian. 1966a. Tension development in highly stretched vertebrate muscle fibers. *J. Physiol.* **184**:143.

———. 1966b. The variation in isometric tension with sarcomere length in vertebrate muscle fibers. *J. Physiol.* **184**:170.

Herlihy, J. T. and R. A. Murphy. 1973. The length-tension relationship of arterial smooth muscle. *Circ. Res.* **33**:275.

Hill, A. V. 1945. On the time required for diffusion and its relation to processes in muscle. *Proc. Roy. Soc. B* **135**:446.

Honeyman, T. H. and F. S. Fay. 1975. Effect of isoproterenol on cyclic-AMP levels and contractility of isolated smooth muscle cells. *Fed. Proc.* **34**:361.

Kelly, R. E. and R. V. Rice. 1969. Ultrastructural studies on the contractile mechanism of smooth muscle. *J. Cell Biol.* **42**:683.

Lane, B. P. 1965. Alterations in the cytological detail of intestinal smooth muscle cells in various stages of contraction. *J. Cell Biol.* **27**:199.

Lundholm, L. and E. Mohme-Lundholm. 1966. Length at inactivated contractile elements, length-tension diagram, active state and tone of vascular smooth muscle. *Acta Physiol. Scand.* **68**:347.

Meiss, R. A. 1971. Some mechanical properties of cat intestinal muscle. *Amer. J. Physiol.* **220**:2000.

Murphy, R. A., J. T. Herlihy and J. Megerman. 1974. Force-generating capacity and contractile protein content of arterial smooth muscle. *J. Gen. Physiol.* **64**:691.

Murray, J. M., P. W. Reed and F. S. Fay. 1975. Effects of the divalent cation ionophore A23187 on isolated smooth muscle cells. *Proc. Nat. Acad. Sci.* **72**: 4459.

Pease, D. C. and S. Molinari. 1960. Electron microscopy of muscular arteries; Pial vessels of the cat and monkey. *J. Ultrastruc. Res.* **3**:447.

Purves, R. D., G. E. Mark and G. B. Burnstock. 1973. The electrical activity of single isolated smooth muscle cells. *Pflügers Arch.* **341**:325.

Purves, R. D., C. E. Hill, J. H. Chamley, G. E. Mark, D. M. Fry and G. B. Burnstock. 1974. Functional autonomic neuromuscular junctions in tissue culture. *Pflügers Arch.* **350**:1.

Rapoport, S. I. 1972. Mechanical properties of the sarcolema and myoplasm in frog muscle as a function of sarcomere length. *J. Gen. Physiol.* **59**:559.

Rosenblueth, J. 1965. Smooth muscle: An ultrastructural basis for the dynamics of its contraction. *Science* **148**:1337.

Sandow, A. and H. Preiser. 1964. Muscular contraction as regulated by the action potential. *Science* **146**:1470.

Singer, J. J. and F. S. Fay. 974. Isolated smooth muscle cells in suspension. I. Detection of contraction. *Fed. Proc.* **33**:435.

Small, J. V. 1974. Contractile units in vertebrate smooth muscle cells. *Nature* **234**:324.

Tomita, T. 1967. Spike propagation in the smooth muscle of the guinea pig taenia coli. *J. Physiol.* **191**:517.

Genetic and Molecular Studies of Nematode Myosin

**Henry F. Epstein, Harriet E. Harris, Frederick H. Schachat,
Edwin A. Suddleson and Janice A. Wolff**

Department of Pharmacology, Stanford University School of Medicine
Stanford, California 94305

The organization of any contractile system requires a series of regulated interactions between genes, structural proteins and movement to occur in a precise manner. The goal of our studies is to understand, at least in outline, some of the principles governing these interactions within the muscle cells of the nematode *Caenorhabditis elegans*. Our approach has been to analyze the consequences of mutations in the *unc-54* I gene upon the movement of animals, the arrangement of contractile filaments within the body wall muscle cells and the molecular properties of myosin.

EXPERIMENTAL PROCEDURES

The conditions for growth of *Caenorhabditis elegans,* N2 strain, the genetic analysis of mutants, and the isolation of N2 and mutant contractile proteins have been described previously (Brenner 1974; Epstein, Waterston and Brenner 1974; Waterston, Epstein and Brenner 1974). The polarized light and electron microscopic methods as applied to nematode muscle have been described elsewhere (Epstein and Thomson 1974; Epstein, Waterston and Brenner 1974).

Plastic petri dishes, 8.5 cm in diameter, were filled with nematode growth medium agar (Brenner 1974). Four-tenths ml of about 10^9 cells $\cdot$ ml^{-1} of *Escherichia coli,* strain OP8, was placed around the circumference of the solid medium and incubated for 24 hours at room temperature. Nematodes, previously washed free of bacteria on Millipore filters (Ward 1973), were placed on a central zone, 2.5 cm in diameter. The number of animals remaining in the center was counted at various times.

The procedure for the purification of native nematode myosin possessing enzymatic and assembly properties characteristic of most other myosins is only outlined here, the details will be published elsewhere (H. E. Harris, J. A. Wolff and H. F. Epstein, in prep.). All work was done at 0–4°C. A

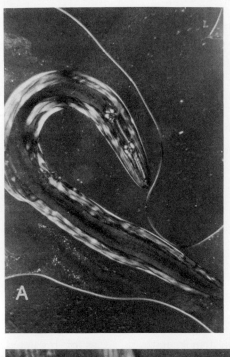

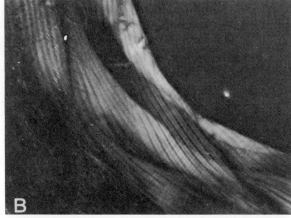

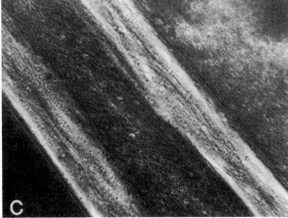

Figure 1 (*See facing page for legend*)

pellet enriched for contractile proteins was sedimented from nematode homogenates in 15 mM sodium phosphate pH 7.0, 50 mM NaCl, 1 mM EDTA, 1 mM PMSF (phenyl methyl sulfonyl fluoride), 5 mM DTT (dithiothreitol). Myosin and paramyosin, as well as actin and tropomyosin, were extracted from this pellet in 20 mM Tris-HCl, 0.6 M KCl, 1 mM DTT and 1 mM PMSF pH 8.0. The extract was centrifuged at 100,000g for 3 hours to give a clear supernatant containing paramyosin and tropomyosin and a pellet containing actin and myosin. Paramyosin was purified from the supernatant by precipitation in 75% ethanol, solubilization by dialysis against 10 mM potassium phosphate, 0.6 M KCl, 2 mM DTT, pH 7.6, and precipitation by dialysis against 10 mM potassium phosphate, 0.1 M KCl, 2 mM DTT, pH 6.0. A second cycle of resolubilization and precipitation was necessary to obtain paramyosin of >90% purity.

The pellet containing actin and myosin was suspended in 20 mM Tris-HCl, 0.6 M KCl, 5 mM ATP, 5 mM MgCl$_2$, 1 mM DTT and 1 mM PMSF pH 8.0, clarified by centrifugation at 100,000g for 15 minutes, and then brought to 0.4 M KI. The solution was placed on a Sepharose-4B column, previously equilibrated with 20 mM Tris-HCl, 0.6 M KCl, 0.2 mM ATP, 1 mM DTT and 1 mM PMSF pH 8.0, and developed in the same buffer. A pre- and postloading wash with 20 mM Tris-HCl, 0.6 M KI, 5 mM ATP, 5 mM MgCl$_2$, 1 mM DTT and 1 mM PMSF pH 8.0 was necessary. Myosin and actin of at least 95% purity were obtained from pooled chromatographic fractions.

Myosin from the E675 strain was chromatographed on a 1×15-cm hydroxylapatite (BioRad) column equilibrated with 5 mM potassium phosphate pH 7.5, 0.6 M KCl, 1 mM PMSF and 1 mM DTT and developed with a 200-ml gradient arising from equal volumes of 5 mM potassium phosphate pH 7.5 and 0.5 M potassium phosphate pH 6.5, with 0.6 M KCl, 1 mM PMSF and 1 mM DTT held constant.

Polyacrylamide slab gel electrophoresis, staining and autoradiography were performed as described previously (Epstein, Waterston and Brenner 1974).

Grids for electron microscopy of filaments were first covered with a Formvar film and then carbon-coated. Cofilaments of myosin and paramyosin were washed on the grids with 0.1 M KCl and then stained with 1% uranyl acetate. Specimens were examined in a Siemens Elmiskop Ia operating at 60 kV, with the sample at liquid nitrogen temperature.

RESULTS

Differences in Morphology and Function between Normal and Mutant Muscles

Two kinds of muscle are predominant in *C. elegans*: pharyngeal and body wall (Fig. 1A). The pharynx is a food pump that is rapidly and continuously

Figure 1
Polarized light micrographs of living nematodes. (*A*) Entire N2 animal about 1 mm long and 50 μm wide. (*B*) N2 body wall sarcomeres, about 2.4 μm between nonbirefringent I bands. (*C*) E675 body wall under identical conditions and magnification as in *B*.

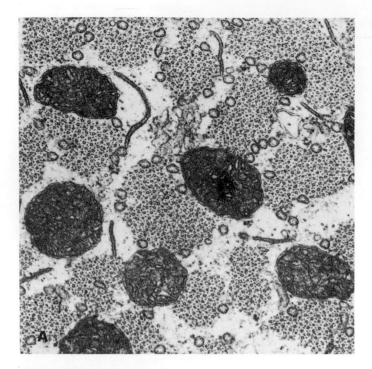

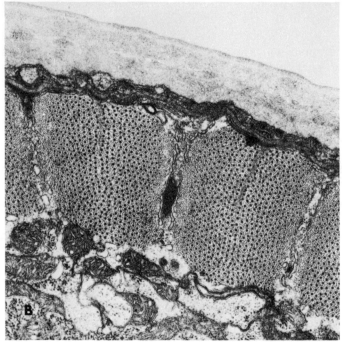

Figure 2
Electron micrographs of nematode muscle. The dimensions of the thick filaments are about 25 nm, and of the thin filaments, about 6–7 nm. (*A*) N2 or E675 pharyngeal muscle. (*B*) N2 body wall muscle. (*C*) E675 body wall muscle. (Reprinted, with permission, from Epstein, Waterston and Brenner 1974.)

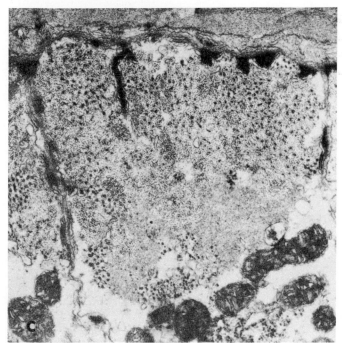

Figure 2 (*continued*)

contracting and relaxing. The body wall muscles are responsible for locomotion and perform work upon demand. There are mutants in at least seven genes with associated body wall muscular defects, but there are none that alter pharyngeal function or structure (Brenner 1974). Probably, pharyngeal function is required for viability on bacterial lawns.

N2 body wall muscles are clearly organized into sarcomeres with distinct A (relatively anisotropic) and I (relatively isotropic) bands (Fig. 1B). These differences in birefringence correspond to regions in which thin filaments (I band) and thin and thick filaments (A band) are found in thin sections of body wall muscle when viewed by electron microscopy (Fig. 2B). The pharynx does not exhibit such alternating patterns of birefringence but it clearly has parallel arrays of thin and thick filaments (Fig. 2A). Both classes of filaments in pharyngeal muscle are more regularly arranged, perhaps in hexagonal lattices, than in the body wall structures. The pharyngeal thick filaments appear to have nonstaining cores, in contrast to most of the body wall thick filaments.

Unc-54 mutant animals move very slowly or have completely paralyzed body wall muscles (Brenner 1974; Epstein, Waterston and Brenner 1974). A quantitative comparison of the movement of N2 (wild type) and E190 (*unc-54* allele) is shown in Figure 3. The rates of leaving the center of a plate in response to the chemotactic gradient of bacterial secretions are first-order for both strains. The calculated apparent rate constant for N2 is 20-fold greater than that for E190. These constants are "apparent" in that they refer to the specific geometry and conditions of our experiments. Pilot experiments suggest that this plate method may be useful as a screening procedure for new

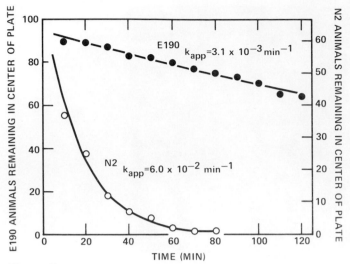

Figure 3

Kinetics of normal and mutant movement. (○, ●) Actual counts of N2 and E190 animals, respectively, at indicated times after placement. The lines and apparent rate constants were calculated from best-fits of the data to first-order rate equations (M. M. Isachsen, E. A. Suddleson and H. F. Epstein, unpubl.).

muscle-defective mutants (M. M. Isachsen, E. A. Suddleson and H. F. Epstein, unpubl.).

The dysfunction of the body wall muscles correlated with disruption of the sarcomeres and the filament arrays within them (Figs. 1C and 2C). In at least two *unc-54* mutants, E190 and E675, the numbers of thick filaments in body wall muscle cells are decreased. In contrast, the fine structure of the pharynges in these mutants appears normal (Fig. 2A). The observed differences between function and structure of normal body walls and pharynges and the localization of mutant effects to the body wall muscles suggest that certain genes may be expressed in only one of these muscle types.

Altered Myosin Heavy Chains in Mutant Body Walls

Despite the extensive disruption of structure and apparent reduction of thick filaments observed by microscopy of mutant body wall muscle, myosin and paramyosin, the major components of the thick filaments, as well as other contractile proteins, can be isolated in quantities comparable to similar N2 preparations (Epstein, Waterston and Brenner 1974; Waterston, Epstein and Brenner 1974). N2 and five homozygous *unc-54* mutant strains are compared in Figure 4. The intense bands in the middle of the gel represent myosin heavy chains (about 210,000 daltons). A faint band of slightly greater mobility (about 206,000 daltons) is also present in all the strains. However, the E675 mutant exhibits a novel major component which migrates still faster (about 203,000 daltons). This polypeptide is very similar to normal myosin

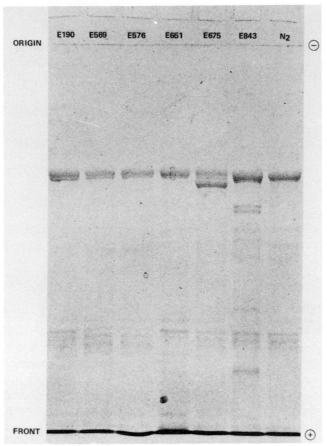

Figure 4
Myosin heavy chains of N2 and *unc-54* strains; 5% acrylamide-sodium
dodecyl sulfate electrophoretogram. (Reprinted, with permission, from Epstein, Waterston and Brenner 1974.)

heavy chain in its distribution of peptides (Epstein, Waterston and Brenner
1974) and its actin-binding and ATPase activities (H. E. Harris, J. A. Wolff
and H. F. Epstein, unpubl.). The retention of these properties and the disruption of thick filaments suggest that the myosin rods have been altered. Hetero-
zygotes (+/E675) synthesize this abnormal heavy chain, but in decreased
amounts relative to normal chains; this finding of codominance is consistent
with the mutation occurring in a structural gene for myosin heavy chains.

These abnormal myosin heavy chains are located in the body walls, but not
the pharynges, of E675 animals. Figure 5 compares single N2 and E675
animals and dissected body walls and pharynges of both strains. In whole
animals, both the 210,000- and the 206,000-dalton components can be seen.
The E675 animal also shows the abnormal 203,000-dalton polypeptide as a
major component. The dissected body walls do not exhibit any 206,000-dalton
component; this polypeptide is restricted to the pharynges, which also synthesize a 210,000-dalton component. The abnormal 203,000-dalton polypeptide

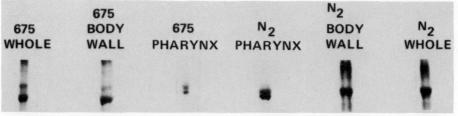

Figure 5
Myosin heavy chains of whole and dissected individual nematodes. No purification
was performed. Animals were grown on ^{35}S-labeled bacteria. Autoradiograms of
5% acrylamide-sodium dodecyl sulfate slab gel separation. (Reprinted, with permission, from Epstein, Waterston and Brenner 1974.)

is observed only in the body walls of E675. This finding correlates with the
location of paralysis and structural disruption within the body wall muscle
cells of the mutant.

Biochemical and Genetic Evidence for Three Myosins

The appearance of two classes of putative myosin heavy chains in pharynges
suggests that the body wall muscle cells may similarly contain two heavy
chains. The presence of normal and abnormal heavy chains in homozygous
E675 mutants could be explained, therefore, by mutation in only one or two
structural genes normally expressed in body wall muscle cells. A minimal
model would have one class of 210,000-dalton heavy chain present in both body
wall and pharyngeal muscles, one class of 206,000-dalton heavy chain synthesized only in pharyngeal muscle, and one class of normally 210,000-dalton
heavy chain expressed only in body wall muscle. As pharyngeal muscle function appears indispensable, only mutants affecting the uniquely body wall
myosin chains would be viable, possibly explaining the failure to find mutations in other genes associated with altered myosins.

In order to examine the hypothesis that several myosins exist, native myosin
from N2 and E675 animals has been purified (Fig. 6). N2 myosin binds actin
in an ATP-sensitive manner and has a Mg^{++}-ATPase which is stimulated by
actin. Such myosin can form short, bipolar filaments. In addition to 210,000-
dalton heavy chains and several light chains in the range of 15–20,000 daltons,
these myosin preparations exhibit a minor component at 206,000 daltons.
Thus the pharyngeal component copurifies with bulk myosin in an extensive
procedure that requires the ability to bind actin in the absence of ATP and
to dissociate from actin in the presence of ATP. For these reasons, the
pharyngeal 206,000-dalton component can be identified as a myosin heavy
chain.

Myosin from E675 can be similarly purified. These preparations exhibit
the same three heavy chains seen in the whole animals and in the partially
purified extracts. When E675 myosin is chromatographed on hydroxylapatite,
partial resolution of myosin-containing heavy chains that are predominantly
210,000 daltons at the leading edge and 203,000 daltons in the trailing

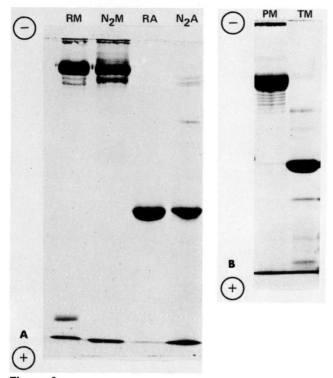

Figure 6

Purified nematode contractile proteins. (*A*) RM, rabbit white skeletal muscle myosin; N₂M, N2 myosin; RA, rabbit white skeletal muscle actin; N₂A, N2 actin. (*B*) PM, N2 paramyosin; TM, N2 tropomyosin. *A* and *B* represent different 10% acrylamide-sodium dodecyl sulfate electrophoretograms, 25 μg of each protein per slab gel slot (H. E. Harris, J. A. Wolff and H. F. Epstein, unpubl.).

shoulder is achieved (Fig. 7). The pharyngeal myosin heavy chain at 206,000 daltons appears only in the first half of the peak. After chromatography on hydroxylapatite, N2 myosin still contains its light chains, and the E675 fractions are quantitatively precipitated by antibodies against the purified N2 protein, suggesting that such myosin is composed of intact molecules. Therefore we propose as a working hypothesis that at least three classes of myosin molecules exist in nematode muscles.

Further experiments are in progress to determine any differences in light-chain composition between these classes of myosin, to test for exchange of polypeptides between myosin molecules in this procedure, and to detect whether any myosin molecules are heterodimeric with respect to heavy chains. The existence of such heterodimers that are stable in vitro would be strong evidence for the existence of two kinds of myosin heavy chains within the same cell. The existence of stable homodimers composed of 203,000-dalton myosin heavy chains and purified on the basis of actin-binding is further evidence for alteration of the carboxyl-terminal, light meromyosin segments of affected myosin molecules in E675 mutants.

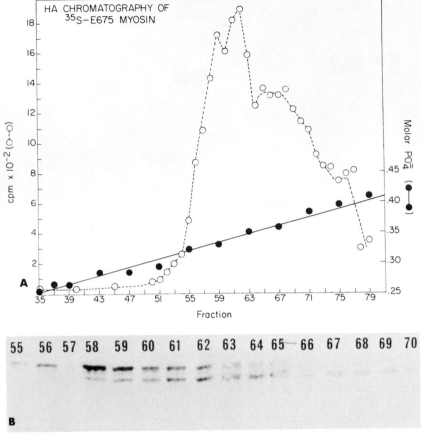

Figure 7
Chromatographic resolution of different myosins. E675 animals were grown
on ^{35}S-labeled bacteria. Myosin was purified as in Fig. 6A. (*A*) Radioactivity
of 50-μl samples of each 1.5-ml fraction indicated. (*B*) Autoradiogram of
5% acrylamide-sodium dodecyl sulfate slab gel separation of peak fractions
as indicated (F. H. Schachat, J. A. Wolff and H. F. Epstein, unpubl.).

Interaction between Myosin and Paramyosin in Thick Filaments

The thick filaments of nematode body wall muscles contain myosin and
paramyosin (Waterston, Epstein and Brenner 1974). As shown in molluscan
muscle, paramyosin comprises the cores of such filaments, whereas myosin
forms a network at their surfaces (Szent-Györgyi, Cohen and Kendrick-Jones
1971). In E675 mutant body wall muscle cells, specific structural alterations
of myosin chains leads to the disruption of thick filaments, despite the presence
of presumably normal paramyosin molecules. This finding suggests that both
intact myosin and paramyosin are necessary for the stability of thick filaments
and the fibrillar lattice in vivo.

This hypothesis, based primarily on genetic and microscopic evidence in
vivo, has led us to study how purified nematode myosin and paramyosin
(Fig. 6) assemble in vitro. When individually dialyzed against 100 mM KCl,
10 mM $MgCl_2$, 0.1 mM PMSF, 0.5 mM DTT, 10 mM potassium phosphate,

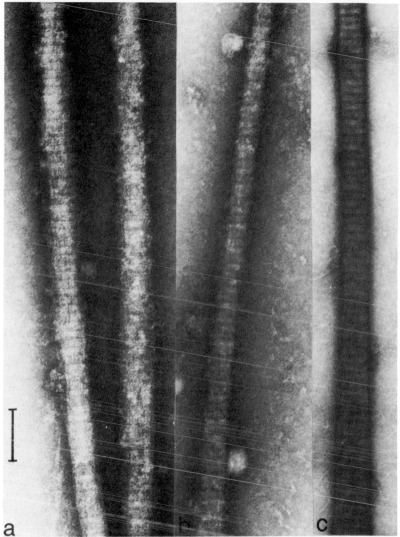

Figure 8
Electron micrographs of reconstituted myosin-paramyosin filaments. Proteins
were purified as in Fig. 6. (*a*) Myosin-paramyosin mixed 1:1 by weight;
(*b*) myosin-paramyosin mixed 1:2 by weight; (*c*) as in *b*, but grids were
washed with 0.3 M KCl. The bar represents 0.1 μM. (H. E. Harris, J. A.
Wolff and H. E. Epstein, unpubl.)

pH 6.5, these proteins form distinct structures. Purified myosin forms predominantly very long filaments (>10 μm in length) of about 10–20 nm in
diameter, with occasional bipolar filaments about 0.2 μm long. All the myosin
structures exhibit headlike protrusions. Purified paramyosin forms smooth,
paracrystalline filaments of varying lengths and diameters. These structures
exhibit prominent axial periodicities, usually 14.5 nm. When a solution containing myosin and paramyosin is treated identically to the individual proteins,

long filaments with protruding structures and underlying 14.5 nm periodicity are observed (Fig. 8). The myosin can be removed by washing with 0.3 M KCl, leaving a paramyosinlike filament. As in the native molluscan filament, myosin forms a cortex about a paramyosin core in the reconstituted nematode structures.

Similar experiments with myosin or paramyosin from mutant strains are in progress. This approach may serve as an assay for more subtle alterations of myosin than in E675 and may lead to a better understanding of the structural interactions between myosin and paramyosin.

FUTURE AIMS

In the ways discussed above, molecular and genetic studies in conjunction with microscopic observations of body wall muscle structure in *C. elegans* may each help us to dissect the physiological relationships that permit the assembly and function of the fibrillar elements. In nematodes, the understanding of how differences in gene expression result in the formation of distinct contractile assemblies within specific cells appears to be an attainable goal.

Acknowledgments

H. F. E. was a Fellow of the National Genetics Foundation (through the Grant Foundation) and the Mellon Foundation; H. E. H. was a Fellow of the Science Research Council of Great Britain; F. H. S. was a Fellow of the Muscular Dystrophy Association, Inc. We thank the National Institute of Child Health and Human Development, through the Adult Development and Aging Branch; the American Heart Association, Inc., through the Santa Clara County chapter; the Muscular Dystrophy Association, Inc.; and the American Cancer Society institutional grant to Stanford for support.

REFERENCES

Brenner, S. 1974. The genetics of *Caenorhabditis elegans*. *Genetics* **77**:71.

Epstein, H. F. and J. N. Thomson. 1974. Temperature-sensitive mutation affecting myofilament assembly in *Caenorhabditis elegans*. *Nature* **250**:579.

Epstein, H. F., R. H. Waterston and S. Brenner. 1974. A mutant affecting the heavy chain of myosin in *Caenorhabditis elegans*. *J. Mol. Biol.* **90**:291.

Szent-Györgyi, A. G., C. Cohen and J. Kendrick-Jones. 1971. Paramyosin and the filaments of molluscan "catch" muscles. II. Native filaments: Isolation and characterization. *J. Mol. Biol.* **56**:239.

Ward, S. 1973. Chemotaxis by the nematode *Caenorhabditis elegans:* Identification of attractants and analysis of the response by use of mutants. *Proc. Nat. Acad. Sci.* **70**:817.

Waterston, R. H., H. F. Epstein and S. Brenner. 1974. Paramyosin of *Caenorhabditis elegans*. *J. Mol. Biol.* **90**:285.

Section 3

THE ASSOCIATION
OF MICROFILAMENTS
AND MICROTUBULES
WITH SURFACE PHENOMENA
RELATED TO CELL MOTILITY

Introductory Remarks:
To Move, or Not to Move,
That Is the Question

Olav Behnke

Anatomy Department C, University of Copenhagen
DK 2200 Copenhagen, Denmark

Cells move, and cytoplasmic components move within cells. The long list that can now be made of cell types in which the presence of actin and myosin (and in several cases also tropomyosin and troponin) has been established makes it increasingly attractive to generalize that the interaction of these proteins provides the molecular basis of motility in numerous biological systems besides muscle.

Because the unique organization of actin and myosin in muscle cells is not found in non-muscle motile cells, it has become customary to use the term "primitive motile systems" to characterize the locomotory systems of these cells. Although it may be justified morphologically (as far as we know their morphology) to characterize them as "primitive," it is now appreciated that functionally they seem to be very complicated systems, and that the term primitive had perhaps better be forgotten. Thus we have learned that in motile non-muscle cells, the proteins of the locomotory system, in contrast to muscle cells, constitute a labile, dynamic system in the sense that the cell can alter and control the state of polymerization, the localization and the organization of these proteins. Actin, for example, may be present in a cell in at least two states: as a recognizable, but ill-defined meshwork or as bundles of filaments that can be readily identified. The one state can apparently convert into the other, but whether the conversion is direct or runs through hitherto unrecognized states is unknown. There is little doubt that the mesh to filament conversion and the reverse can be initiated by a contact-mediated signal originating at and transmitted to the cytoplasm from the cell's periphery. The nature of the signal and how it is transmitted and affected remain unknown, and it is not clear whether the conversion in turn alters the properties of the cell's periphery, locally and universally.

The very lability of the locomotory system and the fact that the system with respect to the above-mentioned states shows an apparent high degree of individuality in different cells make an analysis of the involvement of the

215

system in the various processes to which it has been linked troublesome and tedious.

The recent introduction of fluorescent antibodies against the various protein components of the locomotory system hopefully may develop into a powerful tool in our attempts to understand the system. However, we should not in our enthusiasm forget that the restrictions and limitations of interpretation that must be placed on all observations of fixed cells also apply to studies based on immunofluorescence—perhaps even more so. It is still debatable whether our preparative methods faithfully reveal the proteins of the cellular locomotory system in their precise localization and whether they preserve the proteins in the state they were in at the moment of fixation. Although we know from biochemical studies that important properties of actin and myosin from muscle are shared with actins and myosins from non-muscle cells, there is also evidence that their properties *in the cell* differ, and that therefore their reaction to treatments such as fixation and extraction also may differ. In fact, various procedures may induce depolymerization as well as polymerization.

To clarify these challenging questions we need, as Chairman Mao phrased it, "an enthusiastic, but calm state of mind, and intense, but orderly work."

Organizational Changes
of Actinlike Microfilaments
during Animal Cell Movement

Robert D. Goldman, Jeffery A. Schloss and Judith M. Starger

Department of Biological Sciences, Mellon Institute
Carnegie-Mellon University, Pittsburgh, Pennsylvania 15213

Over the past several years rapid advances have been made in the isolation, physical-chemical characterization and localization of actomyosinlike proteins in many types of non-muscle cells (Pollard and Weihing 1974). The results accumulated so far indicate that these proteins are major elements of the molecular architecture of animal and plant cells and that they bear remarkably similar properties to their analogs found in striated muscle (Pollard and Weihing 1974). These findings have led to the widespread acceptance of the idea that these "contractile" proteins are ubiquitous elements of the protoplasm of eukaryotic cells. Although it is also assumed that these proteins function in many diverse aspects of cell motility, there is very little direct evidence that this is the case. The investigations reported here are aimed at elucidating the functions of these proteins in cultured mammalian cell motility. Since actinlike microfilaments are the only elements of the contractile machinery that can be visualized directly in non-muscle mammalian cells in culture, a study of their structure, organization and distribution during different motile activities should provide new insight into their specific functions.

The Interphase Cell: Stress Fibers and Their Coincidence
with Microfilament Bundles (Mfb)

Many cells in high-density cultures of mouse 3T3 cells, rat embryo cells and human ENSON cells are well spread upon their growth substrates. One of the most distinctive cytoplasmic features of many of the interphase cells when viewed live with phase-contrast, polarized light or Nomarski optics is the presence of stress fibers (Figs. 1–4) (Buckley and Porter 1967; Goldman et al. 1975). Other types of mammalian cells contain few or no obvious stress fibers. These are usually cells that do not become extensively spread when growing on a solid substrate, such as the more fibroblastic BHK-21 cell (Goldman and Follett 1969) or rounded adenovirus-transformed cells

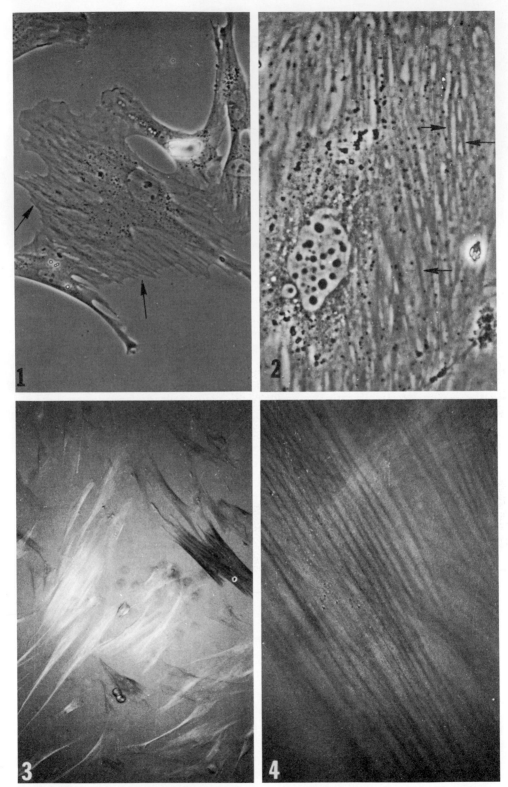

Figures 1–4 (*See facing page for legend*)

(Goldman, Chang and Williams 1975) (Fig. 5). However when BHK-21 cells are treated with colchicine, which causes the disassembly of cytoplasmic microtubules, they tend to flatten out extensively as they lose their fibroblastic shape (Goldman 1971; Goldman and Knipe 1973). In such colchicine-treated cells, distinct stress fibers are easily resolved (Fig. 6) (Goldman 1971; Goldman and Knipe 1973). Therefore it appears that stress fibers are found in a variety of normal cell types.

In most instances, stress fibers in normal cells are located in the region of cell-substrate contact and are usually not found in other regions of the cytoplasm (Buckley and Porter 1967; Goldman 1971; Goldman and Knipe 1973; Goldman et al. 1975). This can be demonstrated most easily in living cells grown on glass cover slips using Nomarski differential interference optics. Differential focusing shows that stress fibers in the cytoplasm are only in focus immediately adjacent to the adhesive surface of the cell (Figs. 7, 8).

Following fixation with glutaraldehyde and processing for electron microscopy (Goldman 1975), the stress fibers are retained in the region of cell-substrate contact (Figs. 9–12). Thin sections of flat-embedded 3T3 cells, rat embryo cells and BHK-21 cells made parallel to the plane of cell-substrate contact contain large numbers of submembraneous microfilament bundles (mfb) that have exactly the same distribution and localization of stress fibers (Goldman et al. 1975; Goldman 1975). These mfb are seen in the cell cortex underlying the plasma membrane in contact with the substrate (e.g., plastic tissue-culture dish) and consist primarily of 4–7-nm microfilaments lined up with their long axes parallel to each other (Figs. 13, 14).

Several light microscopic techniques are available which demonstrate that stress fibers contain actinlike protein. These include indirect immunofluorescence with an antibody preparation directed against actinlike components extracted from SDS gels (Lazarides and Weber 1974; Goldman et al. 1975). When 3T3 cells containing stress fibers are fixed with formaldehyde, extracted with acetone, and then prepared for indirect immunofluorescence, the stress fibers fluoresce intensely above background. By observing the same cell in

Figures 1–4

(*Fig. 1*) Living rat embryo cells. The largest and most extensively spread cell in the field contains stress fibers in the regions indicated (arrows). Other cells in this field are not as well spread, and cytoplasmic stress fibers are not evident. Phase-contrast; magnification, 350×.

(*Fig. 2*) Portion of a living 3T3 cell viewed with phase-contrast optics. This cell was selected because of its extremely flattened configuration. Note stress fibers (arrows). Magnification, 625×.

(*Figs. 3–4*) Field of human ENSON cells (obtained from American Type Culture Association) viewed with polarized light optics. Note birefringent stress fibers seen at low and high magnification. These photomicrographs were taken with Zeiss strain-free polarized light optics with a λ/30 Brace-Koehler compensator set just past the extinction point for maximum contrast. The contrast changed from dark to light or vice versa when the compensator was rotated to one side of extinction or the other. Magnifications: Fig. 3, 125×; Fig. 4, 580×.

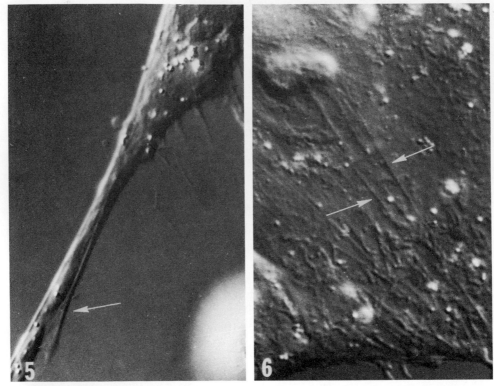

Figures 5–6

(*Fig. 5*) BHK-21 fibroblastlike cell. One stress fiber can be seen along the major cell process (arrow). In many cells of this type, no stress fibers can be seen. Nomarski optics; magnification, 1700×.

(*Fig. 6*) Living BHK-21 cell incubated for 12 hr in culture medium containing 10 μg/ml colchicine. Many stress fibers are easily resolved (arrows). Nomarski optics; magnification, 1850×.

phase-contrast and fluorescence microscopy, a 1:1 relationship between phase-dense stress fibers and fluorescent fibers is seen (Figs. 15, 16) (Goldman et al. 1975).

Similar observations can be made utilizing fluorescein-labeled heavy meromyosin (FlHMM) prepared by conjugating rabbit skeletal muscle HMM with fluorescein isothiocyanate (Aronson 1965; Sanger 1975a,b; Goldman, Schloss and Forer 1976). Cells containing stress fibers are subjected to short-term glycerol extraction at room temperature (Chang and Goldman 1973; Goldman 1975), reacted with FlHMM for 15–30 minutes, followed by extensive washing in standard salt solution (Goldman 1975), and then observed by epifluorescence microscopy. The stress fibers exhibit intense fluorescence, and their general pattern and distribution is similar to that seen following the actin-antibody staining procedure (Fig. 17) (Goldman et al. 1975). Since HMM binds to skeletal muscle actin and non-muscle actinlike microfilaments (Huxley 1963; Ishikawa, Bischoff and Holtzer

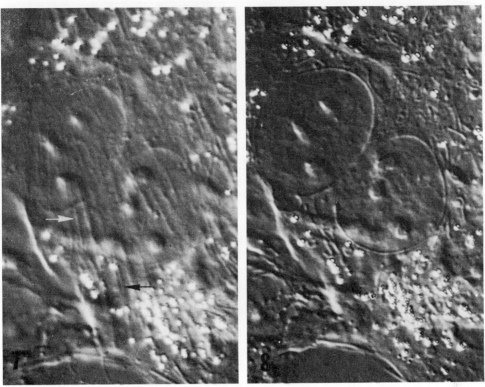

Figures 7–8

Living binucleate cell obtained from a primary rat embryo culture. (*Fig. 7*) An in-focus image in the region of cell-substrate contact containing stress fibers (arrows). (*Fig. 8*) The same cell with nuclei in sharp focus. No stress fibers are seen. Nomarski optics; magnification, 1850×.

1969), the fluorescence of stress fibers following FlHMM treatment also indicates the presence of an actinlike protein.

Although both the antibody and the FlHMM techniques demonstrate that stress fibers contain actin, ultrastructural studies must be utilized in conjunction with these techniques in order to determine the precise organization and molecular arrangement of actin within stress fibers. Therefore cells glycerinated (Goldman 1975) in the same manner as those prepared for the FlHMM assay can be reacted with HMM and used successfully to determine the ultrastructural and ultracytochemical characteristics of stress fibers. Glycerinated cells not treated with HMM contain submembraneous mfb having the distribution and location of stress fibers and that appear essentially identical to mfb seen without glycerination (Goldman 1975). When treated with HMM or FlHMM, the microfilaments within the bundles are "decorated" with HMM. This is recognized primarily by the increased thickness (up to 15 nm), electron density and fuzzy appearance of individual microfilaments within the bundles (Fig. 18). Occasionally arrowhead configurations can be seen (Huxley 1963; Ishikawa, Bischoff and Holtzer 1969; Chang

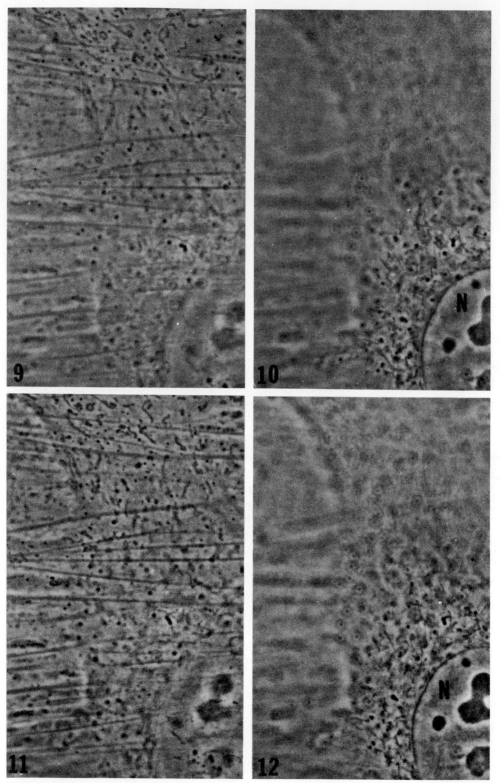

Figures 9–12 (*See facing page for legend*)

and Goldman 1973; Goldman 1975) along microfilaments, especially at the edges of mfb.

Cell Spreading and the Contact-mediated Assembly of Microfilament Bundles

Growing cultures of BHK-21 cells treated with a mixture of trypsin and EDTA (Grand Island Biological) are induced to release their attachments to their growth substrate, round up, and detach. BHK-21 cells treated in this fashion contain no mfb when observed with the electron microscope (Goldman and Knipe 1973; Goldman 1975). Submembraneous mfb form rapidly after cells are allowed to attach and spread on a solid substrate such as a glass cover slip or a plastic petri dish (Goldman 1975). This process of assembly of mfb does not require protein synthesis (Goldman and Knipe 1973), indicating the presence of a pool of precursors to mfb in suspended cells. Since on the basis of their size and interaction with heavy meromyosin microfilaments appear to be similar to the polymerized form of muscle actin (F-actin), one possible source of these precursors is a pool of the monomer form of actin (G-actin). Therefore the precursors to mfb might be G-actin monomers that rapidly assemble into actinlike microfilaments and in turn become organized into mfb when cells become attached to solid substrates. However when the submembraneous cortical region of suspended BHK-21 cells is studied in detail with the electron microscope, a randomly organized meshwork of microfilaments is seen. Some of the microfilaments making up this meshwork appear to insert into the plasma membrane (Goldman 1975). When BHK-21 cells are glycerol-extracted and treated with heavy meromyosin, the meshwork microfilaments are decorated, thus indicating their actinlike nature (Goldman 1975). Within 15–30 minutes following attachment to a plastic dish, mfb are formed exclusively in the cortex underlying the plasma membrane on the adhesive side of the cells. No mfb are seen in regions away from the adhesive surface of the attached cells during the early stages of spreading (Goldman 1975).

We have extended these observations on BHK-21 cells to the contact-inhibited mouse 3T3 cell line. When freshly trypsinized suspensions of 3T3 cells are observed in thin sections prepared for electron microscopy (Goldman 1975), a similar submembraneous meshwork of microfilamentous material is seen just below the plasma membrane (Figs. 19, 20). When suspended

Figures 9–12
(*Fig. 9*) Living rat embryo cell containing stress fibers in the region of cell-substrate contact. (*Fig. 10*) The same cell with the nucleus (N) in focus to demonstrate absence of stress fibers in this region. (*Figs. 11–12*) The same cell following fixation with 1% glutaraldehyde in phosphate-buffered saline (PBS) for 30 min (room temperature) and five washes in PBS to remove the glutaraldehyde. Note that the stress fibers are preserved with regard to their structure and localization towards the substrate (Fig. 11). They are not in focus at the level of the nucleus (Fig. 12). These structures are preserved throughout postfixation with 1% OsO_4, dehydration and flat-embedding. Phase-contrast; magnification, 1650×.

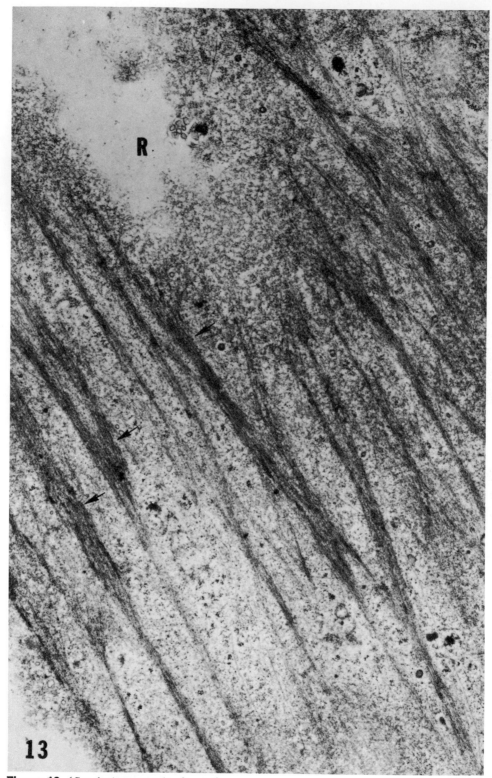

Figure 13 (*See facing page for legend*)

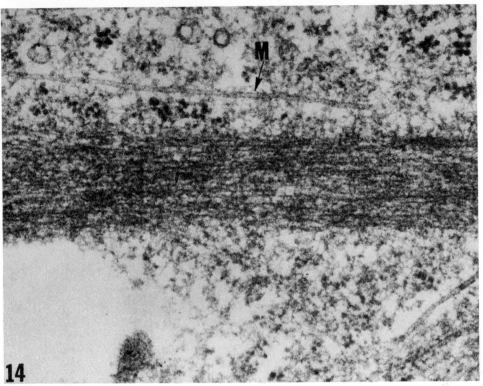

Figure 14
Higher magnification electron micrograph of an mfb showing that it consists of parallel arrays of 4–7-nm microfilaments and some amorphous, electron-dense material (Goldman et al. 1975). A few microtubules (M) are present in the adjacent cytoplasm. Magnification, 69,875×.

cells are extracted with glycerol, treated with rabbit skeletal muscle HMM, and then fixed and embedded for electron microscopy, this region is seen to contain only decorated microfilaments, with occasional arrowhead configurations (Fig. 21) (Goldman 1975).

Suspended 3T3 cells placed in contact with plastic tissue-culture dishes containing growth medium were fixed and prepared for electron microscopy at various time intervals as the cells spread over the plastic substrate. Within 15 minutes following attachment, mfb could be found exclusively in the cell cortex adjacent to the adhesive surface of the cells (Fig. 22). Only the mesh-

Figure 13
Low-magnification electron micrograph of a thin section of a spread 3T3 cell taken just below the plasma membrane at the level of cell-substrate contact. This cell was fixed and flat-embedded as described elsewhere (Goldman 1975). The plane of section passes out of the cell in one region (R), demonstrating that the microfilament bundles (arrows) are just below the plasma membrane. Magnification, 17,340×.

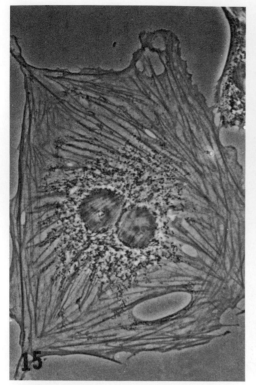

Figures 15–17

(*Figs. 15–16*) A well-spread 3T3 cell that has been fixed with formaldehyde, extracted with acetone, and treated with actin antibody for indirect immunofluorescence (for details, see Goldman et al. 1975). Fig. 15 was taken with phase-contrast optics and Fig. 16 was taken with dark-field fluorescence optics. Note the 1:1 relationship between phase-dense and fluorescent stress fibers. Magnification, 520×. (Reprinted, with permission, from Goldman et al. 1975.)

(*Fig. 17*) Rat embryo cell following short-term glycerination (Goldman 1975) and treatment with FlHMM (Goldman, Schloss and Forer 1976). Note the fluorescent stress fibers. Epifluorescence optics; magnification, 560×.

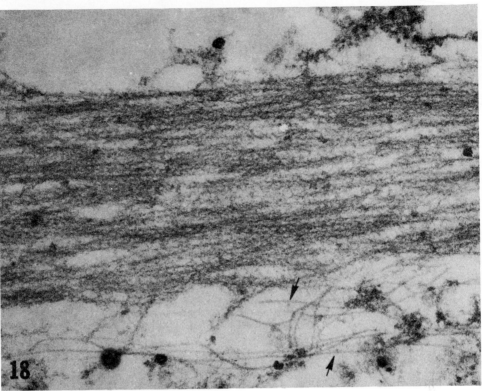

Figure 18
Electron micrograph of a thin section made adjacent to the adhesive surface of a 3T3 cell that has been glycerinated and treated with HMM. A decorated mfb is present as well as a few undecorated 100-Å filaments (arrows). Magnification, 71,750×. (Reprinted, with permission, from Goldman 1975.)

work form of microfilaments could be found in cortical regions not in contact with the substrate during these early stages of cell spreading. These observations were made by examining serial thin sections made parallel to the substrate of flat-embedded cells beginning at the adhesive surface of the cells. As the cells continued to spread over the substrate, more mfb could be found in regions of cell-substrate contact (see Fig. 13). Cells glycerinated and treated with HMM during the spreading process were also fixed and prepared for electron microscopy. Observations of these cells indicated that the individual microfilaments making up these newly formed mfb interacted with HMM (Fig. 23), and that the remainder of the cell cortex not in contact with the substrate contained the more randomly arranged meshwork of decorated microfilaments.

These ultrastructural observations of 3T3 and BHK-21 cells indicate that there are at least two organizational states of microfilaments: meshworks and bundles. In addition, it appears that there is a cell-substrate contact-mediated surface signal that initiates the conversion of microfilament meshworks into mfb (Goldman 1975).

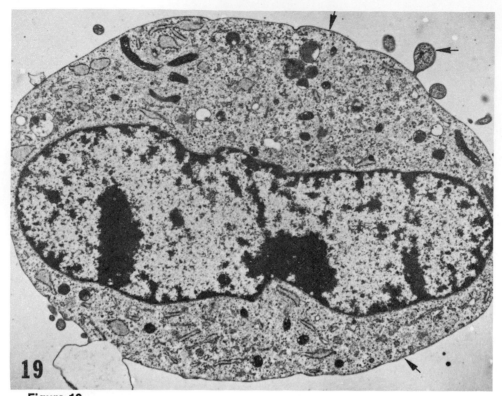

Figure 19

Low-magnification electron micrograph of a suspended 3T3 cell (see Goldman 1975 for EM methods). Arrows point towards the plasma membrane beneath which meshworks of microfilaments are seen (see Fig. 20). Magnification, 6,975×.

Further Ultrastructural Studies on the Nature of the Contact-mediated Assembly of Microfilament Bundles

It has been demonstrated that cultured cells form intermittent close contacts with their substrates (Ambrose 1961; Curtis 1964; Cornell 1969; Brunk et al. 1971; Revel and Wolken 1973; Harris 1973; Revel, Hoch and Ho 1974; Abercrombie and Dunn 1975). When flat-embedded cells are viewed in thin sections made perpendicular to their substrates, the cells are seen to adhere to an electron-dense line that probably represents a film of protein coating the plastic culture dish (Revel and Wolken 1973) (Fig. 24). Much of the lower surface of the cell is not in close contact with this layer of protein. However, intermittent close contacts are seen in which the plasma membrane and the protein film are in close apposition. The cytoplasm adjacent to these close cell-substrate contacts is usually electron-dense (Fig. 24). Similar structures have been described by other investigators (e.g., Brunk et al. 1971) and have been termed plaques by Abercrombie, Heaysman and Pegrum (1971). These electron-dense plaques seen in thin-sectioned material may also correspond to the "attachment sites" described by

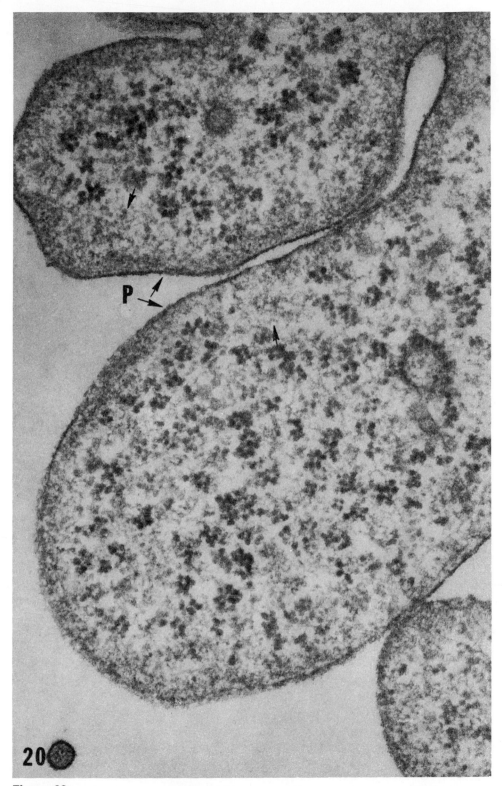

Figure 20
High-magnification electron micrograph of the surface of a suspended 3T3 cell. Note the plasma membrane (P) and meshworks (arrows). Magnification, 75,000×.

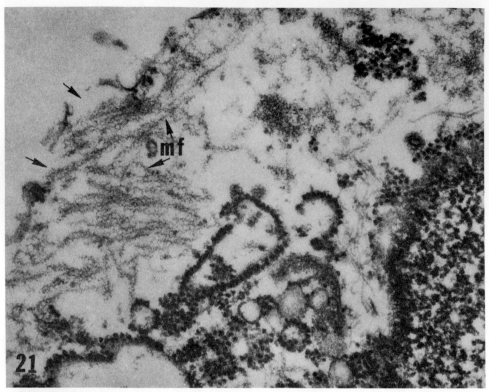

Figure 21
A suspended 3T3 cell that has been glycerinated and treated with HMM (Gold-man 1975). The plasma membrane is missing in several places (arrows), and the submembraneous region contains decorated microfilaments (mf). Magnification, 53,000×.

Revel and Wolken (1973) in whole-cell replicas of BHK-21 cells. The plaques are frequently associated with a submembraneous mfb (Fig. 24; see also Brunk et al. 1971), and in some instances individual microfilaments appear to insert directly into the plaques (R. D. Goldman, J. A. Schloss and J. M. Starger, unpubl. obs.). In thin sections of the cell surface made parallel to the plane of cell-substrate interaction, two major morphological types of electron-dense plaques have been distinguished. Plaques containing a crystal-linelike substructure have been seen in slightly flattened hamster embryo cells that have been transformed by adenovirus type 5 (Goldman, Chang

Figures 22–23
(*Fig. 22*) Electron micrograph of a thin section made parallel to the sub-strate at the level of cell-substrate interaction. This 3T3 cell was fixed and flat-embedded (Goldman 1975) 15 min after attachment to a plastic culture dish. Note submembraneous mfb (arrows). Magnification, 20,500×.

(*Fig. 23*) A 3T3 cell that was glycerinated and treated with HMM (Gold-man 1975) 15 min after attachment to a plastic dish. Mfb are decorated with HMM (arrows). Magnification, 20,500×.

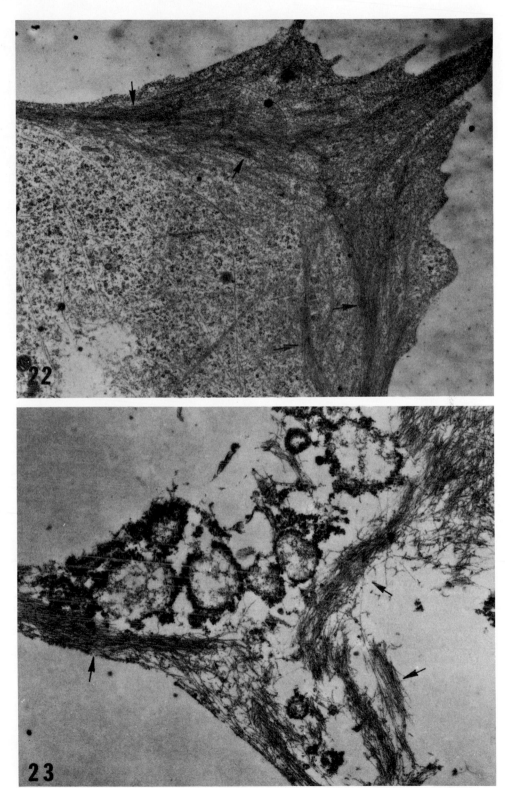

Figure 23 (*See facing page for legend*)

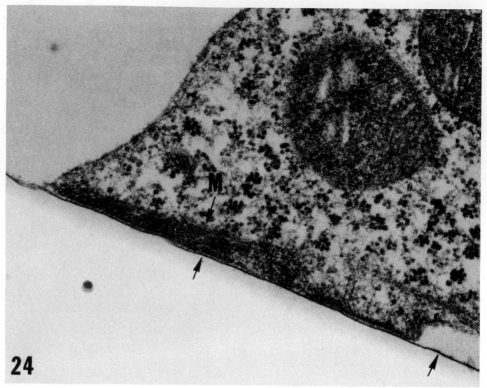

Figure 24

A slightly flattened adenovirus 5-transformed hamster embryo cell that has been fixed, flat-embedded, and sectioned perpendicular to the substrate. The lower cell surface makes intermittent contacts with the protein film (arrows). Note electron-dense region containing a bundle of microfilaments on the cytoplasmic side of the close cell-substrate contact (M). Magnification, 61,875×.

and Williams 1975), BHK-21 cells and human ENSON cells (Figs. 25, 26). Amorphous plaques are seen in monkey BSC-1 cells (R. D. Goldman, J. A. Schloss and J. M. Starger, unpubl. obs.). Mfb are frequently seen in close association with these plaques (Fig. 26). Based on these observations it appears that the electron-dense plaques seen in sections made parallel and perpendicular to the substrate of flat-embedded cells are the same cell-surface structures. Preliminary electron microscopic studies of plaque formation during attachment and spreading of a variety of cell types suggest that plaques may also act as the nucleating sites for the assembly of mfb.

Cell-Cell Contact-mediated Assembly of Microfilament Bundles

Since there is a cell-substrate contact-mediated assembly of mfb in 3T3 cells, we decided to determine whether or not a similar phenomenon took place following the establishment of cell-cell contacts during contact inhibition. The edge of a spreading 3T3 cell frequently contains an actively undulating or ruffling membrane. If one of these edges establishes a contact with a neighboring cell, there is a rapid and immediate inhibition of ruffling (contact inhibition), and in many instances, spikelike projections form either under or over the adjacent cell's surface (Figs. 27–29). In established cultures of 3T3, virtually all cells in contact with neighboring cells contain identical projections (Fig. 30). When viewed by electron microscopy, each spike of cytoplasm is seen to contain a bundle of microfilaments (Fig. 31). When similar contacting pairs of cells are extracted with glycerol (Goldman 1975) and treated with HMM, microfilaments within these bundles are decorated (Fig. 32). Ruffled membrane regions contain no mfb (Fig. 33) but they do contain microfilament meshworks. Heaysman (1973) has also seen groups of oriented microfilaments forming in electron microscope preparations of cells following the establishment of contacts. These observations indicate that there is a cell-cell contact-mediated assembly of mfb and perhaps a cell-cell contact-mediated conversion of microfilament meshworks into mfb.

Cell Locomotion

The locomotion of fibroblasts in vitro has been described in detail by Abercrombie (1961) and Ambrose (1961). The cells move jerkily along a substrate apparently by breaking loose adhesions towards the rear of a cell and establishing new contacts with the substrate at the leading edge. These authors consider that the leading edge is the main locomotory organelle of a fibroblast (see, e.g., Abercrombie 1961). Ambrose (1961) also proposed that there may be a submembraneous contractile system responsible for locomotion. Thus the assembly of submembraneous mfb in the region of cell-substrate contact may be an important factor in determining the mechanism by which a cell migrates across a substrate. The submembraneous bundles may contain the organized form of an actomyosinlike contractile system involved in providing the motive force necessary to overcome the forces of adhesion between cell and substrate during cell translocation.

Preliminary analyses of time-lapse movies and video tapes of locomoting rat embryo fibroblastlike cells indicate that the ruffling leading edge moves slowly over the substrate (a glass cover slip) (see Abercrombie, Heaysman and Pegrum 1970a,b for a detailed description of this type of movement). However, the tail region of a moving rat embryo fibroblast moves intermittently and at a relatively higher rate of speed. In many cases, those latter movements result in total resorption of the tail into the main cell body (Figs. 34–37). These rapid and spontaneous tail movements appear to account for much of the net translocation of fibroblastic cells.

Ultrastructural studies of migrating rat embryo fibroblasts indicate that the tail portion contains a bundle of microfilaments oriented with its long axis parallel to the long axis of the tail (Fig. 38). Frequently a bundle of microfilaments completely fills the tail portion, as visualized in thin sections made along the length of a tail. In glycerinated models of similar cells, heavy meromyosin decorates the tail bundle of microfilaments (R. D. Goldman, J. A. Schloss and J. M. Starger, unpubl. obs.).

These results indicate that actinlike microfilaments are oriented with their long axes along the length of the tail and correspond to the direction in which the tail tends to move, i.e., forward towards the main cell body.

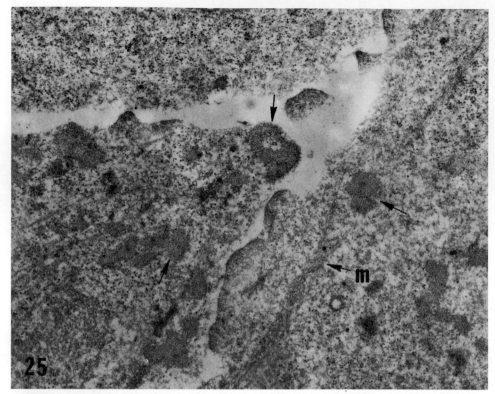

Figures 25–26
Electron-dense plaques (arrows) seen in sections made parallel to the substrate
at the level of cell-substrate interaction in cells similar to those shown in Fig. 24.
Note the juxtaposition of some mfb (m) and plaques. These plaques have a
crystalline appearance. Magnifications: Fig. 25, 23,800×; Fig. 26, 79,500×.

"Contractile" Models of Locomoting Cells

Hoffmann-Berling (1945a,b) was the first to demonstrate that glycerinated
models of cultured cells would respond to exogenous ATP. The response
of interphase cells consisted of rounding up from a flattened position (Hoffmann-Berling 1954a). He concluded that the response was most likely due
to an actomyosinlike contraction. In addition, models of dividing cells were
induced to form cleavage furrows following treatment with ATP (Hoffmann-
Berling 1954b).

In order to determine whether or not locomoting rat embryo cells would
respond to ATP, they were glycerinated using the short-term extraction
procedure described previously (Goldman 1975). Following glycerination,
the cells were treated with 0.05–1.0 mM ATP in a modified standard salt
solution containing 0.05 M KCl, 0.007 M potassium phosphate buffer and
0.015 M $MgCl_2$ (pH 7.0). The response of the cell models was virtually
identical to that seen in living cells; the tail moved forward rapidly toward
the leading edge (Figs. 39–43). Control cells treated with the same solution

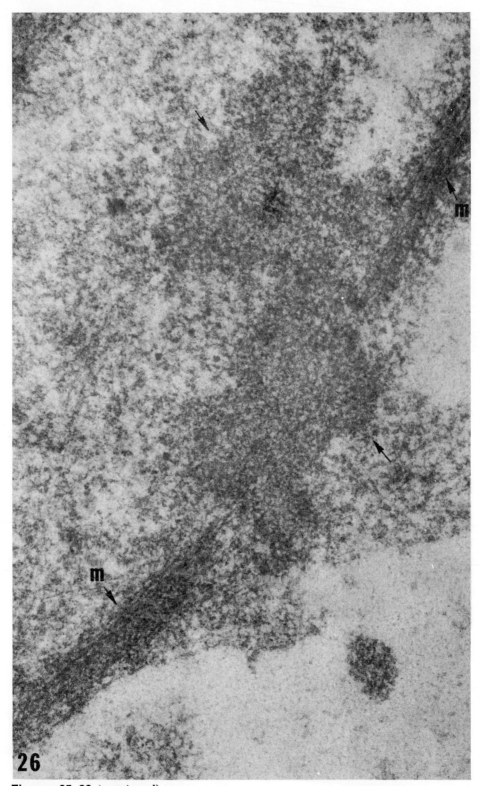

Figures 25–26 (*continued*)

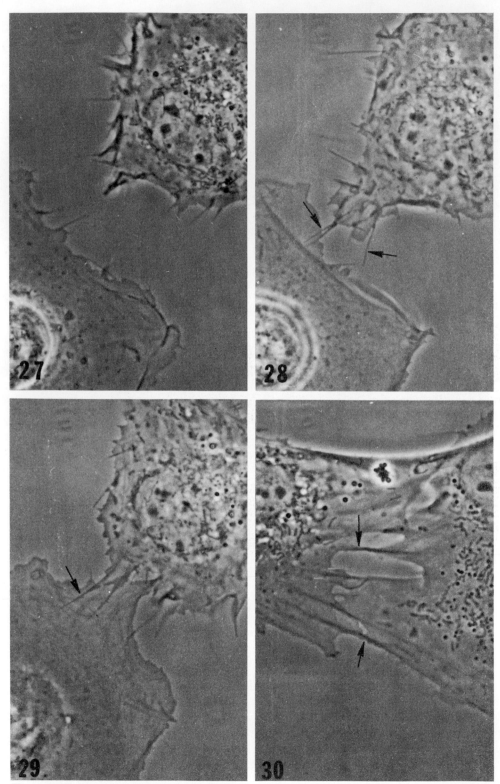

Figures 27–30 (*See facing page for legend*)

236

without ATP show no movements. Furthermore, the response is temperature-sensitive and thus is much faster at 37°C than at 20–22°C.

Similar responses also occur following the addition of ITP and GTP in the same modified standard salt solution, although the movements are slower when compared to the ATP response. However, no movements are induced following the addition of similar concentrations of ADP. Furthermore, SDS gel electrophoresis of the proteins of glycerinated cells indicates that the models contain actinlike and myosinlike proteins (Fig. 44).

DISCUSSION AND SUMMARY

The results of this study demonstrate a close relationship between the organization of submembraneous actinlike microfilaments and certain aspects of cell motility. In many types of interphase cells, microfilaments adjacent to the surface in contact with a solid substrate are arranged in parallel, closely packed arrays to form bundles (mfb), which in living cells are visible as stress fibers. The microfilaments within these bundles appear to be similar to muscle F-actin by several criteria in addition to their size and morphology. They bind rabbit skeletal muscle heavy meromyosin (HMM) (Huxley 1963; Ishikawa, Bischoff and Holtzer 1969; Goldman and Knipe 1973; Goldman 1975), as determined by electron microscopic examination of glycerinated cells that have been treated with HMM. In addition, several fluorescence microscope techniques are available that indicate that actin is a component of stress fibers. These techniques result in stress-fiber fluorescence following treatment with actin antibody (using indirect immunofluorescence) or fluorescein-labeled HMM (FlHMM) (Lazarides and Weber 1974; Goldman et al. 1975; Sanger 1975a,b). Based on these different types of observations, there remains little doubt that stress fibers consisting of mfb contain actin as their major morphological component. Indirect immunofluorescence has also been used to suggest that other musclelike proteins, such as myosin (Weber and Groeschel-Stewart 1974), tropomyosin (Lazarides 1975) and α-actinin, are also components of stress fibers. Therefore it appears likely that at least some submembraneous mfb represent musclelike contractile elements.

Although the rigid appearance of mfb suggests that they are static elements of interphase cells, some insight into their dynamic nature is provided by the finding that microfilament organization is altered in cells that have been detached from their substrates. These suspended cells contain submembraneous microfilament meshworks that may be similar to the microfilament net-

Figures 27–30

(*Figs. 27–29*) Light micrographs of living 3T3 cells spreading over a glass cover slip and contacting one another. As contact occurs, elongated cell processes (arrows) form and surface ruffling ceases. Phase-contrast; magnification, 1400×.

(*Fig. 30*) Micrograph similar to Figs. 27–29, showing established contacts between two 3T3 cells. Note phase-dense projections or fibers (arrows). Phase-contrast; magnification, 1400×.

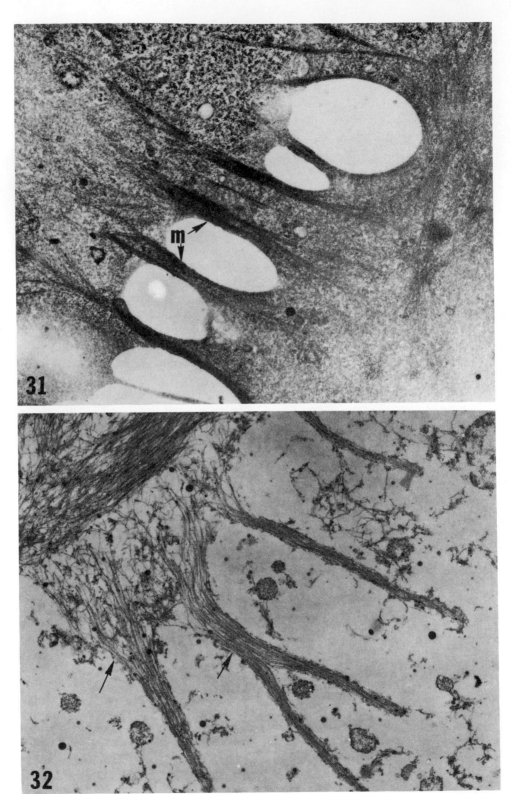

Figures 31–32 (*See facing page for legend*)

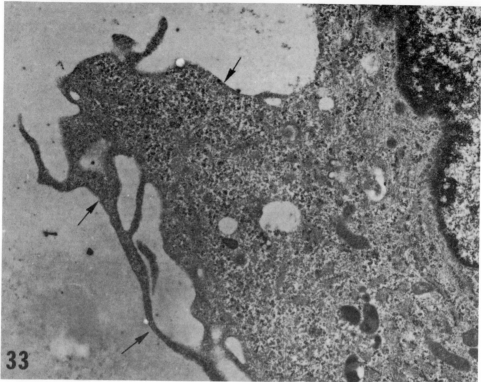

Figures 31–33

(*Fig. 31*) Electron micrograph of a thin section taken parallel to the substrate in a cell-cell contact region similar to that seen in Fig. 30. Note bundles of microfilaments (m) in the spikelike projections. Magnification, 14,000×.

(*Fig. 32*) A region of cell-cell contact selected from a 3T3 cell culture. The cells were glycerinated, treated with HMM (Goldman 1975), and thin-sectioned in a manner similar to that of Fig. 31. Decorated microfilaments are seen (arrows). Magnification, 16,875×.

(*Fig. 33*) Ruffling region of a 3T3 cell not in contact with another cell. Note absence of organized mfb in ruffling region (arrows). Magnification, 10,800×.

works described by Spooner and coworkers (Spooner, Yamada and Wessells 1971). The rapid formation of submembraneous mfb in the cell cortex adjacent to the adhesive surface of spreading cells suggests that there is a cell-substrate contact-mediated conversion of randomly arranged (meshworks) microfilaments into organized mfb. Similar alterations accompany the rounding up of BHK-21 and other cell types entering mitosis; microfilament bundles disappear and only meshworks are seen beneath the plasma membrane (Dickerman 1972; L. H. Dickerman and R. D. Goldman, in prep.) during the early stages of mitosis. Following cytokinesis, the rounded daughter cells begin respreading over the substrate, with simultaneous assembly of mfb (Dickerman 1972; Goldman et al. 1973). These results demonstrate the dynamic nature of the submembraneous system of actinlike microfilaments.

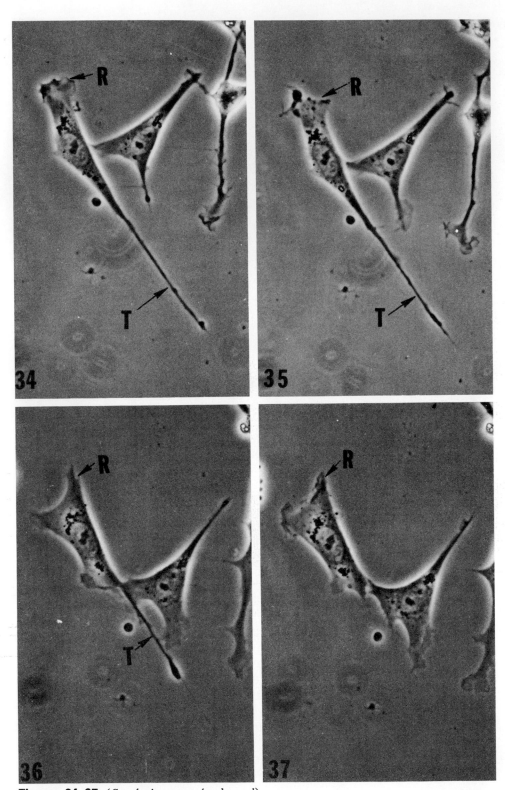

Figures 34–37 (*See facing page for legend*)

240

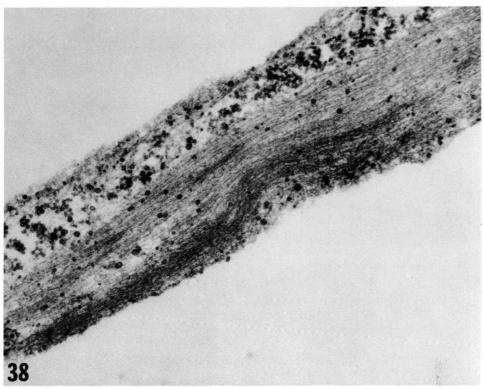

Figure 38

Electron micrograph of a thin section taken parallel to the long axis of the tail of a rat embryo fibroblast possessing the shape of a locomoting cell. Note the mfb, which nearly fills the cytoplasm, running parallel to the direction of movement of the cell. Magnification, 53,625×.

The localization of mfb immediately adjacent to the plasma membrane at the attached surface of normal spread cells suggests that they may provide the motive force necessary to break cell-substrate adhesions during cell locomotion (Goldman et al. 1975). This possibility is supported by the relationship between some mfb and cell-substrate attachment sites or plaques. Preliminary ultrastructural analyses indicate that individual microfilaments may insert directly into the plasma membrane in plaque regions. Thus mfb could exert tension on plaques and in some instances cause the plaque to break loose from the substrate. The resulting loss of adhesion could result in a movement such as that generated when the tail of a moving fibroblast is resorbed into the main cell body.

Figures 34–37

Series of phase-contrast micrographs of a locomoting rat embryo cell showing movement of the cell body and resorption of the tail region (T). Micrographs were taken over a period of 40 min. The leading edge moved about 25 μ in this interval, whereas the tail moved about 115 μ. Ruffled leading edge (R). Magnification, 450×.

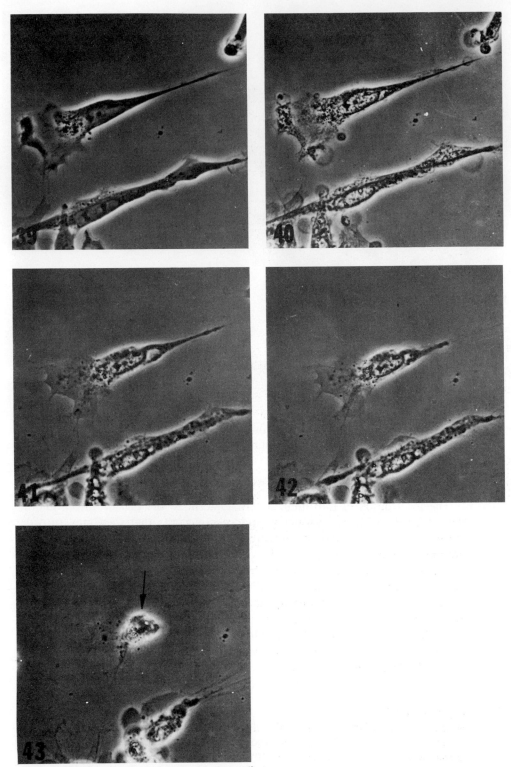

Figures 39–43 (*See facing page for legend*)

242

Figure 44

Preliminary results of SDS-polyacrylamide gel electrophoresis of rat embryo fibroblast cells (w), a glycerol extract of these cells (e) and the cells after extraction (m). Cells grown in plastic petri dishes were extracted with glycerol in a standard salt solution (Goldman 1975). The extracts were pooled, the cell remains scraped from the dishes and sonicated. Live cells were rinsed with phosphate-buffered saline and sonicated. The three samples were concentrated by lyophilization. Protein concentration was determined, and the samples were applied to 5.7% polyacrylamide gels according to the method of Fairbanks (Fairbanks, Steck and Wallach 1971). Bands that comigrate with rabbit skeletal muscle myosin heavy chain and rabbit skeletal muscle actin are labeled m and a, respectively. In these samples, a heavy band in the extract (e) comigrates with serum albumin (arrow) and is thought to originate in the cell growth medium that contains calf serum. Note that very little actin or myosin is present in the extract.

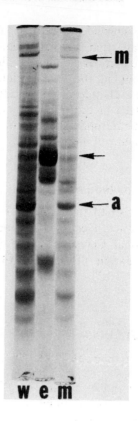

The response of glycerinated models of locomoting cells to Mg^{++}-ATP suggests that the tail movements might be due to actomyosinlike contractile events taking place within the mfb oriented along the long axis of the tail. This idea is also supported by the facts that glycerinated models of cells are enriched for actinlike and myosinlike proteins, and that Mg^{++}-ATP is the appropriate substrate for skeletal muscle actomyosin ATPase activity.

The observations made in these investigations suggest that a dynamic actin-like microfilament system functions in several types of cultured cell motility. The finding that changes in the organization of submembrancous microfila-

Figures 39–43

Phase-contrast micrographs showing a moving rat embryo fibroblast being glycerinated (Goldman 1975) and treated with ATP. Cells were grown on a glass cover slip which was then mounted on a slide (Goldman 1971) for microscopy in such a way that solutions could be perfused over the cells during observation. (*Fig. 39*) Live cell; (*Fig. 40*) after glycerination and standard salt rinse; (*Fig. 41*) 1 min after addition of 1 mM ATP in standard salt solution at room temperature—tail movement has begun; (*Fig. 42*) 2 min after ATP addition; (*Fig. 43*) 8 min after ATP addition—the position of the cell body is unchanged, but the majority of the tail cytoplasm has "contracted" into a refractile mass (arrow) at the base of the tail's original position. Magnification, 520×.

ments coincide with the establishment of cell-substrate and cell-cell contacts indicates that there are signals transmitted from the outer surface of the cell to the cortical cytoplasm. The processing of such signals may be involved in determining the organizational state of submembraneous actinlike microfilaments. Therefore the elucidation of the mechanisms and molecular controls involved in the transmission of these surface signals will undoubtedly lead to a better understanding of the relationship between microfilament organization and cell motility.

Acknowledgments

We wish to thank Anne Bushnell and Anne Goldman for their excellent assistance. Generous support for these studies has been received from The National Cancer Institute (1 RO1 CA17210–01), The National Science Foundation (BMS 73–01446 A01), and The American Cancer Society (VC–110C).

REFERENCES

Abercrombie, M. 1961. The bases of the locomotory behavior of fibroblasts. *Exp. Cell Res.* (Suppl.) **8**:188.

Abercrombie, M. and G. Dunn. 1975. Adhesions of fibroblasts to substratum during contact inhibition observed by interference reflection microscopy. *Exp. Cell Res.* **92**:57.

Abercrombie, M., J. Heaysman and S. Pegrum. 1970a. The locomotion of fibroblasts in culture. I. Movements of the leading edge. *Exp. Cell Res.* **60**:393.

———. 1970b. The locomotion of fibroblasts in culture. II. "Ruffling." *Exp. Cell Res.* **60**:437.

Ambrose, E. J. 1961. The movements of fibrocytes. *Exp. Cell Res.* (Suppl.) **8**:54.

Aronson, J. F. 1965. The use of fluorescein-labeled heavy meromyosin for the cytological demonstration of actin. *J. Cell Biol.* **26**:293.

Brunk, U., J. Ericsson, J. Pontén and B. Westermark. 1971. Specialization of cell surfaces in contact-inhibited human glial-like cells *in vitro. Exp. Cell Res.* **67**:407.

Buckley, I. K. and K. R. Porter. 1967. Cytoplasmic fibrils in living cultured cells. A light and electron microscope study. *Protoplasma* **4**:24.

Chang, C. M. and R. D. Goldman. 1973. The localization of actin-like fibers in cultured neuroblastoma cells as revealed by heavy meromyosin binding. *J. Cell Biol.* **57**:867.

Cornell, R. 1969. Cell-substrate adhesion during cell culture. *Exp. Cell Res.* **58**:289.

Curtis, A. S. G. 1964. The mechanism of adhesion of cells to glass. *J. Cell Biol.* **20**:199.

Dickerman, L. H. 1972. Cytoplasmic fibers and motile behavior of synchronized BHK-21 fibroblasts. *J. Cell Biol.* **55**:60a.

Fairbanks, G., T. L. Steck and D. F. H. Wallach. 1971. Electrophoretic analysis of the major polypeptides of the human erythrocyte membrane. *Biochemistry* **10**:2606.

Goldman, R. D. 1971. The role of three cytoplasmic fibers in BHK-21 cell motility. I. Microtubules and the effects of colchicine. *J. Cell Biol.* **51**:752.

————. 1975. The use of heavy meromyosin binding as an ultrastructural cyto-chemical method for localizing and determining the possible functions of actin-like microfilaments in non-muscle cells. *J. Histochem. Cytochem.* **23**:529.

Goldman, R. D. and E. A. C. Follett. 1969. The structure of the major cell processes of isolated BHK-21 fibroblasts. *Exp. Cell Res.* **57**:263.

Goldman, R. D. and D. Knipe. 1973. Functions of cytoplasmic fibers in non-muscle cell motility. *Cold Spring Harbor Symp. Quant. Biol.* **37**:523.

Goldman, R. D., C-M. Chang and J. Williams. 1975. Properties and behavior of hamster embryo cells transformed by human adenovirus type 5. *Cold Spring Harbor Symp. Quant. Biol.* **39**:601.

Goldman, R. D., J. Schloss and A. Forer. 1976. On the use of myosin fragments to localize actin in the mitotic apparatus: Light and electron optical techniques. *J. Cell Biol.* (in press).

Goldman, R. D., E. Lazarides, R. Pollack and K. Weber. 1975. The distribution of actin in non-muscle cells. *Exp. Cell Res.* **90**:333.

Goldman, R. D., G. Berg, A. Bushnell, C-M. Chang, L. Dickerman, N. Hopkins, M. L. Miller, R. Pollack and E. Wang. 1973. Fibrillar systems in cell motility. In *Locomotion of tissue cells. Ciba Foundation Symposium,* vol. 14, p. 83. Asso-ciated Scientific Publishers, American Elsevier, Amsterdam.

Harris, A. 1973. Location of cellular adhesions to solid substrate. *Devel. Biol.* **35**:97.

Heaysman, J. 1973. In *Locomotion of tissue cells. Ciba Foundation Symposium,* vol. 14, p. 185.

Hoffmann-Berling, H. 1954a. Adenosintriphosphat als Betriebsstoff von Zell-bewegungen. *Biochim. Biophys. Acta* **14**:182.

————. 1954b. Die Glycerin-Wasserextrahierte Telophasezelle als Modell der Zytokinese. *Biochim. Biophys. Acta* **15**:332.

Huxley, H. 1963. Electron microscope studies on the structure of natural and synthetic protein filaments from striated muscle. *J. Mol. Biol.* **7**:281.

Ishikawa, H., R. Bischoff and H. Holtzer. 1969. Formation of arrowhead com-plexes with heavy meromyosin in a variety of cell types. *J. Cell Biol.* **43**:312.

Lazarides, E. 1975. Tropomyosin antibody: The specific localization of tropomyo-sin in non-muscle cells. *J. Cell Biol.* **65**:549.

Lazarides, E. and K. Burridge. 1975. α-Actinin: Immunofluorescent localization of a muscle structural protein in non-muscle cells. *Cell* **6**:289.

Lazarides, E. and K. Weber. 1974. Actin antibody: The specific visualization of actin filaments in non-muscle cells. *Proc. Nat. Acad. Sci.* **71**:2268.

Pollard, T. D. and R. R. Weihing. 1974. Actin and myosin and cell movement. *CRC Crit. Rev. Biochem.* **2**:1.

Revel, J. P. and K. Wolken. 1973. Electron microscope investigations of the underside of cells in culture. *Exp. Cell Res.* **78**:1.

Revel, J. P., P. Hoch and D. Ho. 1974. Adhesion of culture cells to their sub-stratum. *Exp. Cell Res.* **84**:207.

Sanger, J. W. 1975a. Changing patterns of actin localization during cell division. *Proc. Nat. Acad. Sci.* **72**:1913.

————. 1975b. Intracellular localization of actin with fluorescently labelled heavy meromyosin. *Cell Tissue Res.* **161**:431.

Spooner, B. S., K. M. Yamada and N. K. Wessells. 1971. Microfilaments and cell locomotion. *J. Cell Biol.* **49**:595.

Weber, K. and U. Groeschel-Stewart. 1974. Antibody to myosin: The specific localization of myosin-containing filaments in non-muscle cells. *Proc. Nat. Acad. Sci.* **71**:4561.

The Function of Filopodia in Spreading 3T3 Mouse Fibroblasts

Guenter Albrecht-Buehler

Cold Spring Harbor Laboratory, Cold Spring Harbor, New York 11724

It is well known that chemical gradients, gradients of adhesiveness or mechanical obstacles influence the size and direction of the average displacement of erratically moving animal cells or cell processes and thus may be used for cellular navigation. However, we do not know the means by which cells sense these extracellular parameters.

In perhaps a naive way one might expect to find a single animal cells sensory organs that are involved in cellular navigation. At present, the best candidates for such organs are fibrous structures, about 0.2 μm in diameter and varying in length between 5 and 30 μm, that protrude from the cell surface into the environment under various conditions. There is as yet no common terminology for these structures. First described by Porter, Claude and Fullam in 1945 as "fibrous projections," they have since been given various other names, such as "microvilli" (Fisher and Cooper 1967), "microspikes" (Weiss 1961), "filopodia" (Trelstad, Hay and Revel 1967) and "microfibrils" (Gey 1956). We prefer the term "filopodia" because it matches the term "lamellipodia" and does not seem to be associated with a particular cell type.

Filopodia have been observed in primary mesenchyme cells (Gustavson and Wolpert 1961), during the formation of the neural tube in chicken embryos (Revel and Brown 1976), in fibroblasts and epithelial cells in culture (Porter, Claude and Fullam 1945; Gey 1956; Taylor and Robbins 1963; Dahlen and Scheie 1967; Fisher and Cooper 1967; Cornell 1969; Follet and Goldman 1970; Porter, Fonte and Weiss 1970; Witkowski and Brighton 1971; Revel, Hoch and Ho 1974; Buckley 1975), in macrophages (Allison, Davies and de Petris 1971), in amoebocytes (Partridge and Davies 1974), in blood platelets (Zucker 1974) and at the growth cone of neurites (DeRobertis and Setelo 1952; Yamada, Spooner and Wessels 1971; Dunn 1973; Wessels, Spooner and Luduena 1973). The literature contains several suggestions as to the possible sensory or exploratory function of these structures (Gustavson and Wolpert 1961; Dahlen and Scheie 1967; Cornell 1969; Dunn 1973; Trinkaus 1973; Partridge and Davies 1974).

We found spreading 3T3 mouse fibroblasts to be a convenient system for studying the function of filopodia. During interphase, 3T3 cells growing on a solid substratum assume a flat morphology. In the presence of trypsin or EDTA, the cells change their morphology to spheres, which are still anchored to the substratum by radially extending thin "retraction fibers" (Revel, Hoch and Ho 1974). Many of these retraction fibers can still be seen extending from the cell surface after the cells have been shaken off the substratum. After such cells have been allowed to settle on a new solid substratum for 5 to 10 minutes, only very few fibrous extensions of the cells can be observed in light or scanning electron microscopy. Following attachment to the new solid substratum, new fibrous extensions, which morphologically are very similar to the retraction fibers, appear in large numbers all around each cell. We call these fibrous extensions "filopodia." Once the filopodia are attached to the substratum, the cells extend lamellipodia and in this way begin to spread again.

By merely observing these events on a light microscopic level one might suspect that filopodia are the means by which the cells anchor themselves to the substratum. Another function of filopodia is suggested by observing these events at higher levels of resolution. If spreading cells are compared with spread-out cells in scanning electron microscopy or by replica techniques in transmission electron microscopy, it is obvious that the surface of a rounded-up cell consists of a large number of small blebs, thin, short microvilli and long filopodia, all of which seem to disappear as the cell spreads out. Therefore it has been suggested that these surface specializations, including perhaps filopodia, serve as storage areas for surface material required for the flattening of a round cell (Follet and Goldman 1970; Trinkaus and Erickson 1974; Knutton, Sumner and Pasternak 1975).

Without denying these possible functions of filopodia, we should like to emphasize the already mentioned possibility that filopodia serve an exploratory function. A quantitative way of studying filopodia was discovered when we tried to study the dynamics of cell spreading using a pseudosolid substratum. A glass substratum was densely and evenly covered with solid gold particles 0.2–0.4 μm in diameter and particle clusters up to 2 μm in diameter. Since the particles were only loosely attached to the glass, they would have to move if the cells exerted any forces following contact. Freshly suspended 3T3 cells were plated on top of such substrates. We found that filopodia made contact with the particles and removed them before any lamellipodia were extended (Albrecht-Buehler and Goldman 1976). This observation led us to speculate that filopodia may produce a retraction force every time they contact the surface of an object. In the case of the loose particles, this retraction force removed the particles and therefore shortened the attached filopodia. When filopodia are attached to an extended solid substratum or another cell, however, the inability of the retraction force to shorten the attached filopodia may "inform" the cell where in its vicinity firm attachment is possible. We tried to substantiate this speculation by a number of experiments which will be described in this paper.

RESULTS AND DISCUSSION

Morphology of Filopodia in Spreading 3T3 Cells

When freshly plated, still rounded, 3T3 cells are examined in light and scanning electron microscopy 5–10 minutes after plating, hardly any filopodia can be observed. They begin to be produced by the cells between 10 and 25 minutes after plating (see Fig. 1a) and originate predominantly from the cell surface near the substratum. They move rapidly (see below) and eventually attach to the substratum. Subsequently, lamellipodia extend from areas near the base of the attached filopodia towards their tips.

Filopodia vary in length between 5 and 30 μm. Transmission and scanning electron microscopy (Fig. 1b,c,d) show that the thickness of filopodia under normal culture conditions is 0.15–0.20 μm with remarkable uniformity. The true thickness is probably higher than 0.15–0.20 μm. Occasionally, broken filopodia occur in the scanning electron microscope preparations. Judging by the dimensions of the gaps between broken pieces, the filopodia suffer a length shrinkage of 5–10% during preparation. Assuming that the same shrinkage applies to their thickness, their true thickness can be estimated to be 0.17–0.22 μm.

In transmission electron microscopy, attached filopodia appear as membrane-coated, single microfilament bundles with parallel-oriented filaments (Fig. 1b). Free filopodia are difficult to find in thin sections. In accord with reports on filopodia of other cell types, there are very few, if any, microtubules (DeRobertis and Setelo 1952; Taylor and Robbins 1963; Yamada, Spooner and Wessels 1971; Wessels, Spooner and Luduena 1973; Buckley 1975). The microfilament bundles of filopodia at the edge of fully spread cells can extend 5–10 μm into the cell body (Buckley 1975). In the rounded cells described here, the distance of about 3 μm between the plasma membrane and the nuclear membrane does not allow such extensions into the cytoplasm. In thin sections of spreading 3T3 cells, the filopodia microfilaments were often found to bend near the base and follow the direction of the plasma membrane.

In scanning electron microscopy (Fig. 1c,d), many filopodia tips are bulky. We suspect that this indicates the beginning of a retraction caused by the fixation procedure (see below).

Removal of Gold Particles in Spreading 3T3 Cells

Glass cover slips were densely and evenly coated with colloidal gold particles 0.2–0.4 μm in diameter. Such particles are nontoxic for 3T3 cells (Albrecht-Buehler and Yarnell 1972). So far we have not succeeded in producing a similar coat with other particles. If freshly suspended cells are plated on top of such a cover slip in normal medium containing 10% calf serum, the particles disappear around 95% of the cells within 30–40 minutes after plating (Albrecht-Buehler and Goldman 1976) (Fig. 2a).

Observations of living cells reveal that the particles are removed by filopodia in two different ways: (1) a filopodium attaches to a particle and retracts with it to the cell body (Fig. 2b,c); or (2) a filopodium remains

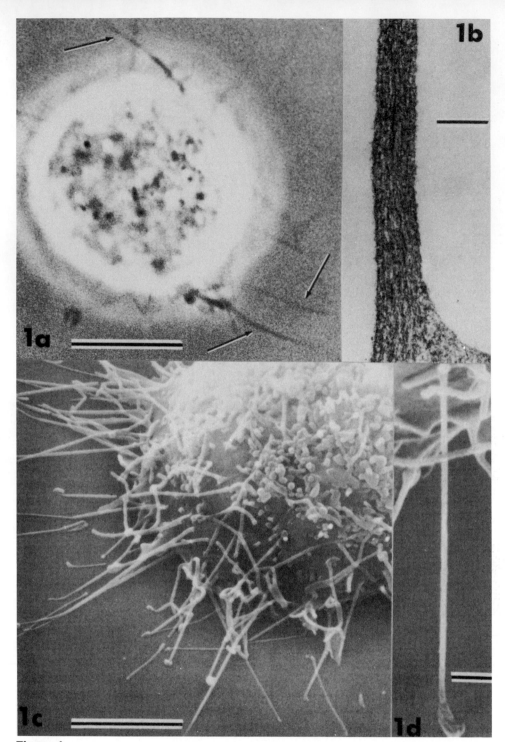

Figure 1

Morphology of filopodia in spreading 3T3 cells. (*a*) Live 3T3 cell in phase-contrast optics 35 min after plating. Arrows point to filopodia. Bar = 10 μm. (*b*) Transmission electron microscope image of an attached filopodium showing the parallel microfilament arrangement. Bar = 0.2 μm. (Photo courtesy of Dr. R. D. Goldman.) (*c*) Scanning electron micrograph of a 3T3 cell 50 min after plating. Note the uniformity in the thickness of filopodia. Bar = 5 μm. (Tilt angle: 45°.) (*d*) Filopodium with bulky tip at higher magnification. Bar = 1 μm.

attached to the substratum and the particle flows centripetally along the surface of the filopodium (Fig. 2d,e). The speed of particle transport is about 0.2 μm/second. In later stages of spreading, particles are also transported along lamellipodia towards the cell body, as was observed earlier (Abercrombie and Heaysman 1954; Harris and Dunn 1972) in fully spread out cells.

The two observed modes of particle removal by filopodia may be two different expressions of a common mechanism: (1) The backflow of particles along a stretched out filopodia may be considered as a special case of the retraction of the filopodia as a whole together with the particles. If two thin filopodia happen to be so close to each other that they cannot be separated in light microscopy, the retraction of one of them with a particle may appear as a flow of the particle along the other filopodium. It may also be that one filopodium can split its microfilament bundle along its long axis and only half of it retracts carrying a particle with it. Or (2) the retraction of a whole filopodium with a particle may be considered as a special case of the backflow mode. If the tip of a filopodium comes loose while a particle flows back along its outside, the forces that produce the backflow may retract the loosened filopodium as a whole. We cannot at present decide between these alternatives.

Most of the back-transported particles become internalized but are exocytozed again 60–120 minutes after plating (Albrecht-Buehler and Goldman 1976). Within 30–40 minutes, the filopodia remove the particles within a circular area 30–40 μm in diameter. While this happens, the cells remain rounded up, which supports the observation with live cells that no lamellipodia are involved in the early stages of particle removal. Most of the cells are found in the center of the cleaned area, indicating that the filopodia that grow out all around the cellular perimeter are equally active.

Approximately 95% of trypsinized cells of unsynchronized 3T3 cell populations remove the gold particles during early spreading. Therefore the effect is independent of the time point in the cell cycle at which the cells were before trypsinization or treatment with EDTA. The phenotype of suspended, freshly plated cells, however, resembles mitotic cells in many ways. Therefore one may suspect that particle removal by filopodia is a characteristic of a mitosislike state of the cytoplasm that is enforced by suspending agents such as trypsin or EDTA.

Since the particles are removed by filopodia in the case of 3T3 cells, the particle removal within 50 minutes after plating can be used as an assay for their retractile function. Table 1 shows the dependencies of filopodium function we have investigated so far. In all cases, the effect of the inhibitory agent is readily reversed by placing the cells back into normal culture conditions. Table 1 shows the pronounced pH and temperature dependence of particle removal. Colchicine does not affect particle removal, indicating that there is no major involvement of microtubules in the removal mechanism.

With the exception of very low temperature (4°C), none of the inhibitory conditions tested could produce cells that were free of filopodia 30 minutes after attachment. On the contrary, in the presence of cytochalasin B, in phosphate-buffered saline as extracellular medium, and at a temperature of 22°C we observed an increased production of filopodia while particle removal was

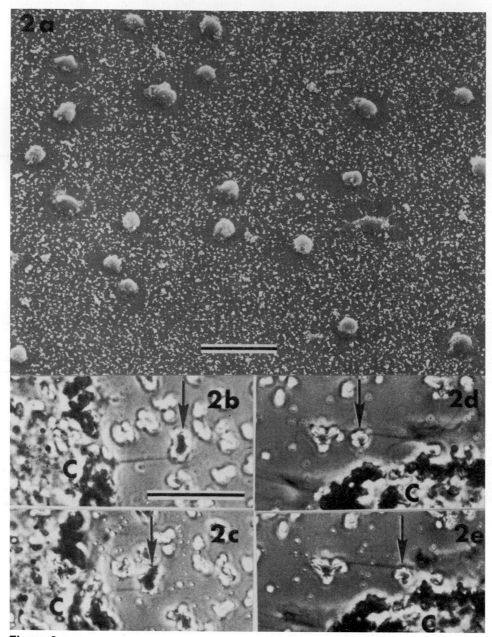

Figure 2

Gold particle removal by filopodia during early spreading of 3T3 cells. (*a*) Scanning electron micrograph of a whole field of cells 50 min after plating. Bar = 50 μm. (Tilt angle: 45°.) (*b,c*) Retraction of a particle-attached filopodium as a whole. C indicates the cell body. Arrow points to the moving cluster of gold particles. Time difference between *b* and *c* is 1 min. Bar in *b* = 10 μm. (*d,e*) Centripetal flow of a particle cluster along a stretched-out filopodium. Time difference between *d* and *e* is 2 min.

252

Table 1
Parametric Dependencies of Particle Removal

Parameter	Inhibition	No inhibition
Temperature	4–26°C	28–37°C
pH	5.8–6.6	7.0–9.0
EGTA	—	[EGTA] = 0–2.5 mM
Calf serum	[serum] = 0%	[serum] = 1–20%
Cytochalasin B	[CB] = 2–10 μg/ml	[CB] = 0–1 μg/ml
Colchicine	—	[colch] = 0–20 μg/ml
dBcAMP[a]	—	[dBcAMP] = 1 mM

[a] Experiment included a 16-hr preincubation of the cells in 0.1 mM dBcAMP.

inhibited. In Figure 3a, which shows the increased production of filopodia in the presence of cytochalasin B, filopodia can be seen beginning to grow out even from the dorsal side of the cell. The inhibition of particle removal by cytochalasin B is accompanied by a loss of the parallel-oriented microfilaments inside the filopodia. Instead, 100-Å thick filaments and an unspecified granulated material appear (Fig. 3b).

Modes of Filopodia Movement

Several distinct modes of movement have been described by Wessels, Spooner and Luduena (1973) for "microspikes" at the growth cone of neurites. We have observed basically identical modes in spreading 3T3 cells.

1. *Outgrowth of filopodia* (Fig. 4a,b). The formation of a filopodium takes only a few seconds.
2. *Lateral waving of filopodia* (Fig. 4c,d). This is also a very fast movement, which rarely leaves the filopodium in the plane of focus. The filopodium remains rigid during the waving.
3. *Retraction of filopodia and possibly related movements.* This mode may start with a sudden backfolding of the filopodium tip (Fig. 4g). Immediately afterwards, a droplet of phase-dense material is formed which moves quickly towards the cell body while increasing in size. The retraction of free filopodia is rare. More often, an attached filopodium suddenly comes loose and then retracts by forming a droplet (Fig. 4e,f). Sometimes the movement of the droplet stops before it reaches the cell body, and the filopodium continues the waving motion until the droplet happens to touch the cell body. Subsequently, the half-retracted filopodium may fuse with the cell body. With cells at later stages of spreading, we observed several filopodia bending at one or more points on the filopodium (Fig. 4h). After a droplet is formed at these points (right arrow in Fig. 4h), the sections between bending points remain rigid and move freely around these points. Often a droplet is formed upon the accidental contact of two filopodia (Fig. 4i). Subsequently, one or both filopodia retract. The

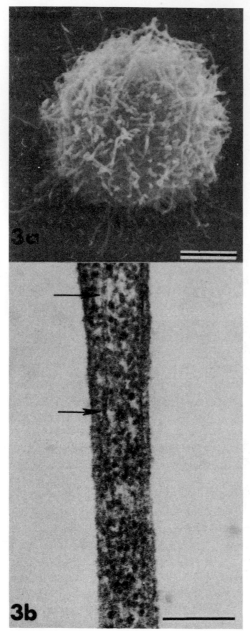

Figure 3
(a) Enhanced production of filopodialike structures by freshly plated 3T3 cells in the presence of 5 μg/ml cytochalasin B. Bar = 5 μm. (b) Loss of parallel-oriented microfilament bundles in the presence of 5 μg/ml cytochalasin B (cf. Fig. 1b). Bar = 0.2 μm. Arrows point to one 100-Å filament. (Photo courtesy of Dr. R. D. Goldman.)

direct observation of filopodium retraction gives the impression of a local disintegration of an otherwise rigid structure.

Tentative Interpretation of Particle Removal

Estimate of Forces Involved in Particle Removal

Particle-attached filopodia obviously produce a retraction force that transports the particles to the cell surface. For this process, the filopodia have to overcome at least four forces: the adhesion between particles and substrate, the friction of the moving particles in the medium, the internal friction of the retracting filopodium, and the weight of the particles (in case one component of their movement is directed against gravity).

The friction of a moving particle in the medium can be calculated by Stokes' law:

$$F_1 = 6\pi\eta r u,$$

where η is the viscosity of the medium (0.01 poise), r, the radius of the particle or particle clusters (0.5–2 μm), and u, the particle speed (0.2 μm/ sec), which yields $F_1 < 8 \times 10^{-10}$ dyne.

The weight of a particle (neglecting buoyancy) is

$$F_2 = \frac{4\pi}{3} r^3 \delta g,$$

where δ is the density of gold (19.3 g/cm^3) and g is the earth's acceleration (981 cm/sec^2), which yields $F_2 < 8 \times 10^{-8}$ dyne.

At most, 10% of this force is required for particle movement, since it occurs mostly parallel to the substratum and at right angles to the direction of gravity.

We measured the adhesion force (F_3) between particles and substratum by removing the particles by centrifugation of particle-coated cover slips at varying g values at 26–37°C. The number of particles that remained attached after each centrifugation was determined by photometric measurements of the light scatter of the cover slip. We found that 50% of the particles were removed at an acceleration of 1550g if the centrifugal force was oriented parallel to the cover slip (sliding force) or at 4700g if it was oriented vertically (detachment force). For particle clusters of 1-μm diameter, these values yield 1.6 $\times$ 10^{-5} dyne for the sliding force and 4.8 $\times$ 10^{-5} dyne for the detachment force. Considering that the breaking of the adhesion force between particle and substratum requires at least a thousand times greater force than is required for the particle back transport, one would expect a rapid snapping back of the particle-attached filopodium once the particle comes loose. We did not observe such a movement and have therefore concluded that the speed-limiting force during particle removal is the internal friction of the retracting filopodium. There is, however, the alternative possibility that the particle-substrate adhesion is not broken by mechanical forces but by a chemical action of the filopodium tip. So far, there is no evidence for such an action of filopodia.

The mechanism of force production is also unknown. Bray (1974) has

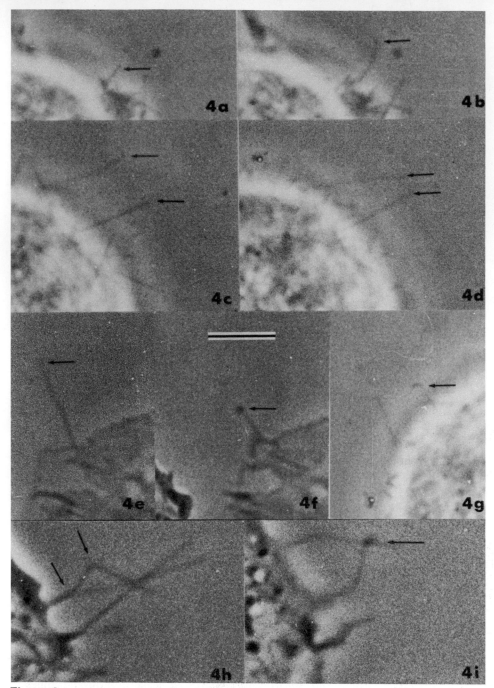

Figure 4

Modes of filopodium movement. Bar = 5 μm. (*a, b*) Outgrowth of a filopodium (arrows). Time difference between *a* and *b* about 5 sec. (*c, d*) Waving of filopodia. The filopodia remain rigid during this mode (arrows). Time difference between *c* and *d* about 5 sec. (*e, f*) Retraction of a filopodium. A droplet is formed at the tip, which then moves toward the cell body. (*g*) Bending at the tip before retraction. (*h*) Breaking of a filopodium into three rigid sections that move freely around the breaking points (arrows). The breaking point at the lower arrow is not a branching point. The impression of branching is due to an optical coincidence with another filopodium. (*i*) Droplet formation (arrow) upon accidental contact of two filopodia, both of which have breaking points.

suggested a model of filopodia movement based on an actomyosin sliding-filament mechanism. However, it is not yet known whether filopodia contain actomyosin. One may try to estimate the maximal force that one filopodium could produce if it were a striated muscle fiber 0.2 μm in diameter producing an optimal tetanic tension of 2.8 Kp/cm^2 (Gordon, Huxley and Julian 1966). Under this assumption, one obtains a maximal value of the force produced by an actomyosin sliding-filament mechanism (F_4) of about 10^{-3} dyne.

However, there is another possible mechanism of force production, which we presently tend to favor in view of the possibility that retracting filopodia "fluidify." Once the material of the filopodium is fluidified, the surface tension of the material would produce a retraction force, F_5:

$$F_5 = 2R\pi\sigma,$$

where R is the radius of the filopodium (0.1 μm) and σ is the surface tension between the fluidified material and the culture medium. Even a small surface tension of the fluidified material of only 30 dynes/cm (benzene-air) would produce a retraction force of 2×10^{-3} dyne.

Interpretation of Particle Retraction

No matter how the retraction force is produced, the suspicion arises that filopodia display the same force if they attach to an *extended* solid substratum. In other words, we assume that the filopodia that removed the gold particles did so because they "mistook" them for an extended solid substratum and produced a retraction force. Considering the importance of anchorage for the proliferation of 3T3 and other cells (Stoker 1968; Pollack et al. 1974), one feels tempted to interpret this reaction of filopodia to contact with a solid object as a test of the firmness of an anchorage point by a tentative pulling. The retraction force seems to be associated with, if not produced by, an at least partial fluidification of the filopodium material, as the backflow of particles along stretched-out filopodia (see Fig. 2c,d) and the "melting" of retracting filopodia suggest.

Possible Exploratory Function of Filopodia

If the retraction force detaches a previously attached filopodium, the filopodium will retract to the cell body (see Fig. 4e,f). If it remains attached, the cell begins to extend a lamellipodium along the attached filopodium. This can be observed during the normal spreading of 3T3 cells (Albrecht-Buehler and Goldman 1976), but perhaps more clearly on a partially gold-coated cover slip. The technique of evaporating metal through the openings of an electron microscope grid, thus obtaining regular metallic squares on a nonmetallic substratum, was introduced by Carter (1967) and Harris (1973). After evaporation of gold, the cover slips have to be sintered in order to prevent detachment of the metal in aqueous media. The sintering happens to produce gold particles of about the same size as the loose particles used in the removal experiments (see Fig. 5c). In this case, however, the particles are too tightly attached to the glass substrate to be removed by filopodia. We observed freshly plated 3T3 cells attached to the glass

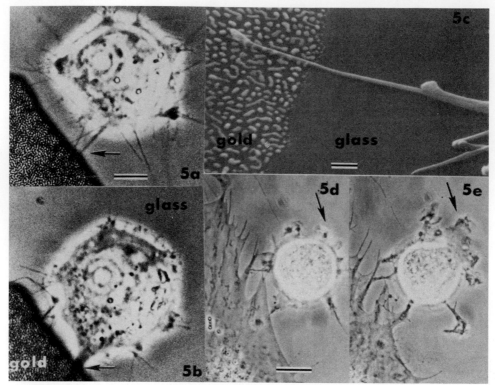

Figure 5

Exploratory function of filopodia. (*a*) Glass-attached 3T3 cell 27 min after plating, showing filopodia contacting a gold-plated area (arrow). Bar = 5 μm. (*b*) The same cell 34 min after plating, showing the first lamellipodia extended towards the gold-plated area. (*c*) Filopodium of spreading 3T3 cell reaching over to a gold-plated area in the presence of 5 μg/ml cytochalasin B. Note the granulated structure of the gold area after the sintering process. Bar = 1 μm. (Tilt angle: 45°.) (*d*) A 3T3 cell 28 min after plating contacting, by filopodia, another spread-out 3T3 cell plated 24 hr before. Bar = 10 μm. (*e*) The same cells shown in *d* but 40 min after plating. The first large lamellipodia have extended away from the spread-out cell (arrows).

area of a so prepared cover slip with a gold area within reach of filopodia. If at least one filopodium attached to the adjacent gold area, presumably producing a retraction force, then 4–6 minutes later a lamellipodium extended towards the gold area, provided the filapodium did not come loose. During this time no lamellipodia extended towards the glass area, even though the 3T3 cells were able to spread on glass and the cell body had no other contact with the gold area than by filopodia (Fig. 5a,b). We take this finding as a strong suggestion for the capacity of filopodia to "detect" areas of a certain, as yet unknown, quality of the substratum in the vicinity of a spreading 3T3 cell and direct lamellipodia towards them (see also Albrecht-Buehler 1976).

In another type of experiment we observed freshly plated 3T3 cells that were in contact with another already spread 3T3 cell by filopodia. We found that the first lamellipodia of the freshly plated cells extended preferentially away from the already spread cell (Albrecht-Buehler 1976). In Figure 5, d and e show an example of this observation. Recalling that the described directionally differentiated extension of lamellipodia following the attachment of filopodia is preceded by a rapid scanning motion of the free filopodia, we suggest that filopodia also have the function of exploring the nonfluid environment of the still rounded cell. The term "exploration" has been used before by Gustavson and Wolpert (1961) in expressing their impression of the action of filopodia in migrating primary mesenchyme cells.

Possible Significance of the Retraction Force for the Exploratory Function of Filopodia

Cytochalasin B (CB) inhibits reversibly the removal of gold particles around a spreading 3T3 cell but not the formation of filopodia. This indicates that the attachment of particles and/or the capacity of filopodia to retract is inhibited by CB. Live cell observations confirm that at least the retractility of filopodia is inhibited by CB.

We found that the preferential extension of lamellipodia towards an adjacent gold area and the lamellar extension away from an already spread out 3T3 cell were both inhibited by CB and recovered rapidly after removal of CB from the culture medium (Albrecht-Buehler 1976) (Figs. 6, 7). The results suggest that the capacity of filopodia of a spreading 3T3 cell to attach and retract may be a necessary first step in the chain of events that eventually leads to the lamellar extension.

Of course there is also the alternative possibility that CB inhibits only the formation of lamellar structures. At present we cannot distinguish between these two possibilities. However, under all conditions where particle removal was inhibited, lamellar extension was inhibited, although filopodia had grown out (see above). This finding may indicate a general link between both phenomena.

Figure 8 summarizes by means of a flow diagram our present view of the basic logic that we assume to underly the described exploratory function of filopodia. Although this diagram emphasizes the role of filopodia in the "exploration" of the nonfluid environment of a spreading cell, we do not intend to exclude the possibility that ruffling lamellipodia, blebs and microvilli perform similar functions. However, filopodia can be studied more conveniently than the other surface extensions of animal cells.

SUMMARY

Filopodia of spreading 3T3 cells appear as single microfilament bundles coated by plasma membrane. They perform a rapid scanning motion before

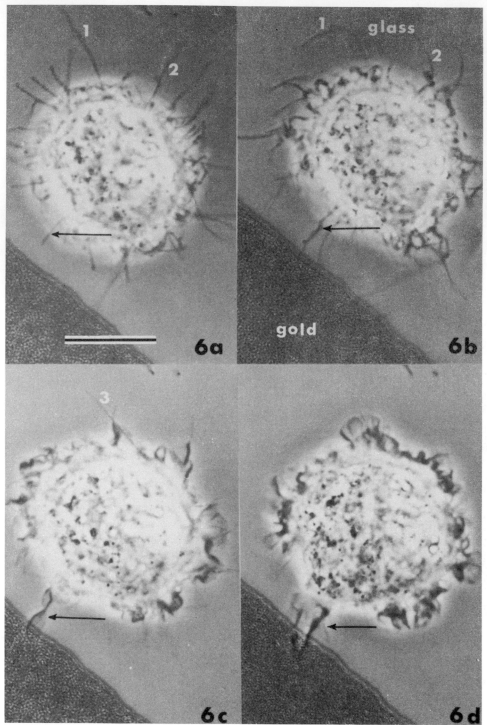

Figure 6

Restoration in live cells of the "recognition" of the gold area after removal of cyto-chalasin B from the medium. Phase-contrast microscopy. Bar in $a = 10$ μm. The cell had been in medium containing 5 μg/ml cytochalasin B for 35 min. (*a*) 24 min after removal of CB. Filopodia are poorly attached (cf. positions of filopodia in *a* and *b*) and thickened (see *c*). A new filopodium grows out (arrow). (*b*) 26 min after removal of CB. The new filopodium has attached to the gold area and assumed a state inter-mediary between filopodium and lamellipodium. Other filopodia are still bending (cf. filopodia 1 and 2 in *a* and *b*). (*c*) 29 min after removal of CB. The first lamellipodium has flown to the gold area, immediately followed by lamellipodia all around the cell. New, thin and rigid filopodia can be seen (spike 3). (*d*) 33 min after removal of CB. Cell is in normal spreading state with large lamellipodium reaching over to the gold area.

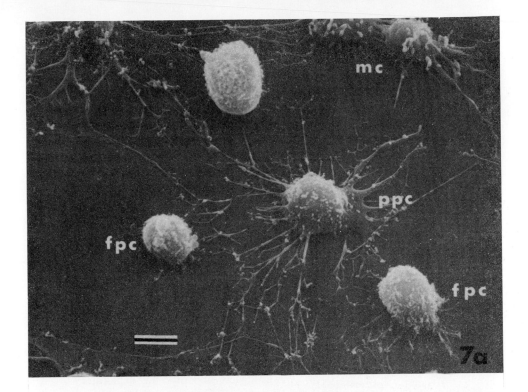

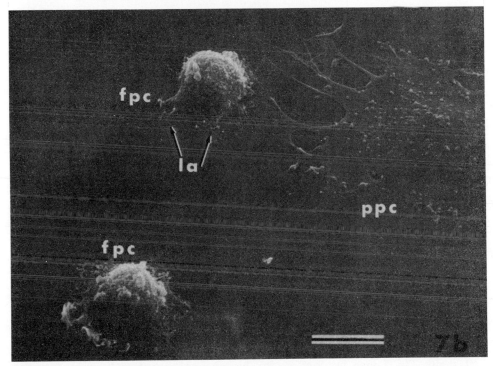

Figure 7

Scanning electron microscopy of the restoration of the "recognition" of another spread-out cell by a spreading cell after inhibition by cytochalasin B. (For live cell observations see Albrecht-Buehler 1976). Bars = 10 μm. (Tilt angle: 45°.) (*a*) Freshly plated cells (fpc) and previously plated cells (ppc) after 35 min in medium that contained 5 μg/ml cytochalasin B. Note the arborized shape of the ppc and the difference between mitotic cells (mc) and fpc's. (*b*) Duplicate of the preparation, transferred into normal medium for 10 min after 35 min incubation in cytochalasin B-containing medium. Lamellipodia (la) of an fpc have flown to the side away from a ppc, whereas the fpc at the lower left-hand corner spread symmetrically.

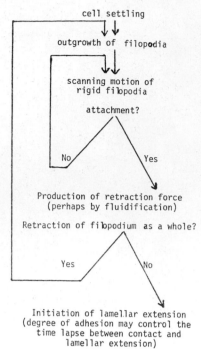

Figure 8

Flow diagram of the basic logic that we suggest underlies the exploratory function of filopodia in 3T3 cells.

they attach to the substratum. If the substratum is densely and evenly covered with colloidal gold particles, the filopodia of a freshly plated 3T3 cell remove the particles in a circular area around the cell before any lamellipodia are extended. If such a cell has settled on a glass substratum and its filopodia have made contact to the glass and to an area of evaporated gold in the glass, lamellipodia preferentially extend towards the gold. If the filopodia of a freshly plated 3T3 cell have made contact to another spread out 3T3 cell and to glass, the first lamellipodia extend preferentially away from the spread out cell. We suggest that the filopodia of freshly plated 3T3 cells produce a retraction force upon contacting an object. If the contact is firm enough so that the retraction force does not shorten the filopodium, then the cell extends a lamellipodium along this filopodium. In this sense, the filopodium of spreading 3T3 cells may be considered as cellular organs that "detect" a certain, as yet unknown, quality of the nonfluid environment of the cell.

Acknowledgments

I thank Dr. James D. Watson for his encouragement and support, and Drs. H. Green (MIT), R. D. Goldman (Mellon Institute), and E. Lazarides (University of Colorado) for valuable suggestions and discussions. I am grateful to Dr. R. E. Pollack for the use of his cell culture facilities and to R. Lancaster and Miss S. Chait for their excellent technical help. The work reported in this paper was partly carried out during the tenure of a Research Training Fellowship awarded by the International Agency for Research on Cancer, and was supported by Grant CA 13106 from the National Cancer Institute and by the Grants GB 38415 and BMS75-09539 from the National Science Foundation.

REFERENCES

Abercrombie, M. and T. E. M. Heaysman. 1954. Observations on the social behaviour of cells in tissue culture. II. Monolayering of fibroblasts. *Exp. Cell Res.* **6**:293.

Albrecht-Buehler, G. 1976. Filopodia in spreading 3T3 cells. Do they have a substrate-exploring function? *J. Cell Biol.* **69**:275.

Albrecht-Buehler, G. and R. D. Goldman. 1976. Microspike-mediated particle transport towards the cell body during early spreading of 3T3 cells. *Exp. Cell Res.* **97**:329.

Albrecht-Buehler, G. and M. M. Yarnell. 1972. A quantitation of movement of marker particles in the plasma membrane of 3T3 mouse fibroblasts. *Exp. Cell Res.* **78**:59.

Allison, A. C., P. Davies and S. de Petris. 1971. Role of contractile microfilaments in movement and endocytosis. *Nature New Biol.* **232**:153.

Bray, D. 1974. Model for membrane movements in the neural growth cone. *Nature* **244**:93.

Buckley, J. K. 1975. Three dimensional fine structure of cultured cells: Possible implications for subcellular motility. *Tissue and Cell* **1**:51.

Carter, S. B. 1967. Haptotaxis and the mechanism of cell motility. *Nature* **213**: 256.

Cornell, R. 1969. In situ observations on the surface projections of mouse embryo fibroblasts. *Exp. Cell Res.* **57**:86.

Dahlen, H. and P. D. Scheie. 1967. Two types of long microextensions from cultivated liver cells. *Exp. Cell. Res.* **53**:670.

DeRobertis, E. and J. R. Setelo. 1952. Electron microscope study of cultured nervous tissue. *Exp. Cell Res.* **3**:433.

Dunn, G. A. 1973. Extension of nerve fibres, their mutual interaction and direction of growth in tissue culture. In *Locomotion of tissue cells. Ciba Foundation Symposium* (ed. M. Abercrombie), vol. 14, p. 211. Associated Scientific Publishers, New York.

Fisher, H. W. and T. W. Cooper. 1967. Electron microscope studies of the microvilli of Hela cells. *J. Cell Biol.* **34**:569.

Follet, E. A. C. and R. D. Goldman. 1970. The occurrence of microvilli during spreading and growth of BHK21/C13 fibroblasts. *Exp. Cell Res.* **59**:124.

Gey, G. O. 1956. Some aspects of the constitution and behaviour of normal and malignant cells maintained in continuous culture. In *The Harvey Lecture Series*, vol. 50, p. 154. Academic Press, New York.

Gordon, A. M., A. F. Huxley and F. J. Julian. 1966. Tension development in highly stretched vertebrate muscle fibres. *J. Physiol.* **184**:143.

Gustavson, T. and L. Wolpert. 1961. Studies on the cellular basis of morphogenesis in the sea urchin embryo. *Exp. Cell Res.* **24**:64.

Harris, A. 1973. Behaviour of cultured cells on substrata of variable adhesiveness. *Exp. Cell Res.* **77**:285.

264   G. Albrecht-Buehler

Harris, A. and G. Dunn. 1972. Centripetal transport of attached particles on both surfaces of moving fibroblasts. *Exp. Cell Res.* **73**:519.

Knutton, S., M. C. B. Sumner and C. A. Pasternak. 1975. Role of microvilli in surface changes of synchronized P815Y mastocytoma cells. *J. Cell Biol.* **66**: 568.

Partridge, T. and P. S. Davies. 1974. Limpet haemocytes. II. The role of spikes in locomotion and spreading. *J. Cell Sci.* **14**:319.

Pollack, R. E., R. Risser, S. Conlon and D. Rifkin. 1974. Plasminogen activator production accompanies loss of anchorage regulation in transformation of primary rat embryo cells by SV40 virus. *Proc. Nat. Acad. Sci.* **71**:4792.

Porter, K. R., A. Claude and E. F. Fullam. 1945. A study of tissue culture cells by electron microscopy. *J. Exp. Med.* **8**:233.

Porter, K. R., V. Fonte and G. Weiss. 1970. A scanning microscope study of the topography of Hela cells. *Cancer Res.* **34**:1385.

Revel, J. P. and S. S. Brown. 1976. Cell junctions in development, with particular reference to the neural tube. *Cold Spring Harbor Symp. Quant. Biol.* **40**:443.

Revel, J. P., P. Hoch and D. Ho. 1974. Adhesion of culture cells to their substratum. *Exp. Cell Res.* **84**:207.

Stoker, M. 1968. Abortive transformation by polyoma virus. *Nature* **218**:234.

Taylor, A. C. and E. Robbins. 1963. Observations on microextensions from the surface of isolated vertebrate cells. *Devel. Biol.* **7**:660.

Trelstad, R. L., E. D. Hay and J. P. Revel. 1967. Cell contact during early morphogenesis in the chick embryo. *Devel. Biol.* **16**:78.

Trinkaus, J. P. 1973. Surface activity and locomotion of Fundulus deep cells during blastula and gastrula stages. *Devel. Biol.* **30**:68.

Trinkaus, J. P. and C. A. Erickson. 1974. Microvilli as a source of reserve surface membrane during cell spreading. *J. Cell Biol.* **63**:351a.

Weiss, P. 1961. From cell to molecule. In *The molecular control of cellular activity* (ed. J. Allen), p. 1. McGraw-Hill, New York.

Wessels, N. K., B. S. Spooner and M. A. Luduena. 1973. Surface movements, microfilaments and cell locomotion. In *Locomotion of tissue cells. Ciba Foundation Symposium* (ed. M. Abercrombie), vol. 14, p. 53. Associated Scientific Publishers, New York.

Witkowski, J. A. and W. D. Brighton. 1971. Stages of spreading of human diploid cells on glass surfaces. *Exp. Cell Res.* **68**:372.

Yamada, K., B. S. Spooner and N. K. Wessels. 1971. Ultrastructure and function of growth cones and axons of cultured nerve cells. *J. Cell Biol.* **49**:614.

Zucker, M. B. 1974. The value of the blood platelet in hemostasis and as a research tool. *Trans. N.Y. Acad. Sci.* (Ser. II) **36**:561.

Actin in Dividing Cells: Evidence for Its Role in Cleavage but Not Mitosis

Thomas E. Schroeder

Friday Harbor Laboratories, University of Washington
Friday Harbor, Washington 98250

Cell division in animal cells has two well-known aspects, mitosis and cell cleavage. In the simple operational sense it is the mitotic apparatus (MA) that accomplishes chromosome motion and the contractile ring (CR) that causes cleavage constriction. The movements of chromosomes and deformation of the cell during cleavage are expressions of mechanical forces generated by or applied within the MA and the CR. There is ample evidence that the forces of mitosis are totally separate and distinct from the forces of cleavage, even though the two processes are efficiently coupled in the cell. The search for the separate mechanisms of mitosis and cell cleavage involves attempts to identify the relevant structures, biochemical substances and physiological events that produce the motive force in each case. A proper understanding of those mechanisms will emerge from consistent and clear patterns of congruence between the many facets of the events: the cell's structural organization, the functional biochemistry of its parts, their physiological requirements, and biomechanical interrelations between primary forces and resulting movements of cell parts. We are approaching, but have not yet reached, a sophisticated understanding of cell division in these various aspects.

I believe that there are a few general principles operating during cell division and that some valuable guidelines for approaching the problem can be extracted from them.

1. All cells possess, as a common denominator of their shared ancestry, the same essential mechanism for mitotic chromosome motion. Likewise all animal cells cleave by utilizing one common mechanism. In each case, the primary force-producing system has probably undergone rather little alteration during evolution—although it is correspondingly reasonable to expect evolutionary modifications of accessory or modulating aspects of the systems.

2. Equal in importance with the primary force-producing mechanisms of either mitosis or cleavage are biomechanical "governors" which guarantee that the motive forces are delivered in the proper place, direction and

265

time. These regulatory agencies may be so intimately associated with the motive agencies as to be difficult to distinguish or they may be physically quite removed.

3. The primary physical processes are active, vectorial mechanical forces of the familiar tensile or elastic type. However, accessories to the system may apply secondary, passive forces that redirect the primary forces and thereby introduce complex relations between forces and observed movements.

4. Once begun, chromosome motion and cleavage constriction proceed to completion under very favorable thermodynamic conditions. Natural selection would quickly eliminate irresolute, unreliable or imprecise mechanisms of cell reproduction.

These are all a priori assertions. But if such principles do operate in cells, then we would do well to incorporate them into our thinking and in the direction and practice of our investigations. Correct or not, they point out the importance of (1) giving careful attention to the *precise localization* (in both space and time) of each structure, substance and event under study; (2) accurately assessing the normal *sequence of events* in dividing cells; and (3) recognizing the extent of natural diversity and threads of similarities among dividing cells of different kinds—in other words, it is important to adopt a broad *comparative perspective* in interpreting our findings.

Lately we have been faced with circumstantial evidence that places actin at or near the scenes of both chromosome motion and cleavage constriction. Is the presence of actin causal or incidental? Furthermore, what are the criteria for deciding such questions? My purpose in this article is to evaluate and discuss the possible functional roles of actin in the MA and CR during cell division. To what extent have methods for detecting and localizing actin advanced our understanding of cell cleavage or mitosis?

Actin in the Contractile Ring

The contractile ring (CR) is an operational term, but one with increasingly definite structural and functional overtones. It refers to the specialized cytoplasm underlying the bottom of the cleavage furrow and is thought to account for the known tensile forces expressed there. These forces measure 10^{-3} to 10^{-2} dynes (Rappaport 1967; Hiramoto 1975). Based on manipulations of living cells and light microscopic observations, there is no doubt that the cortical cytoplasm of a cleavage furrow behaves uniquely. However, unique features of structural organization are revealed only by electron microscopy.

The CR is fundamentally a well-defined array of microfilaments (see Fig. 1; and, for example, Schroeder 1975). Important features of the contractile ring include: (1) its specific location beneath the plasmalemma of the concave portion of the cleavage furrow; (2) the circumferential orientation of its constituent microfilaments; (3) its transient existence, which coincides precisely with the brief period of cell constriction (often a period of about 10 min); and (4) its occurrence in many quite different cell types (Figs. 2–6).

The hypothesis that the CR is directly responsible for generating the force of cleavage constriction is supported by a considerable amount and variety of evidence, which falls into two general categories. The distinctive geometrical

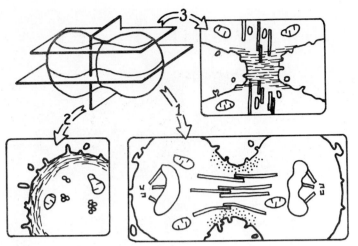

Figure 1
Diagram showing three critical planes of section through a cleaving cell, illustrating different views of the relationships between the contractile ring and mitotic apparatus. Plane 1 is a mid-longitudinal section; mitotic microtubules are seen lengthwise, and microfilaments beneath the cleavage furrow are cut across (dots). Plane 2 is an equatorial section in which microfilaments are curving lines and microtubules are cross-sectioned profiles. Plane 3 grazes the cleavage furrow, exposing microfilaments in the contractile ring as parallel lines and microtubules coursing perpendicularly.

and temporal features of its overall structure, as already mentioned, correlate very consistently with the development of tensile force and cell deformation. In addition, experimental applications of selective inhibitors of furrowing result in a corresponding loss of normal CR structure. These inhibitors include hydrostatic pressure (Tilney and Marsland 1969) and cytochalasin B (Schroeder 1969, 1970, 1972). Mechanical disruption of the subfurrow cytoplasm (the putative CR) also results in rapid relaxation of tension in the cleavage furrow and subsequent failure of cleavage constriction (Rappaport 1966).

As with other instances of cell contractility, it is reasonable to turn to muscle biology in the search for clues to the molecular aspects of cell cleavage. From the outset (Schroeder 1968), CR microfilaments were suspected of being composed of actin (the tension-bearing elements in muscle), although the suspicion was supported only by dimensional similarities. Experiments with heavy meromyosin (HMM) have verified that suspicion, but the confidence behind that conclusion depends largely on an appreciation of the stringent conditions of the observations.

The pioneering work of Ishikawa, Bischoff and Holtzer (1969), using HMM as an ultrastructural probe for actin in non-muscle cells, revealed that HMM-decorated actin filaments occur widely in some cells. In some dividing cells, the entire cortex is filled with bundles of actin whose existence prior to extraction and exposure to HMM is questionable. Therefore to apply the method to the CR successfully, it was imperative to know that the specific microfilaments under examination actually belonged to the CR in every instance.

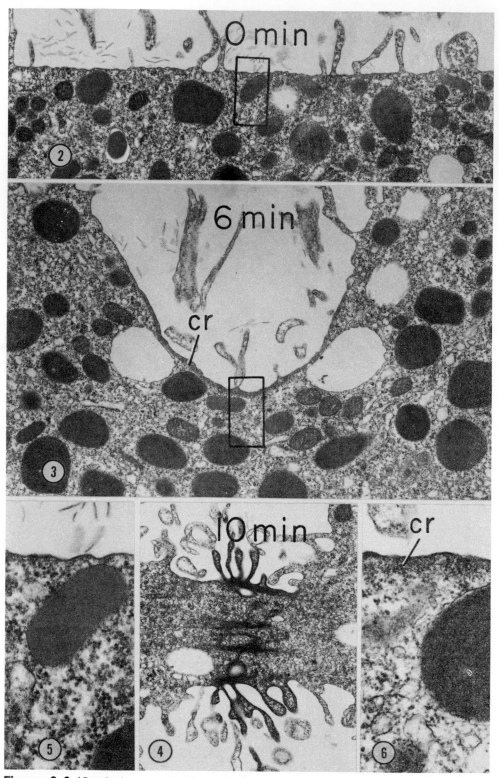

Figures 2–6 (*See facing page for legend*)

As the original method (Ishikawa, Bischoff and Holtzer 1969) did not preserve microtubules, the extraction and incubation method was abbreviated and performed at room temperature. Consequently, stem bodies (Stemkorper) of microtubules were retained as unequivocal landmarks of the equator of a cleaving cell, even if other geometrical cues were lost. Circumferentially aligned microfilaments occurring at the exact cell equator (judged by stem bodies) were attributed to the CR. In this way CR microfilaments in mammalian cells were found to combine with HMM under conditions that have been accepted as specific for actin (Schroeder 1973; see also Fig. 11). Other evidence that the CR contains actin has been demonstrated in amphibian eggs (Perry, John and Thomas 1971) and insect spermatocytes (Forer and Behnke 1972b).

Taken together, the structural, biochemical and biomechanical data form a consistent pattern that suggests a direct functional role of actin polymers in causing cleavage constriction. Physiological studies indicate that cleavage contraction requires ATP (Hoffmann-Berling 1954; Kinoshita and Hoffmann-Berling 1964) and ionic calcium (Baker and Warner 1972; Timourian, Clothier and Watchmaker 1972; Schroeder and Strickland 1974; Arnold 1975). In general, aspects of cell cleavage are reminiscent of muscle contraction, and it would not be surprising if myosin and other "muscle" proteins were functionally significant in the CR and that the cell possesses a special mechanism for providing the presumptive cleavage furrow with elevated free calcium.

Actin in the Mitotic Apparatus

There is considerable excitement (and rightly so) surrounding recent evidence for the presence of actin in the vicinity of the MA. The relevant data arise from three independent and sophisticated methods that have been applied to at least four cell types. However in their fine details, which are directly per-

Figures 2–6

(*Fig. 2*) Longitudinal section of *Arbacia* sea urchin egg at the end of anaphase, moments before cleavage begins. The field of view is precisely at the equator, which, although no contractile ring is yet visible, is the presumptive site of furrowing. Rectangle is Fig. 5. 13,000×.

(*Fig. 3*) Six minutes later the furrow is halfway through the egg. This longitudinal section shows the bottom of the cleavage furrow and the contractile ring (cr) in its characteristic location. This part of the cell corresponds exactly with that shown in Figs. 2 and 4. Rectangle is Fig. 6. 13,000×.

(*Fig. 4*) Cleavage constriction is complete about 10 min after it begins. Microtubules of the stem bodies are bunched together; the contractile ring is no longer evident; and the equatorial surface is irregularly folded and underlain by a curious density. 13,000×.

(*Fig. 5*) Higher magnification of part of Fig. 2, showing that no contractile ring exists at this stage. Compare with Fig. 6. 46,500×.

(*Fig. 6*) Part of Fig. 3, showing the contractile ring (cr) as it appears throughout cleavage. It resides immediately beneath the plasmalemma as an exclusion layer about 0.2 μm thick. 46,500×.

tinent to the mechanism of mitosis, the data are seriously contradictory and tend to nullify one another.

As a basis for discussion I have chosen to include here data from my own experiments which bear on this issue. These studies involved electron microscopic examination of HMM-treated HeLa cells at various stages of division, and I have attempted to apply the same standards of cell selection and localization used for the CR and mentioned above. In fact, the CR and MA studies were conducted in parallel. The conditions for HMM decoration were considered successful on the grounds that HMM-decorated filaments were present in the cortex at most mitotic stages.

Particular attention has been paid to the spindle (chromosome-to-pole) regions during metaphase, anaphase and telophase and to the interzone (between chromosomes) at anaphase and telophase. My evidence for actin in the MA is primarily negative in character, which makes it somewhat awkward to present.

At all stages there definitely are some HMM-decorated filaments scattered throughout the cytoplasm. A few—but only a few—solitary filaments can be seen in chromosome-to-pole regions; Figure 7 is typical in that it contains no evidence whatsoever of decorated filaments in the chromosome-to-pole region. The *best* evidence from my experiments of HMM-decorated filaments in the spindle is shown in Figure 8; in this metaphase cell, a few decorated filaments lie disoriented *in the vicinity* of the spindle fibers, and one or two are possibly associated with spindle microtubules. I have never seen bundles of filaments in this region, nor any HMM-decorated filaments in the areas of kinetochores or centrioles, regardless of mitotic stage.

At anaphase and telophase, HMM-decorated filaments are practically never seen in chromosome-to-pole regions in my preparations, even though highly decorated bundles lie in the cortex very nearby. On the other hand, somewhat more significant numbers of HMM-decorated filaments occur in the interzone, but these are mostly solitary and oriented at random (Fig. 7). Some are oriented parallel to interzone microtubules (Fig. 7), and this is *no more* apparent than in Figure 11, which I include here as my most positive evidence for actin bundles in the vicinity of the mitotic apparatus.

In Figure 11 (lower asterisk), a small bundle of HMM-decorated filaments appears to merge with a small sheaf of interzone microtubules. Although very intriguing, this image is unrepresentative, if not unique, among my observations.

Preparations that have been glycerinated but have not been treated with HMM (Fig. 9) also display fine, solitary microfilaments in the MA, indicating that in these cases HMM treatment is not necessarily revealing microfilaments not ordinarily present. Undecorated microfilaments are occasionally, though infrequently, in close association with microtubules.

These findings agree substantially with previous reports of HMM-decorated filaments in the MAs of insect spermatocytes (Gawadi 1971, 1974; Forer and Behnke 1972a) and mammalian cells (Hinkley and Telser 1974). It is relevant to my interpretations that these collective data concur in showing that HMM-decorated filaments in MAs are few in number, usually solitary, and present between chromosomes and poles predominantly at metaphase

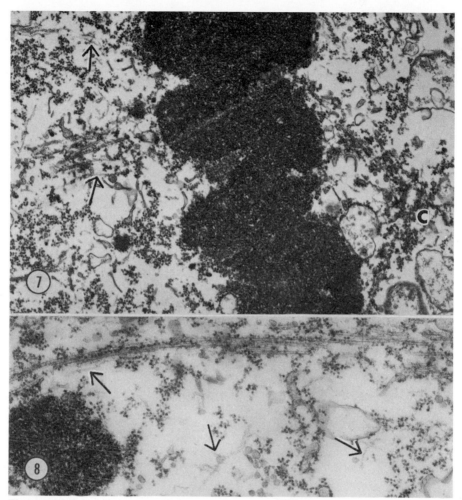

Figures 7–8

(*Fig. 7*) Typically, HMM-treated HeLa cells display little or no HMM-decorated actin in the region between chromosomes (dark) and the poles (c, for centriole). This is a longitudinally sectioned cell late in anaphase. HMM-decorated filaments are more frequently seen in the interzone, sometimes parallel to interzone microtubules (arrows), but never in large numbers. 31,850×.

(*Fig. 8*) As shown here, even at metaphase HMM-decorated filaments are few in number in the chromosome-to-pole regions, and many of these filaments are randomly oriented and solitary; one is seen coursing parallel to spindle microtubules (arrows). 35,770×.

and rarely thereafter and almost exclusively in the interzone during and after anaphase.

In short, HMM-decorated filaments in the MA either remain fixed in place during chromosome motion or are preferentially located *behind* the moving chromosomes. Their distribution therefore seems inconsistent with the idea of

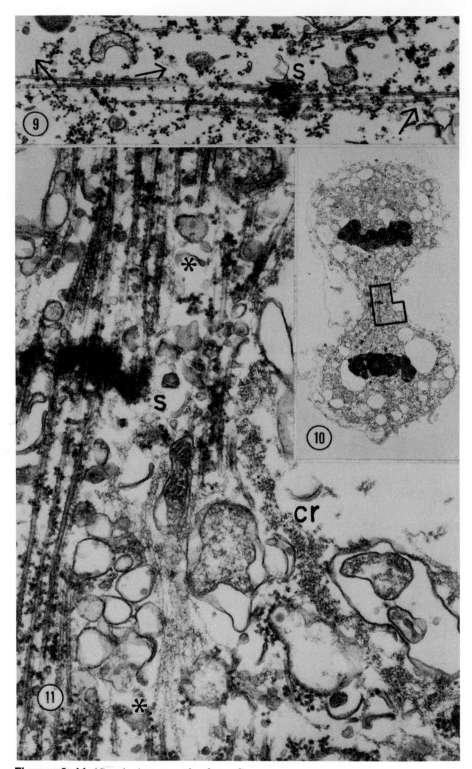

Figures 9–11 (*See facing page for legend*)

272

an actin-powered "tractor" capable of pulling chromosomes along microtubule paths from the metaphase plate to the poles.

The two light microscopic methods for detecting actin in dividing cells utilize fluorescent probes, either active HMM (Sanger 1975) or actin antibody (McIntosh et al., this volume). According to each of these methods, abundant fluorescence (hence actin) accompanies chromosome-to-pole fibers at all stages. Moreover, the regions of the kinetochores and centrioles are likewise distinctly fluorescent. Curiously, in these preparations there is little or no fluorescence in the cell cortex or interzone, where actin has been demonstrated by electron microscopy. To summarize the fluorescence data, virtually all of the actin associated with the MA appears *in front* of moving chromosomes.

The sharp contradictions between the fluorescence microscopic and electron microscopic methods for demonstrating actin in the MA are disturbing. There are good reasons to believe that each of these methods potentially can reveal all cytoplasmic actin, regardless of its polymerization state; yet the glaring disharmony in the results raises grave doubts that this is happening. Inconsistencies in the fundamental pattern of actin localization in the MA underscore the prematurity of claims that actin exerts a motile function in chromosome motion.

Cell Division in Muscle Cells

As a final note on the roles of actin in mitosis and cell cleavage it is perhaps valuable to consider instances in which cell division proceeds in cells known to be rich in actin and other contractile proteins. Many cells could be chosen, for these proteins have been biochemically identified in abundance in sea urchin eggs, amoebae and many other cell types (Pollard and Weihing 1974). For dramatic purposes, I can think of no better example than cell division in differentiated muscle cells.

Both cardiac and smooth muscle cells in developing chick embryos undergo mitosis and cell cleavage even after active myofibrils have been organized.

Figures 9–11
(*Fig. 9*) Longitudinal sections of telophase HeLa cells. Cell not exposed to HMM, showing a stem body (S) for reference and the presence of a few undecorated microfilaments (arrows) associated with interzone microtubules. 24,000×.

(*Fig. 10*) Entire telophase cell; enclosed area is shown enlarged in Fig. 11. 3,300×.

(*Fig. 11*) Interzone region from Fig. 10, confirmed by presence of stem body (S). This micrograph illustrates the *most suggestive* evidence from this study favoring the idea that HMM-decorated microfilaments are associated with interzone microtubules (asterisks). The contractile ring (cr) is also well decorated. Although this image of actin in the mitotic apparatus is intriguing, the small numbers of microfilaments, their exclusive presence *behind* moving chromosomes and the irregularity of the results suggest that actin may not be a significant agent in chromosome motion. 47,000×.

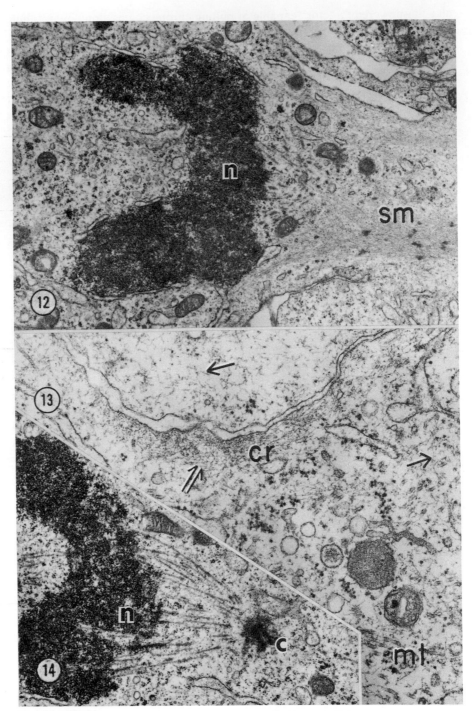

Figures 12–14

Longitudinal sections through smooth muscle cells of 7-day embryonic chicken gizzard, preserved by routine methods.

(*Fig. 12*) This early telophase cell demonstrates that myofibrils of these differentiating muscle cells already contain well-formed smooth myofibrils (sm), which are restricted to poleward regions of these elongate cells. Part of the telophase spindle can be seen emanating from the telophase nucleus (n). 34,300×.

In addition to actin, one may safely assume that such cells possess entire complements of contractile, regulatory and accessory substances needed for muscle contraction. Using routine ultrastructure methods I have begun to examine the question of whether the MA or CR is altered under such circumstances.

In the case of smooth muscle cells from the developing gizzard, smooth myofibrillar elements are segregated away from the region of the MA during division (Fig. 12). Throughout the cytoplasm, including the domain of the MA, there are short, solitary, unaligned strands that strongly resemble actin-containing microfilaments. Nevertheless, there is no preferential concentration, alignment or association of these putative actin microfilaments in any portion of the MA at any stage of division (Figs. 12 and 14). I include this as additional evidence against a significant role for actin in chromosome motion.

With regard to the CR in smooth muscle cells, its microfilaments appear normal in size, number and distribution (Fig. 13). There is no difficulty in distinguishing CR microfilaments from the solitary cytoplasmic ones, although they appear to be the same size. Some of the non-CR filaments accumulate in front of the advancing cleavage furrow, beneath the CR (Fig. 13); however, I believe that they are probably swept into that position by the act of furrow constriction.

CONCLUSIONS

In this article I have reiterated my conclusion that the actin microfilaments of the CR are actively involved in the mechanism of cell cleavage. A more rigorous treatment of the dynamics of the CR is presented elsewhere (Schroeder 1975), but the central factors in this judgment are the following: (1) the repeated confirmation in various cell types of (2) aligned microfilamentous structures occurring (3) in the appropriate location and (4) time which coincide with (5) known forces of constriction. These microfilaments resemble actin polymers in (6) dimensions and (7) by virtue of their specific interaction with HMM.

In the vast literature on non-muscle motility, there are relatively few instances where the specific involvement of a discrete population of microfilaments is unambiguously implicated in a natural contractile event. Ascidian

Figures 12–14 (*continued*)

(*Fig. 13*) The cleavage furrow of this cell displays a typical contractile ring (cr) despite an abundance of solitary actinlike filaments scattered throughout the cytoplasm (single arrows), some of which appear to be bunched together by the advancing furrow (double arrow). Microtubules of the mitotic apparatus are also present (mt), without any obvious association to the solitary actin filaments. 40,670×.

(*Fig. 14*) In this late anaphase cell, the entire chromosome-to-pole region is shown. Microtubules emanate from the nucleus (n) to the pole (c, for centriole), yet there is no clear association between myofibrillar elements or the solitary actin filaments and the microtubules, which might be expected if actin were functionally involved in chromosome motion. 17,000×.

tail retraction, as studied by Cloney (1966, 1969; and others), and the contractile ring are among these few cases and are characterized by contractile behavior that is "simple," i.e., linear shortening or circular constriction.

There is a disarmingly larger number of cellular phenomena that clearly involve actin, but the involvement is very often inconclusive with regard to contraction as such (i.e., shortening). Often there is little correspondence between the location, directionality and timing of microfilaments, on the one hand, and any easily specifiable contractile behavior, on the other. Sometimes the contractile behavior is not "simple" in its geometry or apparent mechanics. In other instances, actin appears to be involved in events that are patently noncontractile; it may form bundles that mechanically resist deformation (a "passive" role akin, perhaps, to rigor in muscle) or it may actually propel cellular extensions in the act of polymerizing. These roles for actin alert us to mechanisms quite different from the "contractile" role it serves in muscle.

In the case of mitosis, I am pessimistic that actin is functionally involved. Because mitotic chromosome motion is so essential to cell reproduction, and considering the sophistication of present methods for identifying actin in cells, I would expect that if actin were involved in force production that its location would be easily and reproducibly visualized. Since that is not the situation with the mitotic apparatus, the very scantiness and inconsistency of the evidence suggest that the entire matter is a tantalizing red herring.

Of course, the stories of actin in the MA or CR do not end here. Evidence for biochemical or biophysical interactions between actin and microtubules may yet reopen the question of actin's role in mitosis. With regard to cell cleavage it is well to consider the rapidity and specific location in which actin microfilaments appear at the equator during late anaphase. It is known that the mitotic apparatus supplies essential information that stimulates the furrowing process (Rappaport 1971), but the nature of the information is unknown. From what pool of actin subunits are CR microfilaments assembled? What is the trigger for their contraction? How do they become aligned? What is the nature of microfilament attachment to the membrane? What other components coexist with the actin in the CR?

The multiplicity of structural and functional states of actin is now widely acknowledged. How are actin-associated proteins, nucleotides and specific ions implicated in these transformations? The contractile ring is a choice example of a transient array of actin polymers that elicits a demonstrable contraction in specific coordination with other well-defined cellular events— chromosome motion. Understanding how a cell regulates and distributes its actin appears to be central to many issues of non-muscle motility.

Acknowledgment

Previously unpublished work reported here was supported by National Institutes of Health Grant GM 19464.

REFERENCES

Arnold, J. M. 1975. An effect of calcium in cytokinesis as demonstrated with ionophore A23187. *Cytobiologia* **11**:1.

Baker, P. F. and A. E. Warner. 1972. Intracellular calcium and cell cleavage in early embryos of *Xenopus laevis*. *J. Cell Biol.* **53**:579.

Cloney, R. A. 1966. Cytoplasmic filaments and cell movements: Epidermal cells during ascidian metamorphosis. *J. Ultrastruc. Res.* **14**:300.

————. 1969. Cytoplasmic filaments and morphogenesis: The role of the notochord in ascidian metamorphosis. *Z. Zellforsch.* **100**:31.

Forer, A. and O. Behnke. 1972a. An actin-like component in spermatocytes of a crane fly (*Nephrotoma suturalis* Loew). The spindle. *Chromosoma* **39**:145.

————. 1972b. An actin-like component in spermatocytes of a crane fly (*Nephrotoma suturalis* Loew). The cell cortex. *Chromosoma* **39**:175.

Gawadi, N. 1971. Actin in the mitotic spindle. *Nature* **234**:410.

————. 1974. Characterization and distribution of microfilaments in dividing locust testis cells. *Cytobios* **10**:17.

Hinkley, R. and A. Telser. 1974. Heavy meromyosin-binding filaments in the mitotic apparatus of mammalian cells. *Exp. Cell Res.* **86**:161.

Hiramoto, Y. 1975. Force exerted by the cleavage furrow of sea urchin eggs. *Devel. Growth Diff.* **17**:27.

Hoffmann-Berling, H. 1954. Die glyzerin-wasserextrahierte Telophasezelle als Modell der Zytokinese. *Biochim. Biophys. Acta* **15**:332.

Ishikawa, H., R. Bischoff and H. Holtzer. 1969. Formation of arrowhead complexes with heavy meromyosin in a variety of cell types. *J. Cell Biol.* **43**:312.

Kinoshita, S. and H. Hoffmann-Berling. 1964. Lokale Kontraktion als Ursache der Plasmateilung von Fibroblasten. *Biochim. Biophys. Acta* **79**:98.

Perry, M. M., H. A. John and N. S. T. Thomas. 1971. Actin-like filaments in the cleavage furrow of newt eggs. *Exp. Cell Res.* **65**:249.

Pollard, T. D. and R. R. Weihing. 1974. Actin and myosin and cell movement. *Crit. Rev. Biochem.* **2**:1.

Rappaport, R. 1966. Experiments concerning the cleavage furrow in invertebrate eggs. *J. Exp. Zool.* **161**:1.

————. 1967. Cell division: Direct measurement of maximum tension exerted by furrow of echinoderm eggs. *Science* **156**:1241.

————. 1971. Cytokinesis in animal cells. *Int. Rev. Cytol.* **31**:169.

Sanger, J. W. 1975. Presence of actin during chromosomal movement. *Proc. Nat. Acad. Sci.* **72**:2451.

Schroeder, T. E. 1968. Cytokinesis: Filaments in the cleavage furrow. *Exp. Cell Res.* **53**:272.

————. 1969. The role of "contractile ring" filaments in dividing *Arbacia* eggs. *Biol. Bull.* **137**:413.

————. 1970. The contractile ring. I. Fine structure of dividing mammalian (HeLa) cells and the effects of cytochalasin B. *Z. Zellforsch.* **109**:431.

————. 1972. The contractile ring. II. Determining its brief existence, volumetric changes, and vital role in cleaving *Arbacia* eggs. *J. Cell Biol.* **53**:419.

————. 1973. Actin in dividing cells: Contractile ring filaments bind heavy meromyosin. *Proc. Nat. Acad. Sci.* **70**:1688.

————. 1975. Dynamics of the contractile ring. In *Molecules and cell movement* (ed. S. Inoue and R. E. Stephens), pp. 334–352. Raven Press, New York.

Schroeder, T. E. and D. L. Strickland. 1974. Ionophore A23187, calcium and contractility in frog eggs. *Exp. Cell Res.* **83**:139.

Tilney, L. G. and D. Marsland. 1969. A fine structural analysis of cleavage induction and furrowing in the eggs of *Arbacia punctulata*. *J. Cell Biol.* **42**:170.

Timourian, H., G. Clothier and G. Watchmaker. 1972. Cleavage furrow: Calcium as determinant of site. *Exp. Cell Res.* **75**:296.

Effects of Colcemid on Morphogenetic Processes and Locomotion of Fibroblasts

J. M. Vasiliev and I. M. Gelfand

Institute of Experimental and Clinical Oncology, Academy of Medical Sciences
and Laboratory of Mathematical Biology, Moscow State University
Moscow, USSR

The aim of this paper is to review the results of experiments in which we investigated the effects of colcemid on the spreading and polarization of fibroblasts in culture. Normal spreading and polarization include numerous local reactions of the formation and attachment of cytoplasmic processes. Results of our experiments indicate that colcemid does not inhibit these local reactions but prevents their integration, which is essential for the coordinated course of spreading and polarization.

Cultures of fibroblasts have traditionally been used in the investigations of mechanisms controlling cell shape, cell attachment to various surfaces and cell locomotion. The advantages in using fibroblasts are that they are able to exist in culture in at least three states (Fig. 1), each different from the others both morphologically and functionally, and that they are able to transit easily and in a relatively short time from one state into another. These three different states of fibroblasts are:

1. A suspended spherical state, in which the cell is unspread and is not attached to any surface.
2. A radially spread state, in which the cell has discoid shape and is firmly attached to the substrate. All the peripheral edge of the cell in this state is active, that is, continuously forming cytoplasmic processes.
3. A polarized state, in which the cell is attached to the substrate and has the peripheral edge divided into a few active and nonactive zones. Only the polarized cells are able to move directionally on the substrate.

Cell transition from the suspended state into the radially spread one will be referred to as spreading. Transition from the suspended or radially spread state into the polarized state will be referred to as polarization.

The experimental data summarized below show that colcemid drastically inhibits both these transitions. We assume, as a working hypothesis, that normal spreading and polarization, as well as cell locomotion on the substrate, result from various combinations of a few basic cellular morpho-

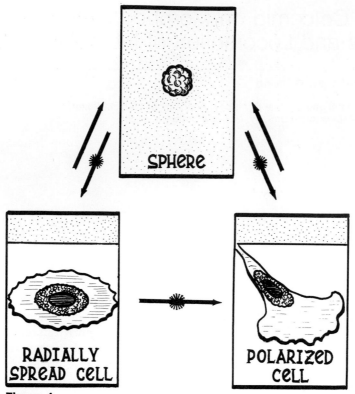

Figure 1
Scheme of three possible states of fibroblasts in culture. Arrows with asterisks designate transitions affected by colcemid; arrows without asterisks, transitions unaffected by colcemid.

genetic reactions. Two of these reactions are local in character; that is, only small parts of the cell participate in each reaction of this type. The main local morphogenetic reaction is the so-called reaction of active attachment, which consists of three stages: protrusion of a cytoplasmic process, attachment of this process to some other cellular or noncellular surface, and development of tension within the attached process. Another local morphogenetic reaction is contact inhibition of movement: inhibition of the formation of cytoplasmic processes in the area of cell-cell contacts. The analysis of the effects of colcemid to be presented further on shows that this drug does not inhibit either of these two local reactions. This analysis suggests that there exists a special colcemid-sensitive morphogenetic reaction or group of reactions. This reaction is global in character, that is, involving the whole cell. It controls the distribution of sites of formation of cytoplasmic processes as well as the distribution of sites of contraction and detachment of these processes.

Secondary cultures of mouse embryo fibroblastlike cells (hereafter referred to as fibroblasts) were used in all the experiments summarized in this paper. The main methods used were time-lapse microcinematography, scanning electron microscopy and transmission electron microscopy. Details of the experimental procedures have been published previously (Vasiliev et al. 1969, 1970, 1975; Domnina et al. 1972; Guelstein et al. 1973). All the effects of colcemid described below were observed in the experiments in which this drug (obtained from Ciba, Switzerland) was added to the culture medium in concentrations ranging from 0.05–0.1 μg/ml. These effects were reversible: they disappeared after transfer of cultures into medium without the drug. Effects similar to those produced by colcemid were observed in experiments with vinblastine sulfate (Richter, Hungary; 0.01–0.05 μg/ml) and colchicine (Merck, GFR; 0.1 μg/ml). Effective concentrations of all three anti-tubulins were near to those that produce complete mitotic block in cultures of mouse embryo fibroblasts (Guelstein and Stavrovskaja 1972). These data indicate that the observed alterations of spreading and polarization described below are due to the specific effect of anti-tubulins on microtubules. Although this suggestion is very probable, it cannot be strictly proven at present. Therefore we will use the neutral term "colcemid-sensitive structures."

SPREADING OF FIBROBLASTS

Normal Course of Spreading

Spreading begins when suspended fibroblasts contact an adhesive substrate. Approximately 80–90% of the fibroblasts accomplish the transition into the radially spread state in about 40–60 minutes after seeding on a flat glass surface. The morphology of spreading (Figs. 2, 3) has been described by several authors (Taylor 1961; Witkowski and Brighton 1971; Domnina et al. 1972; Vasiliev and Gelfand 1973; Rajaraman et al. 1974; Bragina, Vasiliev and Gelfand 1976). At the first stage of spreading, the lower surface of the spherical cell is attached to the surface of the substrate. This attachment is accompanied by the flattening of the lower surface; no other morphological changes are visible at this stage. The protrusion of cytoplasmic processes from the cell body and their attachment to the substrate is characteristic of the second stage of spreading. These processes are of two main morphological types: long and narrow filopodia and flattened lamellar processes. Filopodia are most common at the beginning of the second stage of spreading, whereas lamellar processes predominate at later phases. Specialized substrate attachment sites connected by bundles of microfilaments are often seen at the ends of both types of processes; the morphology of these sites is similar to that of "plaques" seen at the anterior edge of moving fibroblasts (Abercrombie, Heaysman and Pegrum 1971). The bundles of microfilaments connected with these attachment sites are often present in the cytoplasm of the processes. It is important to note that in the course of spreading, first "plaques" and microfilament bundles always appear in the attached cytoplasmic processes. They are not developed in other parts of the spreading cell (Bragina, Vasiliev and Gelfand 1976). The third, final stage of spreading is characterized by the formation of a circular rim of lamellar cytoplasm, that is, a thin cytoplasmic zone containing numerous attachment sites and bundles of microfilaments. Formation of lamellar processes is continued at the peripheral edge of the lamellar cytoplasm; attachment of these processes leads to a gradual increase of the area of lamellar cytoplasm. At this stage, the central

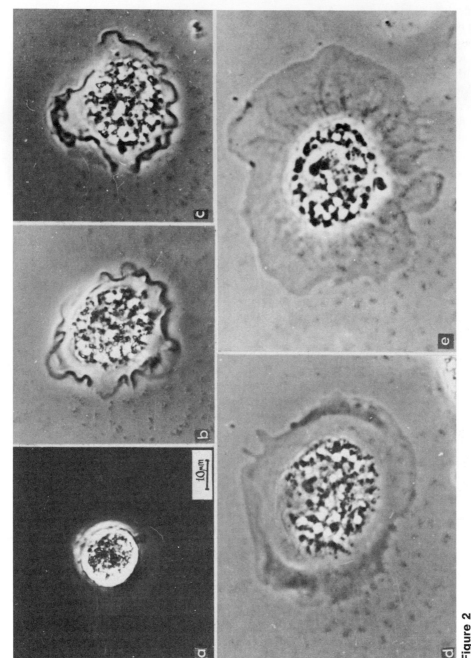

Figure 2
Series of photographs of a living cell spreading on glass in control medium. Time after seeding (in min): (*a*) 10, (*b*) 20, (*c*) 30, (*d*) 50, (*e*) 100.

part of the cell is flattened. Scanning electron microscopy of these cells shows that their upper surfaces become relatively smooth except for a few micro-villi and ruffles.

The results from these morphological investigations of spreading, as well as other experimental data (Di Pasquale and Bell 1974; Vasiliev et al. 1975), suggest that only certain parts of the cell surface have adhesive properties in the sense that they are able to initiate firm attachments with various other cellular or noncellular surfaces. These adhesive properties are characteristic only for the surface of cytoplasmic outgrowths actively protruded by the cell. The ability to develop bundles of microfilaments in their cytoplasm seems to be another property that is characteristic only for these outgrowths. We suggested (Vasiliev et al. 1975) that new cell-cell and cell-substrate attachments are formed in the course of a special morphogenetic reaction, i.e., a reaction of active attachment, that has three main phases. In the first phase, a cytoplasmic outgrowth (filopodium or lamellipodium) is protruded by the cell. If this outgrowth contacts some other surface, e.g., that of the substrate, then the second phase of the reaction, the formation of specific attachment sites, may take place. The third phase is the development of tension in the attached outgrowth; this tension is morphologically correlated with the formation of bundles of microfilaments. Spreading may be regarded as a large series of such reactions. As a result of these reactions, a rim of lamellar cytoplasm is formed from the attached processes at the cell periphery; a system of microfilament bundles and of attachment sites is developed; and the tension of the attached processes leads to the flattening of the cell body.

Effects of Colcemid

In experiments made in collaboration with O. Y. Ivanova and L. B. Margolis we found that mouse fibroblasts were unable to perform rapid and organized spreading in colcemid-containing medium. The time needed for the transition from the spherical into the radially spread state increased considerably: during the first hour after seeding, only about 10% of the cells in colcemid-containing medium reached the radially spread state, in contrast to about 80% in controls. In the following hours, the percent of spread cells gradually increased, but only at 6–8 hours did 60–80% of the cells reach the well-spread state. A decreased rate of spreading in the presence of anti-tubulins had been observed earlier by Kolodny (1972).

Morphological examination (Figs. 4, 5) of the intermediate stages of spreading in colcemid-containing medium revealed several characteristic abnormalities:

(1) *Abnormal distribution and shape of cytoplasmic outgrowths.* The spreading cells formed outgrowths very actively in colcemid-containing medium. However, in contrast to controls, filopodia at the early stages of spreading were not evenly distributed among the various radial directions but instead were concentrated in one or two sectors (Fig. 5a). Often cytoplasmic outgrowths had altered size: instead of numerous relatively small filopodia or lamellipodia, colcemid-treated cells formed only one (Figs. 4a,

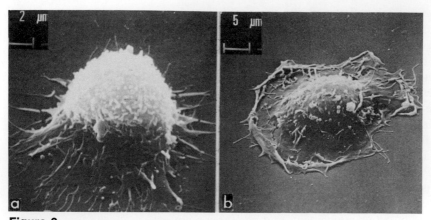

Figure 3
Scanning electron micrographs of critical-point-dried cells spreading on glass
in control medium. (*a*) Lamellar processes and filopodia around the unspread cell body; (*b,c*) radially spread cells with the peripheral lamellar
cytoplasm and flattened central part. Time after seeding (in min): (*a*) 10,
(*b*) 20, (*c*) 40. Tilt angle, 45°; voltage, 20 kv.

5b,c) or two large outgrowths 2–5 μm in diameter; the length of these outgrowths could reach 10–15 μm. At more advanced phases of spreading, the
number of attached large processes often reached three or four. The ends of
these processes became flattened and transformed into areas of lamellar cytoplasm. However, parts of the cell edge between the attached processes remained unspread for a long time (Fig. 5f).

(2) *Lack of correlation between the spreading of various cell parts.* At the
intermediate stages of spreading one sees the cell divided into distinct parts
in such a way that one larger part is relatively well spread and has a relatively
smooth surface, whereas the other remains unspread and retains a spherical
shape with a bulbous or microvillous surface (Fig. 5d). Usually one such
"sphere" was seen, but some cells had two or three spheres connected with
various areas of the well-spread part of the cell. These unspread spheres
could persist for 1–3 hours, undergoing during that time several contractions
and expansions.

(3) *Instability of spreading.* This feature was especially characteristic for
the cells spreading in colcemid-containing medium. Attached cytoplasmic
processes of these cells often retracted and were detached from the substrate
(Fig. 4). The cells that were already relatively well-spread often detached
several processes and had to begin spreading again from an almost spherical
state. These unsuccessful attempts of spreading were often repeated several
times until one of the attempts was successful, that is, led to the transition
of the cell into a stable, well-spread state. Well-spread cells in colcemid-
containing medium (Figs. 4, 8b) had a number of similarities with control
radially spread cells. These cells had the peripheral rim of lamellar cytoplasm
and the flattened central part. Scanning electron microscopy showed that
the upper surface of such cells was relatively smooth. The whole external

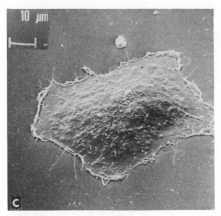

Figure 3 (*continued*)

edge of these cells was active. However, in contrast to control radially spread cells, the cytoplasmic processes formed at this edge varied considerably in size and their distribution along the edge was less uniform. As a result, colcemid-treated, well-spread cells often had an irregular polygonal contour rather than the rounded shape; the position of the nucleus underwent continuous random fluctuations. Well-spread cells could remain in this state for many days in colcemid-containing medium.

POLARIZATION OF FIBROBLASTS

Polarized fibroblasts have various shapes: fanlike, spindle-shaped, stellate, etc. A general feature of all polarized cells, distinguishing them from radially spread ones, is the division of their edges into several active and stable parts, that is, into parts continuously protruding and not protruding cytoplasmic outgrowths. Usually the number of active edge areas of a single cell varies from two to four. Areas of lamellar cytoplasm are usually located immediately behind the active parts of the edge. In polarized cells, in contrast to radially spread ones, these areas are distributed asymmetrically with regard to the center of the nucleus. Often, but not always, these cells have an anterior-posterior axis of symmetry. Directionally moving polarized cells (see Fig. 8a) usually have a wide active edge part and a large area of lamellar cytoplasm located near one (anterior) pole; another small active edge area and a small area of lamellar cytoplasm are often located near the other (posterior) pole. Polarized cells have a complex system of microfilament bundles that are often accompanied by microtubules (Buckley and Porter 1967; Goldman and Follet 1969; Spooner, Yamada and Wessels 1971). If a cell has an anterior-posterior axis, most of these structures are oriented in parallel with that axis. Reactions of active attachment continually going at the active edges of polarized cells lead to the formation of new zones of lamellar cytoplasm with new sites of attachment and new microfilament bundles near these edges.

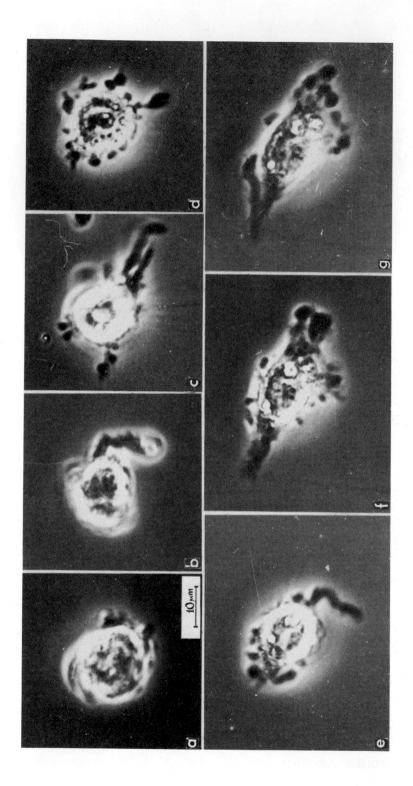

286

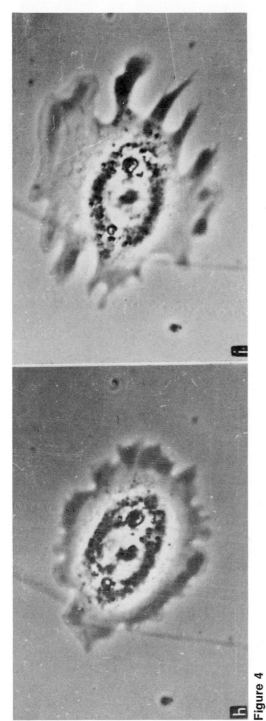

Figure 4

Series of photographs of a living cell spreading in colcemid-containing medium. An unspread cell (*a*) forms a large bended process (*b*). (*c,d,e*) This process is retracted, and several other processes are formed and later retracted. Attachment of several long processes finally leads to successful spreading (*f,g,h,i*). Time after seeding (in min): (*a*) 30, (*b*) 40, (*c*) 60, (*d*) 90, (*e*) 120, (*f*) 130, (*g*) 140, (*h*) 160, (*i*) 200.

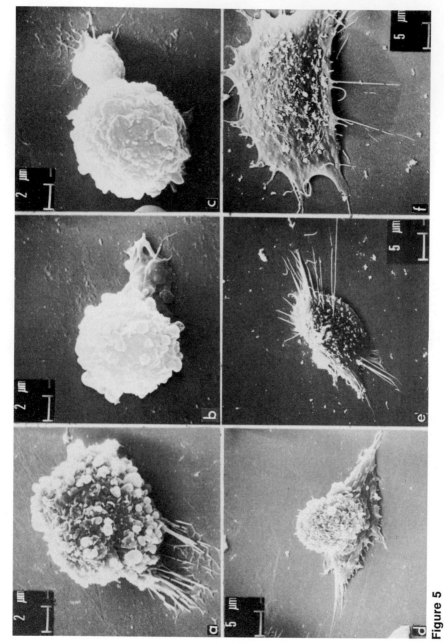

Figure 5

Scanning electron micrographs of cells spreading on glass in colcemid-containing medium. (*a*) An unspread cell with a group of filopodia attached to the substrate; (*b*,*c*) unspread cell with single large process; (*d*) a cell attached by two large processes; the central part of the cell is unspread; (*e*) a cell that after a long time on the substrate is attached by several filopodial processes; it is also possible that this is a stage of retraction after unsuccessful spreading; (*f*) a flattened cell attached by several processes; unattached parts of the edge are still seen between the processes. Time after seeding (in min) : (*a*,*b*,*c*,*d*) 40, (*e*) 360, (*f*) 60.

Polarization on Isotropic Substrate

Polarization takes place on all types of adhesive substrates, although, as shown by Weiss (1961), both the shape and orientation of polarized cells strongly depend on the structure of the substrate. On a flat isotropic substrate, e.g., glass or plastic, radially spread mouse fibroblasts are spontaneously polarized at 3–6 hours after seeding. It is not clear which external factor, if any, effects this polarization. It is possible that hidden anisotropy of the substrate surface is of some importance. It would be interesting to find out whether polarization will take more time on specially prepared flat surfaces that have an increased degree of smoothness.

Colcemid completely prevents polarization on isotropic substrates (Vasiliev et al. 1970; Goldman 1971; Gail and Boone 1971). Polarized cells transferred into colcemid-containing medium become nonpolarized and morphologically indistinguishable from cells that had spread in the presence of colcemid (see above).

Polarization on Narrow Strips of Substrate

When the cells are forced to spread on narrow strips of adhesive substrate surrounded by nonadhesive substrate, radial spreading becomes impossible. In these conditions, suspended cells transit directly into the polarized state without the intermediate transition into the radially spread state. In experiments made in collaboration with O. Y. Ivanova, the glass was covered with nonadhesive lipid films and 20–30 μm wide strips were then scratched in this film (see Ivanova and Margolis 1973 for description of the method). Cells seeded on this substrate and attached to the strips formed outgrowths in all directions. Naturally, only the outgrowths oriented along the strips attached themselves to the substrate. After about 30–90 minutes, the formation of outgrowths in the direction perpendicular to the strip was stopped; cellular edges contacting the edges of the lipid film became nonactive (Fig. 6). Thus polarization, that is, the division of the edge into active and stable parts, took place very early on this substrate. Formation and attachment of outgrowths at two opposite active edges led to the gradual elongation of the polarized cell.

Colcemid had the effect of disorganizing spreading on the strips. Certain manifestations of this disorganization (abnormal shape of outgrowths, instability of spreading) were similar to the effects observed on isotropic substrates as described above. In addition, two particular effects of colcemid were observed: (a) Retraction of the attached processes often led to complete detachment of cells from adhesive strips of substrate. At 15 hours after seeding, the relative number of attached cells per unit area of adhesive strip in colcemid-containing medium was found to be only 30–40% of that in control cultures on the same substrate. At the same time, the number of cells that attached to isotropic glass was similar in both media. Obviously the difference is due to the fact that the area of the substrate around each cell convenient for attachment is much smaller on the strips than on the isotropic glass. Thus on the strips, detachment of a few attachment sites may lead to complete separation of a cell from the substrate, and the disorganized

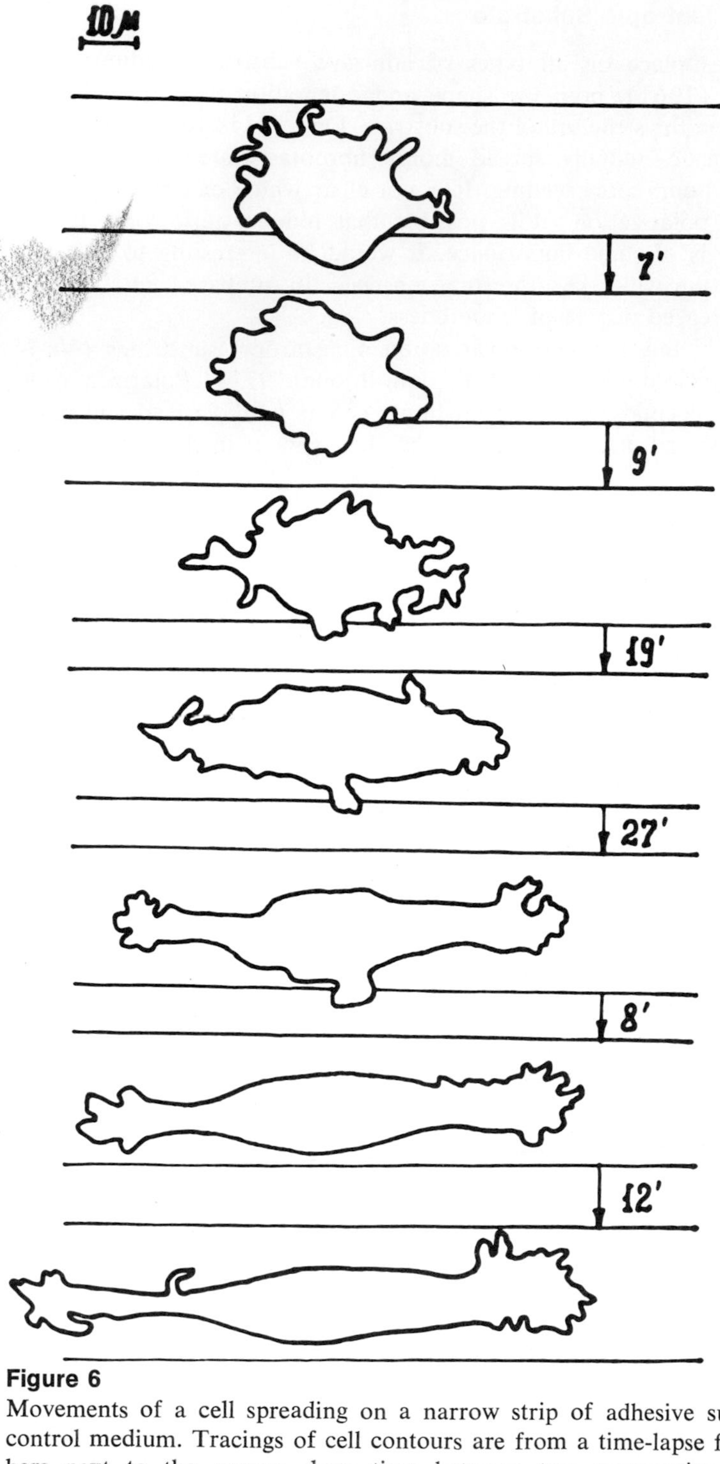

Figure 6

Movements of a cell spreading on a narrow strip of adhesive substrate in control medium. Tracings of cell contours are from a time-lapse film. Numbers next to the arrows show time between two consecutive drawings. Parallel lines show edges of the strip. Notice that at the beginning of the spreading, the cell forms outgrowths in all directions; but later, cellular edges parallel to those of the strip are stabilized.

cell often cannot form outgrowths that would find appropriate sites on the substrate for new attachments.

(b) Cellular edges contacting lipid film are not stabilized, that is, do not become nonactive. Even when cells on the strips achieve some degree of elongation, they are not polarized; their edge is not divided into active and nonactive zones (Fig. 7). These cells continue to make outgrowths in all directions for many hours.

Other External Factors Affecting Polarization

Two other extracellular factors that may determine the orientation of polarized cells should be mentioned. One of these factors is the geometrical structure of the substrate surface. For instance, cells seeded on chemically homogeneous plastic surfaces with cylindrical prominences orient themselves in parallel with the axes of the cylinders (Rovensky, Slavnaja and Vasiliev 1971). Quantitative examination of this phenomenon has shown that the degree of orientation on cylindrical substrates grows with increasing curvature of the cylinders (Samoilov et al. 1975). Colcemid prevented and reversed polarization on these substrates (Rovensky, Slavnaja and Vasiliev 1971).

Another factor affecting cell orientation is contact with another cell. In the area of such contact, formation of protrusions is immediately stopped by contact inhibition (Abercrombie 1970). Then formation and attachment of protrusions at the other parts of the edge gradually changes the orientation of the cell and the direction of cell movements. The part of the edge that has been in contact with the other cell often remains nonactive even after the break of this contact. In colcemid-containing medium, protrusion of outgrowths is also locally inhibited in the area of cell-cell contact. However, it is immediately resumed when contact is broken (Vasiliev et al. 1970). To sum up, colcemid inhibits and reverses all the varieties of polarization of fibroblasts controlled by various extracellular factors.

MORPHOLOGY OF COLCEMID-TREATED FIBROBLASTS AFTER ADDITIONAL INCUBATION WITH CYTOCHALASIN B

The distribution of intracellular structures in fibroblasts incubated with antitubulins has not yet been studied in detail. Of course, microtubules usually disappear in these cells. Wessels, Spooner and Luedena (1973) observed that microfilament bundles often acquire tangential orientation in relation to cellular edges. In order to gain more information about the distribution of intracellular polarized structures in colcemid-incubated fibroblasts we have examined the morphology of these cells after additional treatment with cytochalasin B. It is well known that cytochalasin induces so called arborization of normal polarized fibroblasts: lamellar areas of cytoplasm disappear and are replaced by a system of cytoplasmic cords of various diameters (Carter 1967; Sanger and Holtzer 1972). The nature of the structures forming these cords is not yet quite clear. Transmission electron microscopy often

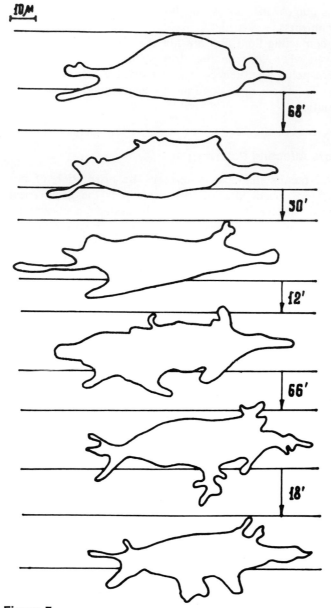

Figure 7
Movements of a cell spreading on a narrow strip of adhesive substrate in colcemid-containing medium. Stabilization of cellular edges does not take place; the cell continues to form outgrowths in all directions during the entire observation period.

reveals microfilament bundles in these cords (Goldman and Knipe 1973; Vasiliev, Gelfand and Tint 1975). However, it would be premature to insist that cords are always formed around preexisting microfilament bundles. As mentioned by Goldman and Knipe (1973), in certain conditions it is difficult to find the bundles within the cords. For the aims of our experiments it is sufficient to assume that the distribution of cytochalasin-induced cords corresponds to the distribution of some intracellular structures connecting various parts of the cell.

In experiments done in collaboration with T. M. Svitkina we have examined the distribution of cytochalasin-induced cords in normal polarized fibroblasts and in well-spread cells that had lost their polarization in colcemid-containing medium. An examination of the effects of colcemid on radially spread cells is in progress and will be reported later. At 24 hours after seeding, fibroblasts were transferred into colcemid-containing medium and incubated in this medium for 6 hours. Control fibroblasts were incubated in the same medium without colcemid. Then 10 μg/ml of cytochalasin B (ICI, England) was added to the cultures kept in colcemid-containing and in control medium. Cells were incubated with cytochalasin for 1 hour, fixed in 2% glutaraldehyde, dehydrated in a series of alcohols and amyl acetate, critical-point-dried from CO_2, coated with silver, and examined by scanning electron microscopy (Cambridge, Stereoscan S-4). Control cytochalasin-treated cells (Fig. 8c) had a system of cytoplasmic cords that sprouted from the central part of the cell and terminated at the periphery. It was usual for these cords to branch into ones of smaller diameter. Connections between the terminal branches of various cords at the cell periphery were rare.

Colcemid-treated cells (Fig. 8d) had a system of small lamella and cords at their periphery; cords crossed each other and were connected with each other at various angles. These peripheral cords and lamellas formed a half-circle or an almost complete circle. The "central" part of the cell, containing the nucleus, was located either in one sector of the circle or near the geometric center of the circle. In the latter case, the central part was connected with the periphery by several cords of various diameters. Thus normal polarized cells have a system of radial, cytochalasin-resistant structures connecting the cell center with its periphery. After incubation with colcemid, this system disappears and is replaced by a circumferential system of structures connecting peripheral sites with each other.

GENERAL DISCUSSION OF THE EFFECTS OF COLCEMID

Morphogenetic Processes Sensitive and Insensitive to Colcemid

Spreading, polarization and movement of polarized cells may be regarded as complex processes, each of them being comprised of a series of more simple, basic morphogenetic reactions. As mentioned above, the reaction of active attachment seems to be the main local morphogenetic reaction and is repeated many times in the course of spreading and polarization. Contact inhibition of movement (Abercrombie 1970) may be regarded as another basic local morphogenetic reaction. This second reaction is not involved in radial spreading but may be important in some cases for the determination of the direction of orientation of polarized fibroblasts. We will now discuss the above-described effects of colcemid on fibroblasts with the aim of determining which basic morphogenetic reactions are affected by this drug. First of all, can the effects of colcemid on spreading and polarization be regarded as a result of inhibition of the local reactions mentioned above? Protrusion

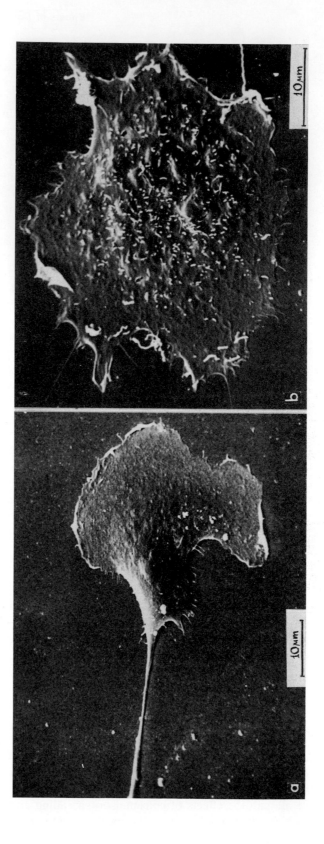

294

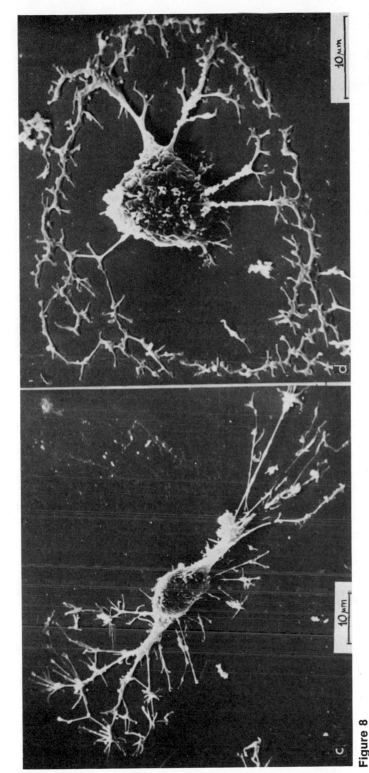

Figure 8

Scanning electron micrographs of well-spread polarized cells treated with colcemid and cytochalasin B. Polarized mouse fibroblast (*a*) control, (*b*) treated with colcemid—loss of polarization; (*c*) treated with cytochalasin B—note the system of branching, substrate-attached radial cords connecting the central part of the cell with the peripheral areas; (*d*) treated with colcemid and cytochalasin B—the radial system of cords is absent, but a circle of substrate-attached cords and small lamellae is seen at the cell periphery; this circle is connected with the central part of the cell by a few cords.

of cytoplasmic outgrowths is the first stage of the active attachment reaction. This stage is not inhibited by colcemid; formation of outgrowths is observed in the course of spreading and at the cell edge after spreading. The ability of outgrowths to form attachments to the substrate is not inactivated by colcemid: in spite of the disorganization of spreading, a suspended cell is able finally to achieve the well-spread state with numerous attachment sites. A number of results indicate that the third stage of the active attachment reaction, that is, development of tension in the attached processes, is also not inhibited by colcemid. This is confirmed by active contractions of cellular outgrowths observed in the course of spreading. The flattened shape and relatively smooth upper surface of well-spread fibroblasts in colcemid-containing medium also suggest that intracellular tension forces are active in these cells. After detachment from the substrate by trypsin or EDTA, colcemid-treated cells, like normal ones, immediately contract and assume the spherical shape (Vasiliev, Gelfand and Tint 1975). Thus none of the manifestations of the active attachment reaction are inhibited by colcemid. As mentioned above, another local reaction, contact inhibition of movement, is also active in the presence of colcemid; the formation of processes is stopped by contacts.

These considerations show that the effects of colcemid on spreading and polarization cannot be described in terms of the inhibition of local morphogenetic reactions. These effects seem to be more complicated and more interesting. Functional components of morphogenesis affected by this drug have some special features different from those of the other known morphogenetic reactions. Analysis of the effects of colcemid seems to be the only approach to characterization of this special morphogenetic reaction or group of reactions.

In contrast to local morphogenetic reactions, this colcemid-sensitive reaction is global, that is, its action involves the whole cell, not just some small part of it. One of the manifestations of this reaction is the regulation of the distribution of sites of formation of cytoplasmic outgrowths. In the course of spreading, colcemid considerably increases fluctuations of the size of outgrowths and of their distribution around the cell body.

Protrusion of an outgrowth at some point of the cellular surface obviously involves transport of some cytoplasmic components to that point. One may say that in the course of normal spreading, the intensity of this transport in various radial directions seems to be equalized, whereas colcemid pervents this equalization. As a result, two large outgrowths are formed or numerous small outgrowths may be concentrated in some areas of the cell surface, whereas in other areas, outgrowths are not formed at all. A distinctive feature of the polarized state is the stable differentiation of the cellular edge into active and nonactive parts. Therefore prevention of polarization by colcemid can also be described in terms of disorganization of the normal distribution of sites of formation of cytoplasmic processes.

Random retraction of the attached cytoplasmic processes is another characteristic feature of colcemid-treated cells. At the intermediate stages of spreading, retractions often lead to partial detachment of cells from the substrate. Possibly these retractions are due to disorganization of the distribution of tensions within the cell; random fluctuations of tensions occasionally lead to violent retraction of some part of the attached edge. Disorganization of the distribution of tensions between the peripheral and central parts of the cell may also explain the lack of coordination in the spreading of various cell parts. When a colcemid-treated cell finally achieves the well-spread state, its shape becomes more stable, but fluctuations of the tensions within the cell probably continue. This is indicated by frequent detachments of some parts of the cellular edge, as well as by random movements of the position of the nucleus.

To summarize, the colcemid-sensitive morphogenetic reaction controls the distribution of the sites of formation and retraction of cytoplasmic processes. In other words, this global morphogenetic reaction integrates local reactions of active attachment going on in various parts of the cell. This global reaction may be called a "stabilization reaction," the term "stabilization" implying the prevention of excessive fluctuations of local activities. One can distinguish two variants of stabilization: (1) a reaction responsible for the stabilization of activity within the active edge, that is, for the equalization of the formation and retraction of the outgrowths at various points on the active edge; this variant is essential for organized spreading, and (2) a reaction responsible for the stabilization of the differentiation of active and non-active parts of the edge; this variant is essential for polarization.

Obviously, the main functional characteristics of both these variants are rather similar and possibly may be regarded as different manifestations of the same morphogenetic reaction.

Radial Organization of Colcemid-sensitive Structures—A Hypothesis

In the previous section we discussed some functional aspects of the effects of colcemid. Now let us examine the possible pattern of organization of intracellular structures responsible for these effects.

Effects of colcemid are global in character, and therefore it is probable that colcemid-sensitive structures form a united system connecting various parts of the cell. We suggest that this system has radial organization, that is, consists of radial elements going from the central part of the cell into various peripheral areas; radial elements are somehow connected with each other in the central part of the cell. One should add that in a polarized cell, elements of the system go along only a few of all the possible radial directions. This suggestion concerning the radial organization of the colcemid-sensitive system is in good agreement with the results of experiments with cytochalasin: the radial system of cords revealed by this agent in a normal polarized cell disappears after incubation with colcemid.

The colcemid-sensitive system may be responsible for the radial organization of various intracellular processes. Two such processes should be specifically mentioned: (1) the intracellular transport of substances and/or organelles and (2) the distribution of intracellular tensions. Colcemid-treated cells retain their ability for transport and for the development of tensions; however, their radial organization is lost. This suggestion is in good agreement with experiments that have shown the inhibition of secretion (see

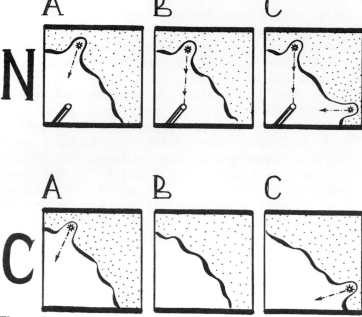

Figure 9

Two series of schematic drawings showing the time sequences of the alterations of the shape of cellular active edges and of the hypothetical distribution of tensions near these edges in control (N) and in colcemid-containing (C) medium. Active edges are shown by wavy lines and attachment sites by asterisks. Arrows show the direction of tensions from the cell acting on the attachment sites in the attached cellular processes; of course, equal, but directionally opposite, tractional forces from the attached processes act on other parts of the cell.

review in Allison 1973) and randomization of movements of intracellular organelles (Bhisey and Freed 1971b) by anti-tubulins. Besides transport and tension, the radial system may control the directions of other polarized intracellular processes, e.g., directional movements of surface components (Edelman, Yahara and Wang 1973; Ukena et al. 1974). The distribution of the sites of formation of cytoplasmic outgrowths may also be a function controlled by the radial system. This control may be related to the control of intracellular transport; as mentioned before, some directional movements of cytoplasmatic components are obviously involved in the protrusion of outgrowths. One may suggest that the formation of outgrowths in normal cells is possible only in those parts of the surface that are located near peripheral terminations of the elements of the radial system.

In the normal cell, each attached outgrowth probably interacts with an element of the radial system (Fig. 9) in such a way that the local tension developed in this outgrowth is transmitted into the central part of the cell and is balanced by the tensions from other peripheral areas (Fig. 10). In the course of normal spreading, such an interaction of individual outgrowths with the elements of the radial system leads to the coordinated stretching

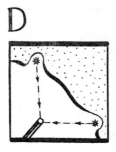

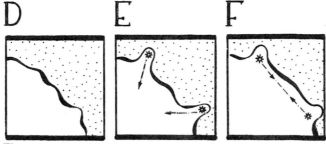

Figure 9 (*continued*)

In a control cell, the tension developed within the attached process (*A,B*) acts on an element of the radial colcemid-sensitive system (shown as a tubule) in the cytoplasm. The second process, which is attached later, also interacts with this radial element (*C,D*).

In a colcemid-incubated cell, radial elements are absent. The unbalanced tension acting on the attached processes leads to detachment and retraction of these processes (*A–D*). Occasionally, forces from two simultaneously attached processes balance each other (*E,F*).

of the cell in all radial directions. In a moving polarized cell, most radial elements are directed from the center to the anterior edge. Interaction of these elements with the attached outgrowths leads to transmission of tension through the center into the tail part of the cell. From time to time this transmitted tension leads to the detachment of the tail part from the substrate.

In the course of spreading in colcemid-containing medium, an attached outgrowth cannot interact with a destroyed radial system. Tensions created by various outgrowths randomly combine with each other, and the sum of these tensions becomes very unstable. Equilibrium of the tensions developed in various parts of the cell can be established only by chance in the course of such random fluctuations. We have not yet made detailed model calculations, but it seems probable that in these conditions, the establishment of equilibrium most often would be a result of the balancing of tensions between the adjacent peripheral outgrowths; organized transmission of tensions to the central part of the cell will not take place. For the establishment of such an equilibrium in the course of spreading, several outgrowths should be simultaneously attached in "convenient" positions (Fig. 9). Obviously, the probability of this event is not high. This explains why the spreading cell

Figure 10

Hypothetical distribution of tension forces (black lines) acting between the attachment sites (asterisks) in a control polarized cell (*left*) and in a well-spread, colcemid-treated cell (*right*). In the control cell, the tensions are distributed radially and partially balance each other in the central part of the cell. In the colcemid-treated cell, tangential orientation of tensions predominates. This is a result of the balancing of tensions between the attached processes (see Fig. 9E,F).

in colcemid-containing medium usually reaches the well-spread state only after several unsuccessful attempts. These considerations also suggest that well-spread cells in the presence of colcemid should have a tangential orientation of the tensions acting between the adjacent peripheral attachment sites (Fig. 10). The distribution of certain intracellular structures, e.g., the microfilament bundles, may be correlated with the distribution of tensions. The experiments with cytochalasin described above confirm the suggestion that in the presence of colcemid well-spread cells develop circumferential structures that connect peripheral attachment sites with each other.

Another important problem is understanding how extracellular factors may determine the orientation of the radial system in the course of polarization. Factors affecting orientation either inhibit formation of the outgrowths (contact inhibition) or alter the probability of the attachment of these outgrowths (structure of the substrate) in certain parts of the cellular edge. As a result, the distribution of tensions within the cell may be altered. For instance during spreading on narrow strips of adhesive substrate, cytoplasmic processes obviously are able to attach themselves and to develop tension only in the direction of the strip. It is probable that this predominant direction of tensions somehow determines the orientation of colcemid-sensitive elements and thereby the direction of cell polarization. A similar mechanism may act in the case of cell-cell contact. The formation of outgrowths in the area of contact is locally stopped by contact inhibition but is continued in other parts of the edge. As a result, the distribution of tensions and the orientation of colcemid-sensitive elements may be changed. This leads to the alteration of the direction of cell movement and to the stabilization of the nonactive state of that part of the edge that has been temporarily stopped by the contact.

Thus the interrelationships between the local reactions of active attachment and the global radial colcemid-sensitive system are reciprocal. The orientation of radial elements may be changed by the tensions developed by attached outgrowths. On the other hand, once the radial system has acquired a certain orientation it would stabilize the distribution of the tensions and the sites of further formation of outgrowths.

It would be easy to describe a plausible morphological model of the colcemid-sensitive system as consisting of long radial cytoskeletal elements (microtubules) connected by some central structure (centriole ?) and interacting with tension-producing structures (bundles of microfilaments ?). However, little is known about the central connections of microtubules, or about the nature of the interactions of microtubules with microfilament bundles and with cortical microfilaments, or the mechanisms of formation and attachment of cytoplasmic outgrowths, etc. Therefore too detailed speculations about the organization of the colcemid-sensitive system seem at present premature.

Role of Colcemid-sensitive Stabilization Reactions in the Orientation of Movements of Various Cell Types

As discussed above, colcemid-sensitive reactions of fibroblasts are responsible for the stabilization of the effects of extracellular factors on cell shape and on the direction of cell movements. In a similar way, the colcemid-sensitive system in many other types of cells may be a structure "memorizing" the effects of various extracellular factors during morphogenetic processes accompanied by oriented movements or directional cellular growth. In particular, colcemid-sensitive structures were found to be essential for directional translocation not only of fibroblasts but also of macrophages (Bhisey and Freed 1971a) and granulocytes (Bandmann, Rydgren and Norberg 1974). A common feature of all these cells is they are moving individually without forming stable cell-cell contacts. In epithelial sheets, where all the cells are firmly linked with each other, a colcemid-insensitive mechanism is responsible for the orientation of cellular movements. Microcinematographic observations of the cells in these sheets show that all cell edges in contact with other cells are not active; formation of outgrowths is observed only at the free edge of marginal cells. Migration of these cells on the substrate in wounded cultures was found to be insensitive to colcemid (Vasiliev et al. 1975). Thus in this system, polarization of the cellular edge is a result of the firmness of local cell-cell contacts and of the efficiency of contact inhibition. In this case, additional global colcemid-sensitive stabilization is not required for directional translocation.

Colcemid-sensitive Reactions of Transformed Fibroblasts

In conclusion we will comment briefly on the state of colcemid-sensitive morphogenetic reactions in fibroblasts morphologically transformed by oncogenic viruses and other agents. Several types of observations with different strains of transformed fibroblasts (J. M. Vasiliev and I. M. Gelfand, unpubl.) indicate that these cells retain an ability for colcemid-sensitive polarization on various substrates:

1. Transformed fibroblasts are able to migrate directionally on glass, although the rate of this migration may be reduced compared to that in their normal ancestors. Microcinematographic observations show that the edges of these cells are divided into stable and active parts, although the active edges of transformed fibroblasts are often narrower than those of normal cells. This polarization and directional migration are prevented by colcemid.
2. Transformed fibroblasts are able to become polarized on narrow strips of adhesive substrate. The average length reached on these strips by transformed fibroblasts of the L strain was found to be similar to that reached by normal mouse fibroblasts (experiments by O. Y. Ivanova and L. B. Margolis).
3. It was found that transformed fibroblasts were able to become polarized and to orient themselves on cylindrical surfaces. The maximal radius of curvature of the substrate surface that could still induce orientation of transformed cells was usually smaller than that which could induce orientation of normal cells (experiments of V. I. Samoilov, Y. A. Rovensky and I. L. Slavnaja).

Thus, at present, there are no data suggesting that transformed cells are deficient in the ability to perform colcemid-sensitive stabilization reactions. A number of facts reviewed and discussed elsewhere (Vasiliev and Gelfand 1976) suggest that abnormalities in the shape and locomotory behavior of transformed cells may be satisfactorily explained by a deficiency in another morphogenetic reaction, namely, in the reaction of active attachment.

REFERENCES

Abercrombie, M. 1970. Contact inhibition in tissue culture. *In Vitro* **6**:128.

Abercrombie, M., J. E. M. Heaysman and S. M. Pegrum. 1971. The locomotion of fibroblasts in culture. IV. The electron microscopy of the leading lamella. *Exp. Cell Res.* **67**:359.

Allison, A. C. 1973. The role of microfilaments and microtubules in cell movement, endocytosis and exocytosis. In *Locomotion of tissue cells. Ciba Foundation Symposium* (ed. M. Abercrombie), vol. 14, p. 109. Associated Scientific Publishers, New York.

Bandmann, U., L. Rydgren and B. Norberg. 1974. The difference between random movement and chemotaxis. Effects of antitubulins on neutrophil granulocyte function. *Exp. Cell Res.* **88**:63.

Bhisey, A. N. and J. J. Freed. 1971a. Ameboid movement induced in cultured macrophages by colchicine or vinblastine. *Exp. Cell Res.* **64**:419.

―――. 1971b. Altered movement of endosomes in colchicine-treated cultured macrophages. *Exp. Cell Res.* **64**:430.

Bragina, E. E., L. M. Vasiliev and I. M. Gelfand. 1976. Formation of bundles of microfilaments during spreading of fibroblasts on the substrate. *Exp. Cell Res.* (in press).

Buckley, J. K. and K. R. Porter. 1967. Cytoplasmic fibrils in living cultured cells. *Protoplasma* **64**:349.

Carter, S. B. 1967. Effects of cytochalasin on mammalian cells. *Nature* **213**:261.

Di Pasquale, A. and P. B. Bell. 1974. The upper cell surface: Its inability to support active cell movement in culture. *J. Cell Biol.* **62**:198.

Domnina, L. V., O. Y. Ivanova, L. B. Margolis, L. V. Olshevskaja, Y. A. Rovensky, Ju. M. Vasiliev and I. M. Gelfand. 1972. Defective formation of the lamellar cytoplasm by neoplastic fibroblasts. *Proc. Nat. Acad. Sci.* **69**:248.

Edelman, G. M., I. Yahara and J. L. Wang, 1973. Receptor mobility and receptor cytoplasmic interactions. *Proc. Nat. Acad. Sci.* **70**:1442.

Gail, M. H. and C. W. Boone. 1971. Effect of colcemid on fibroblast motility. *Exp. Cell Res.* **65**:221.

Goldman, R. D. 1971. The role of three cytoplasmic fibers in BHK-21 motility. I. Microtubules and the effects of colchicine. *J. Cell Biol.* **51**:752.

Goldman, R. D. and E. A. C. Follet. 1969. The structure of major cell processes of isolated BHK-21 fibroblasts. *Exp. Cell Res.* **57**:263.

Goldman, R. D. and D. M. Knipe. 1973. The functions of cytoplasmic fibers in non-muscle cell motility. *Cold Spring Harbor Symp. Quant. Biol.* **37**:523.

Guelstein, V. I. and A. A. Stavrovskaja. 1972. The effect of colcemid on mitosis, migration and DNA synthesis of mouse fibroblasts in vitro (*in Russian*). Bull. Exp. Biol. Med. **73**:94.

Guelstein, V. I., O. Y. Ivanova, L. B. Margolis, J. M. Vasiliev and I. M. Gelfand. 1973. Contact inhibition of movement in cultures of transformed cells. *Proc. Nat. Acad. Sci.* **70**:2011.

Ivanova, O. Y. and L. B. Margolis. 1973. The use of phospholipid membranes for preparation of cell cultures of given shape. *Nature* **242**:200.

Kolodny, G. M. 1972. Effect of various inhibitors on readhesion of trypsinized cells in culture. *Exp. Cell Res.* **70**:196.

Rajaraman, R., D. E. Rounds, S. P. S. Yen and A. Rembaum. 1974. A scanning electron microscope study of cell adhesion and spreading in vitro. *Exp. Cell Res.* **88**:327.

Rovensky, Y. A., I. L. Slavnaja and J. M. Vasiliev. 1971. Behaviour of fibroblast-like cells on grooved surfaces. *Exp. Cell Res.* **65**:193.

Samoilov, V. I., M. S. Slovatchevsky, Y. A. Rovensky and I. L. Slavnaja. 1975. Shape and orientation of nuclei in embryo fibroblast-like cells on substrate with regular relief (*in Russian*). *Cytologija* **17**:453.

Sanger, J. W. and H. Holtzer. 1972. Cytochalasin B. Effects on cell morphology, cell adhesion and mucopolysaccharide synthesis. *Proc. Nat. Acad. Sci.* **69**:253.

Spooner, B. S., K. M. Yamada and H. K. Wessels. 1971. Microfilaments and cell locomotion. *J. Cell Biol.* **49**:595.

Taylor, A. C. 1961. Attachment and spreading of cells in culture. *Exp. Cell Res.* (Suppl.) **8**:154.

Ukena, T. E., J. Z. Borysenko, M. J. Karnovsky and R. D. Berlin. 1974. Effects of colchicine, cytochalasin B and 2-deoxyglucose on the topographical organization of surface-bound concanavalin A in normal and transformed fibroblasts. *J. Cell Biol.* **61**:70.

Vasiliev, J. M. and I. M. Gelfand. 1973. Interactions of normal and neoplastic fibroblasts with the substratum. In *Locomotion of tissue cells. Ciba Foundation Symposium* (ed. M. Abercrombie), vol. 14, p. 311. Associated Scientific Publishers, New York.

————. 1976. Morphogenetic reactions and locomotory behaviour of transformed cells in culture. In *Proceedings of Symposium on the fundamental aspects of metastasis* (ed. L. Weiss). Roswell Park Memorial Institute, Buffalo, New York. (In press.)

Vasiliev, J. M., I. M. Gelfand and I. S. Tint. 1975. Processes causing cell shape alteration after cell detachment from the substratum (*in Russian*). *Cytologija* **17**:633.

Vasiliev, J. M., I. M. Gelfand, L. V. Domnina and R. I. Rappoport. 1969. Wound healing processes in cell cultures. *Exp. Cell Res.* **54**:83.

Vasiliev, J. M., I. M. Gelfand, L. V. Domnina, O. S. Zacharova and A. V. Ljubimov. 1975. Contact inhibition of phagocytosis in epithelial sheets: Alterations of cell surface properties induced by cell-cell contacts. *Proc. Nat. Acad. Sci.* **72**:719.

Vasiliev, Ju. M., I. M. Gelfand, L. V. Domnina, O. Y. Ivanova, S. G. Komm and L. V. Olshevskaja. 1970. Effect of the colcemid on the locomotory behaviour of fibroblasts. *J. Embryol. Exp. Morphol.* **24**:625.

Weiss, P. 1961. Guiding principles in cell locomotion and cell aggregation. *Exp. Cell Res.* (Suppl.) **8**:260.

Wessels, N. K., B. S. Spooner and M. A. Luduena. 1973. Surface movements, microfilaments and cell locomotion. In *Locomotion of tissue cells. Ciba Foundation Symposium* (ed. M. Abercrombie), vol. 14, p. 53. Associated Scientific Publishers, New York.

Witkowski, J. A. and W. D. Brighton. 1971. Stages of spreading of human diploid cells on glass surfaces. *Exp. Cell Res.* **68**:372.

Surface-modulating Assemblies in Mammalian Cells

Gerald M. Edelman, John L. Wang and Ichiro Yahara

The Rockefeller University, New York, New York 10021

In this article we propose to review various lines of evidence supporting the hypothesis (Edelman, Yahara and Wang 1973) that the motion and distribution of cell surface receptors are under the control of a submembraneous macromolecular assembly. This surface-modulating assembly (SMA) appears to be a metastable, tripartite structure consisting of certain surface receptors, microfilaments (together with associated myosin) and microtubules. There is evidence suggesting that interactions with any or all of these components can affect the behavior of this assembly and alter cell surface receptor movement and cell movement, as well as regulatory functions in growth control. For these reasons it is attractive to consider the larger hypothesis that the SMA serves as a mediator of receptor–cytoplasmic interactions and therefore as a central signal regulator coordinating these cellular functions.

Most of the evidence supporting the existence of the SMA comes from perturbation experiments using external ligands to reveal various cellular states in lymphocytes. The evidence is therefore dynamic rather than structural. Nevertheless, there is increasing evidence supporting the existence of similar structures in other cells, and the dynamic evidence is consistent with the presently available structural information. After reviewing the known properties of cells that suggest the existence of the SMA, we shall consider the critical evidence that must be developed to establish its exact structure and function.

Modulation of Cell Surface Receptors

The initial observation (Yahara and Edelman 1972; Yahara and Edelman 1973a) providing direct evidence that the movement of cell surface receptors can be modulated was based on the previous demonstration by Taylor et al. (1971) that cross-linking of lymphocyte surface receptors by divalent antibodies can result in clustering of these receptors in a diffusion-limited nucleation event (Fig. 1). This event, known as patch formation, is metabolism-

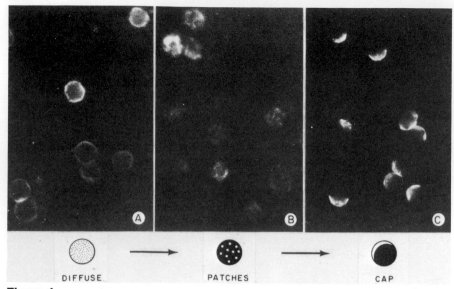

Figure 1

Labeling patterns of surface immunoglobulin molecules on lymphocytes after binding with fluorescein-conjugated anti-immunoglobulin antibodies. (*A*) Diffuse labeling pattern observed at 4°C or in the presence of Con A; (*B*) patchy distribution observed in the presence of 10 mM NaN$_3$; (*C*) cap formation after incubation at 21°C.

independent. Once formed, however, patches can be gathered at one pole of the cell by an active process, called capping, that depends upon cellular metabolism. Both patch and cap formation can be revealed by fluorescence microscopy, as shown in Figure 1 for the immunoglobulin receptors on lymphocytes. It has been shown that under different conditions lectins as well as antibodies can induce patch and cap formation in different cells (Unanue, Perkins and Karnovsky 1972; Loor 1974; de Petris 1974).

Lateral mobility of cell surface receptors within the plane of the membrane (Frye and Edidin 1970) is necessary for patch formation. For this reason we attempted to inhibit patch formation in order to reveal conditions that might lead to inhibition of this motion. It was found (Yahara and Edelman 1972; Edelman, Yahara and Wang 1973) that the addition to lymphocytes of concanavalin A (Con A) or various other lectins results in the inhibition of patching and capping for a variety of receptors, including the glycoprotein receptors that bind the lectins themselves (Table 1). Wheat germ agglutinin, which does not share binding sites with Con A on the lymphocyte surface (Yahara and Edelman 1975a; Sela, Wang and Edelman 1975), did not induce modulation.

Recent experiments using polymorphonuclear leukocytes (Ryan, Borysenko and Karnovsky 1974; Oliver, Zurier and Berlin 1975), leukemic basophils (Carson and Metzger 1974) and normal and transformed fibroblasts (Ukena et al. 1974; Yahara and Edelman, unpubl.) indicate that the phenomenon of surface modulation is not restricted to lymphocytes. In addition, we have examined a variety of mouse tissues, including thyroid, thymus, kidney, testis

Table 1

Tissues, Cell Types and Receptors Tested for Modulation Effects

Tissue/cell	Receptor
Splenic lymphocytes	IgG and IgM
Thymocytes	Fc receptors
Thyroid	θ-antigens
Kidney	H-2 antigens
Testis (spermatogonia)	β_2-microglobulin
Ovary (cumulus cells)	cell surface carbohydrates
Polymorphonuclear leukocytes	(including receptors for
Cultured fibroblasts	lectins and anti-carbohydrate
	antibodies)

(spermatogonia) and ovary (cumulus cells), and have obtained similar results. It appears, therefore, that modulation, or restriction of receptor mobility, is a general phenomenon that does not depend strictly on the specificity of surface receptors or upon cell type (Table 1).

Modulation Is Reversible and Occurs at the Level of Single Receptors

Modulation requires the binding of Con A to cell surface carbohydrates, for if α-methyl-D-mannoside (αMM) is added to remove Con A, patching and capping can again occur (Yahara and Edelman 1972). This indicates that the effect of Con A-binding is reversible and does not result from permanent metabolic inhibition or killing of the cell.

Electron microscopic observations (Yahara and Edelman 1975b) have shown that the addition of Con A immobilizes individual receptors; in this case, the glycoprotein receptors for the lectin detected using ferritin-labeled Con A were found to remain in a diffuse distribution (Fig. 2). Moreover, by using freeze-fracture techniques in conjunction with electron microscopy it was shown that the random distribution of intramembranous particles is not

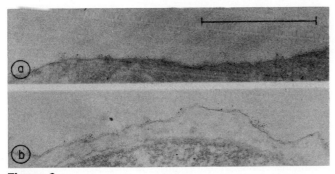

Figure 2

Distribution of Con A receptors on cell surfaces after labeling with ferritin-conjugated Con A. Untreated lymphocytes (*a*) and colchicine-treated cells (*b*) were incubated with ferritin-Con A. The colchicine-treated cells show a patchy distribution of ferritin-Con A, whereas the untreated cells show a diffuse distribution. Bar = 1 μm.

correlated with either the redistribution of receptors or the modulation of receptor mobility. Finally, preliminary spin-labeling experiments suggest that no extensive alteration occurs in the fluidity of membrane lipids under conditions of receptor modulation.

Modulation Is Induced by Cross-linkage of Certain Receptors

Experiments comparing the effect of tetravalent Con A with divalent succinyl-Con A and acetyl-Con A (Gunther et al. 1973; Yahara and Edelman 1973a) suggest that the induction of modulation requires the cross-linkage of certain glycoproteins at the cell surface (Table 2). The possibility that charge effects also play a role has not been excluded. Indeed, like charges on divalent Con A derivatives may act to repel their complexes with the cell surface, and, as shown in Table 2, the binding of divalent derivatives does not result in modulation. Cross-linking of these cell-bound succinyl-Con A molecules by divalent antibodies to Con A or their $(Fab')_2$ fragments again results in modulation. In corroboration of the cross-linking hypothesis, univalent fragments of antibodies to Con A had little or no effect.

Under Perturbing Conditions, Modulation Involves the Entire Cell Surface and Is a Propagated Phenomenon

Lectins are external perturbants that induce modulation. So far, no selectivity or partial modulation appears to have been observed after lectin-binding; i.e., all of the surface receptors are fixed in their distribution and motion. The possibility of partial or selective modulation has been raised, however,

Table 2
Modulation of Receptor Mobility by Con A and Antimitotic Drugs

Treatment	Cap-forming cells (%)
Control	87
Con A	4
Succinyl-Con A[a]	86
Succinyl-Con A + anti-Con A[b]	53
Succinyl-Con A + Fab' anti-Con A	80
Con A + colchicine[c]	30
Con A + vinblastine	51
Con A + vincristine	15
Con A + podophyllotoxin	10
Con A + lumicolchicine	1

In order to test for the inhibition of immunoglobulin receptor cap formation by Con A, the percent of cap-forming cells obtained with fluorescein-conjugated anti-immunoglobulin (100 μg/ml) was measured in the presence of Con A (100 μg/ml).

[a] The concentration of succinyl-Con A was 50 μg/ml.

[b] The concentration of succinyl-Con A was 50 μg/ml; the concentration of the antibody reagents was 100 μg/ml.

[c] The concentrations of the various drugs were as follows: colchicine, vinblastine and vincristine, 10^{-4} M; lumicolchicine, 5×10^{-4} M; podophyllotoxin, 10^{-3} M.

by the work of Berlin and coworkers (Ukena and Berlin 1972; Oliver, Ukena and Berlin 1974).

Modulation does not appear to be the result of simple cross-linking of all of the receptors at the cell surface. First, the effect of modulation is observed at relatively low doses of Con A, representing occupancy of less than 10–15% of all surface glycoproteins capable of binding the lectin (Edelman, Yahara and Wang 1973). In addition, the use of Con A coupled to the solid surfaces of nylon fibers and latex beads indicates that binding and cross-linkage of only a small local population of receptors is sufficient to cause total modulation (Rutishauser, Yahara and Edelman 1974; Yahara and Edelman 1975a). This suggests that the initial signal inducing such a global cellular change must be capable of propagation to the cellular structures responsible for the phenomenon.

Implication of Submembraneous Structures in Surface Modulation

The observations summarized above suggest strongly that surface modulation results from interactions of receptors with cytoplasmic structures that serve as anchorage points for the receptors. In addition, the finding that capping requires metabolism (Taylor et al. 1971) but does not depend upon translocation of shape changes of the cell (Rutishauser, Yahara and Edelman 1974; Unanue and Karnovsky 1974) implies strongly that the patched receptors must be in contact with submembraneous structures that mediate their systematic movement. This interpretation requires that cell surface receptors penetrate the lipid bilayer, as in the case of erythrocytes (Bretscher 1971). This is a particularly attractive possibility because it suggests a means by which receptor–cytoplasmic interactions may take place. More direct evidence to support this idea comes from an examination of the effects of drugs that are known to affect cytoplasmic fibrillar structures.

Early observations on the effect of temperature on modulation (Yahara and Edelman 1973a) led us to suspect that temperature-sensitive structures such as microtubules (Tilney and Porter 1967) may be involved directly or indirectly in receptor anchorage. In accord with this hypothesis, colchicine and the Vinca alkaloids were found to reverse partially the restriction of receptor mobility by Con A (Yahara and Edelman 1973b; Table 2). It is important to note that the colchicine effects were also observed at low concentrations (10^{-6} M), although the use of these low doses lengthened the time required for reversal of modulation. Moreover, the drug effects were partially reversible; i.e., removal of colchicine after washing the cells restored the modulation. It was found that colchicine did not bind to Con A, nor did it alter the cell binding properties of the lectin. In addition, electron micrographs indicated that microtubules in lymphocytes were disrupted by these agents. Finally, lumicolchicine, the photo-inactivated derivative of colchicine, had no effect on the modulation of receptor mobility by Con A. All of these data provide strong evidence implicating microtubular assemblies in the reversible anchorage of cell surface receptors.

Colchicine and Vinca alkaloids have no effect on capping itself, and thus the mechanism of movement of patches to form caps in lymphocytes does not appear to depend upon the state of microtubular assemblies. We conclude, therefore, that another structure must interact with the receptors in order to carry out cap formation. Because of the obvious role of microfilaments in cell movement, these structures appear to be the best candidates. Support for this proposal has come from several observations. Cytochalasin B, which disrupts microfilaments (Wessells et al. 1971), has been found to inhibit capping of preformed patches on both cells in suspension (Taylor et al. 1971) and on sessile lymphocytes bound to nylon fibers (Rutishauser, Yahara and Edelman 1974). Moreover, de Petris (1974) has found that simultaneous addition of colchicine and cytochalasin B prevents capping.

Relationship of Patching and Capping and Their Modulation to Cell Movement

Since it appears that microfilaments and microtubules are involved in capping and modulatory inhibition of patching, it is important to determine the relationship of these phenomena to cell movement. The use of nylon fibers with covalently coupled antibody or lectin ligands permitted an initial study of this problem (Rutishauser, Yahara and Edelman 1974). Bone marrow-derived (B) lymphocytes bound to fibers by their antibody receptors underwent morphogenetic movements as well as patching and capping (Fig. 3). It was observed that the caps were always opposite the point of attachment of the cells to the fibers, raising the possibility that global cellular movements determine either the occurrence or the position of the caps. Addition of

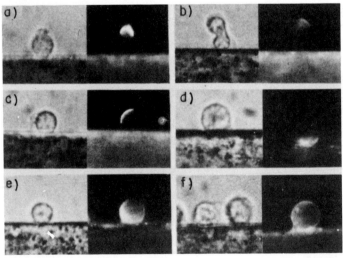

Figure 3
Various types of cells bound to nylon fibers and stained with fluorescein-conjugated anti-immunoglobulin antibodies. Dinitrophenylated bovine serum albumin-derivatized fibers were used in *a–e;* Con A-derivatized fibers were used in *f.* (*a, b*) Morphologically changed cells showing distal caps; (*c, d*) round cells showing caps with random orientations; (*e*) a round cell showing diffuse staining in the presence of Con A; (*f*) two cells bound to a Con A-derivatized fiber; one (*right*) shows diffuse staining pattern, whereas the other (*left*) shows no staining with the fluorescent reagent.

colchicine induced the fiber-bound cells to round up and stop their characteristic shape changes. Nevertheless, these cells formed caps, the positions of which were random with respect to the point of attachment to the fiber.

These observations should be compared to the finding that cells bound on lectin-coated Sepharose beads formed antibody-induced caps oriented mainly towards the point of attachment to the bead (Loor 1974). It has also been reported (Ryan, Borysenko and Karnovsky 1974) that in polymorphonuclear leukocytes, cap formation was concentrated predominantly at the tail end of motile cells, whereas in sessile cells, the caps were central. Regardless of the different conditions, all of these results suggest the conclusion that although cap formation can occur in the absence of global cell movements and cell locomotion (Edidin and Weiss 1972; Rutishauser, Yahara and Edelman 1974; Unanue and Karnovsky 1974; Ryan, Borysenko and Karnovsky 1974), the direction or position of the cap may depend upon such movements.

It was also found that cell movement, as well as patch and cap formation, in cells bound to nylon fibers can be inhibited by addition of soluble Con A or by attaching the cells to fibers derivatized with the lectin (Fig. 3). This inhibition of capping can also be reversed by the addition of αMM or colchicine. Strong corroboration of these local effects was obtained (Yahara and Edelman 1975a) using latex beads or platelets coated with Con A. In this case, addition of colchicine led to the capping of the particles themselves (Fig. 4), despite the fact that their attachment to a small area of the cell surface was sufficient to restrict all receptor mobility in the absence of the drug.

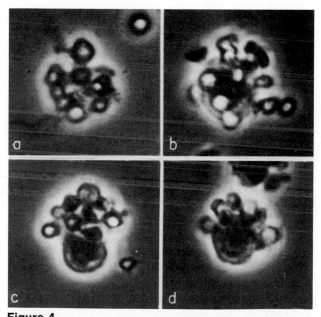

Figure 4
(a, b) Random distribution of Con A platelets bound to splenic lymphocytes.
(c, d) Polar distribution of Con A platelets observed when colchicine-treated lymphocytes were incubated with these particles.

*Evidence for Structures in Lymphocytes Containing Myosin, Actin and
Colchicine Binding Proteins*

All of the preceding conclusions were based on the assumption that muscle-
related proteins and microtubular proteins (or at least colchicine binding proteins) exist in lymphocytes. A number of recent findings indicate that this is
indeed the case.

As reported previously in an electron microscopic study (Yahara and
Edelman 1975b), microtubular and microfilamentous structures can be seen
in lymphocytes. Moreover, colchicine binding proteins (CBP) were detected
in extracts of mouse splenic lymphoctyes (Table 3). When the extraction and
fractionation were performed at 4°C, about 90% of the total colchicine
binding activity was found in the supernatant fraction. In addition, the colchicine binding activity was highly labile, as reported for isolated tubulin.
The lifetime ($t_{1/2} = 38.5$ min at 37°C, pH 6.8 without GTP) of the activity
in lymphocyte extracts was similar to that found in mouse brain extracts.

We have also observed that pretreatment of lymphocytes with mitogenic
doses of Con A increased the colchicine binding activity of the extracts. In
contrast, treatment of cells with modulating doses of Con A resulted in a
decrease in the colchicine binding activity (Table 3). Although detailed
analyses of these observations remain to be performed, we tentatively suggest
that external ligands are capable of altering the state of the CBP and possibly
the microtubules.

In addition to these studies we have also examined the localization of actin,
myosin and α-actinin in lymphocytes by means of indirect immunofluorescence
techniques using antibodies directed against these proteins. In contrast to the
fibrillar staining patterns of fibroblasts labeled with anti-actin (Lazarides and
Weber 1974), lymphocytes exhibited diffuse labeling with anti-actin antibodies (Fig. 5a). The staining patterns were similar to those obtained with
fibroblasts that were not attached to the substratum (Lazarides 1975).

The staining patterns of lymphocytes with antibodies to myosin or α-actinin
were similar to those found with anti-actin antibodies, except that stronger

Table 3
Effect of Con A on the Colchicine Binding Activity of Lymphocytes

Conditions	[^3H]colchicine bound (cpm)/ mg protein	Ratio[a]
Mouse splenic lymphocytes		
control	2952	—
10 µg/ml Con A	4664	1.6 ± 0.3
100 µg/ml Con A	1918	0.6 ± 0.1
100 µg/ml Con A + αMM	3365	1.1 ± 0.2
Mouse brain	60,150	—

Cells were incubated with various reagents for 15 min at 37°C. The cells were then
homogenized and centrifuged at 100,000g. Colchicine binding activity in the supernatant
fraction was assayed using the filter assay described by Weisenberg, Borisy and Taylor
(1968).
[a] [^3H]colchicine bound in the presence of cells and added reagents/[^3H]colchicine
bound in the presence of cells alone.

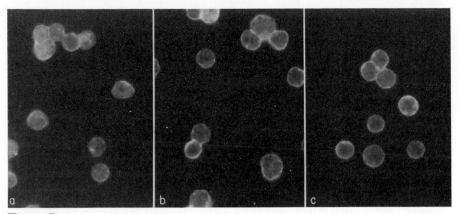

Figure 5

Fixed and dehydrated lymphocytes stained with (*a*) anti-actin antibody, (*b*) anti-myosin antibody and (*c*) anti-α-actinin antibody. The cells were fixed with formaldehyde and dehydrated with methanol-acetone on microscope slides. They were then incubated with the respective antisera and then labeled with fluorescein-conjugated goat anti-rabbit-Ig antibody. The antisera were obtained from rabbits immunized with SDS-striated muscle actin, uterus myosin and striated muscle α-actinin, respectively.

staining was observed on the cell periphery (Fig. 5b, c). The anti-myosin antibodies were produced against mouse uterine myosin, which had a purity of greater than 90% as determined by SDS-polyacrylamide gel electrophoresis. In agreement with the observation of Gwynn et al. (1974) and Willingham, Ostlund and Pastan (1974), these antibodies could stain the cell surface of lymphocytes. We have found, however, by means of radioimmunoprecipitation assays (Sela, Wang and Edelman 1975), that the cell surface components binding the anti-myosin antibodies have, in the presence of SDS and β-mercaptoethanol, molecular weights of 95,000, 70,000, 60,000 and 23,000. Very little of this material has a molecular weight corresponding to myosin (MW 200,000).

Adsorption of the anti-myosin sera with myosin coupled to Sepharose beads decreased strongly the binding of these antibodies to lymphocyte cytoplasmic myosin and, to a lesser extent, to lymphocyte surface components. On the other hand, adsorption of the sera with intact mouse splenic lymphocytes did not decrease significantly the reactivity of the anti-myosin antibodies to either purified uterine myosin or lymphocyte cytoplasmic proteins with molecular weights of about 200,000. These results suggest two possibilities: (a) that myosin does not exist on the lymphocyte surface, and the cross-reactivity of antibodies with surface components is due to contaminating antigens in the myosin preparation used for immunization; or that (b) only parts of the myosin polypeptide chains are exposed on the cell surface and these are fragmented in situ or during preparation of cell lysates. Although we cannot decide between these two possibilities, it is pertinent that induction of capping of surface receptors by anti-Ig, anti-H-2 or Con A (using cells pretreated with colchicine) resulted in the cocapping of surface components reactive with anti-myosin antibodies, whereas the distribution of myosin inside the cell was not affected. In similar experiments using antibodies directed against actin and α-actinin, the staining patterns were not altered by capping on the cell surface.

A Model for the Surface-modulating Assembly

It has been suggested previously (Edelman, Yahara and Wang 1973; Yahara and Edelman 1975b) that surface modulation is mediated by a tripartite assembly (Fig. 6): (1) certain surface receptors that bind to external ligands; (2) microfilaments and associated proteins that are responsible for systematic movement of the receptors; and (3) microtubules that are responsible for reversible anchorage of receptors.

At present, the most stringent test of the structural features of this model of the SMA is to demonstrate that the proposed interactions occur. According to the present serial model (Fig. 6), there are at least three levels of interaction: receptor-microfilament, microfilament-microtubule and microtubule-tubulin subunits. Of course, alternative arrangements are possible. Obvious examples are parallel or alternating models in which both microfilaments and tubulin subunits can interact with receptors. The serial model, however, seems simpler, particularly in view of the fact that assembled microtubules are not often found just beneath the cell membrane. But whichever model is chosen, a major difficulty remains: How can all cell surface receptors interact at their

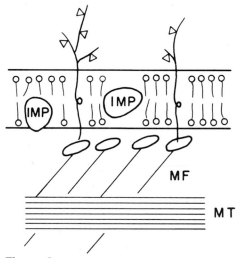

Figure 6
A model for the organization of various components in the hypothesized surface-modulating assembly. The model assumes that microfilaments (MF) interact with receptors, possibly via a myosinlike structure, and that the microfilaments in turn interact with microtubules (MT). The interactions among the various components are assumed to involve reversible association-dissociation reactions. It is assumed that intramembraneous particles (IMP) are not involved in these equilibria.

base with a cytoplasmic structure? What is required is some type of adaptor molecule capable of interacting with all of the receptors to mediate their interaction with the fibrillar components.

Functionally, the SMA has three major properties: (1) the interactions among the components are metastable and reversible; (2) receptors can exist in either free or anchored states; and (3) both external perturbations (such as cross-linking) and internal alterations can affect the state and behavior of the entire SMA. This provides a signal pathway in both directions: from the outside world via the cell surface into the cytoplasm as well as from the cytoplasm to alter the conditions of the cell surface.

These properties of the SMA also make it an appropriate candidate for coordinating and regulating various surface states in major cellular functions such as global cellular movement, cell–cell interactions via surface molecules and the commitment to cell division and proliferation. For these reasons we have examined the possible role of microtubular components as regulators of the mitogenic stimulation of lymphocytes to undergo DNA synthesis and proliferation. The results suggest that cellular commitment to mitosis can in fact be regulated by altering the state of the SMA or its microtubular components.

Role of Microtubules and SMA States in Lymphocyte Mitogenesis

Lymphocytes form a resting population of cells that can become stimulated to maturation and cell division as a result of alterations of their surface states. One of the most useful classes of agents for inducing activation consists of lectins such as Con A. These mitogenic lectins act first at the cell surface (Greaves and Bauminger 1972; Andersson et al. 1972), and it is therefore relevant to ask several key questions about the relationship between SMA states and the activating signal: Are patching and capping necessary for mitogenesis? Do modulation effects alter the mitogenic response? Do components of the SMA play a role in stimulation?

Analysis of the Dose-Response Curve of Stimulation

Mitogens show characteristic unimodal dose-response curves, i.e., concentrations significantly below or above the optimum show no stimulation. For example, the curve obtained after stimulation of mouse splenic lymphocytes by native Con A shows a rising and a falling limb, with a maximum at a concentration of about 5 μg/ml (Fig. 7a) (Gunther et al. 1973). In contrast, the dose-response curve for dimeric succinyl-Con A, which is just as mitogenic as the native lectin, showed no falling limb over a tenfold concentration range beyond the optimal dose. Similar results were obtained with human lymphocytes (Fig. 7b, c), except that the maximal response to the succinyl derivative was lower than that observed with the native lectin (Wang, McClain and Edelman 1975).

Correlation of these effects of Con A and succinyl-Con A on lymphocyte stimulation with their different effects on the modulation of cell surface receptors suggests that neither capping nor the inhibition of receptor mobility is strictly required for the mitogenic stimulation of lymphocytes. In addition, the

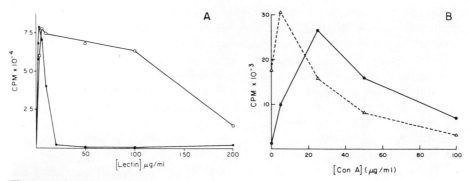

Figure 7

(*A*) Dose-response curve showing the incorporation of [³H]thymidine after stimulation of mouse splenic lymphocytes by Con A (•——•) and succinyl-Con A (o——o). (*B*) The effect of the phorbol ester, TPA, on the stimulation of human lymphocytes by Con A.

data suggest that the rising and falling portions of the dose-response curve reflect two independent events—the stimulation and inhibition, respectively, of lymphocyte proliferation. This conclusion is strongly supported by experiments using Con A and succinyl-Con A with another type of mitogen, 12-0-tetradecanoyl-phorbol-13-0-acetate (TPA).

It has been shown that TPA is independently mitogenic for bovine (Mastro and Mueller 1974) and human (Wang, McClain and Edelman 1975) lymphocytes. The addition of TPA to cultures containing suboptimal doses of Con A (Fig. 7b) or succinyl-Con A (Fig. 7c) greatly enhanced the response of the cells. However, the phorbol ester led to decreased responses when added to cultures containing doses of Con A that alone would be optimally mitogenic, whereas the addition of TPA to cultures containing any dose of succinyl-Con A only enhanced the positive response. No inhibition was seen even at very high doses of the lectin derivative (Fig. 7c).

All of these observations support the conclusion that the typical unimodal dose-response curve can be dissected into stimulatory and inhibitory portions that can be manipulated independently. The fact that high doses of Con A inhibited mitogenesis as well as cell surface receptor mobility, whereas succinyl-Con A had neither effect, suggests the possibility that the high-dose unresponsiveness of Con A-treated cultures may be correlated with the modulation of receptor mobility. It is of particular significance that the effect of low doses of TPA in reducing the cellular response to mitogenic stimulation was lectin-dependent and was observed if, and only if, concentrations of the lectin higher than the optimally mitogenic dose could by themselves deliver the inhibitory signal. The recent experiments of Oliver, Zurier and Berlin (1975) demonstrating a role for cyclic GMP in the modulation of receptor mobility are in accord with this hypothesis, particularly because the action of TPA is thought to be mediated by the cyclic nucleotide (Goldberg et al. 1974). Therefore the possibility arises that components of the SMA responsible for the inhibition of receptor mobility may also be involved in the regulation of various signals involved in cell proliferation.

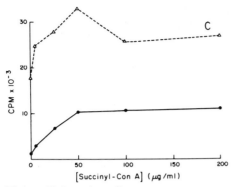

Figure 7 (*continued*)
(*C*) The effect of TPA on the stimulation of human lymphocytes by succinyl-
Con A. (●——●) Cultures containing no TPA; (△– – – –△) cultures con-
taining 100 nM TPA.

Inhibition of Mitogenic Stimulation of Lymphocytes by Colchicine

To test whether microtubular states in the SMA might be involved in cell
activation, the effect of colchicine on Con A-stimulated mitogenesis was
analyzed. It has been shown previously that colchicine inhibits the incorporation of [³H]thymidine in mouse splenic lymphocytes stimulated by Con A
(Edelman, Yahara and Wang 1973) and human peripheral lymphocytes
activated by phytohemagglutinin (Medrano, Piras and Mordoh 1974). Recent
studies (Wang, Gunther and Edelman 1975) indicate that colchicine inhibits
stimulation early, as shown by a decrease in both the level of [³H]thymidine
incorporated and the percentage of lymphocytes transformed into blast cells.
A detailed analysis indicated that the inhibition could not be accounted for
by decreased cell viability, metaphase arrest, blockage of thymidine transport
or inhibition of DNA synthesis in cells already in the S phase of the cell cycle.
Moreover, lumicolchicine had no effect on the same system (Wang, Gunther
and Edelman 1975).

Kinetic data have now been obtained indicating that colchicine blocks
stimulation early in the sequence of events following addition of the mitogen,
and that the time of inhibition may be correlated with the kinetics of cellular
commitment to lectin activation. We have found (Gunther, Wang and Edelman 1974) that when the competitive inhibitor αMM is used to remove Con A
from the cell surface at various times after lectin addition, the rising level of
[³H]thymidine incorporation is proportional to the number of cells committed
to mitogen activation (Fig. 8a). A strikingly similar curve was obtained when
colchicine was introduced into cultures at various times after the addition of
Con A (Fig. 8b). The extent of inhibition decreased the later the colchicine
was added, and no inhibition was observed when the drug was added 30 hours
after the lectin. Analysis of the kinetic curve on a cell-by-cell basis has shown
that the increasing level of [³H]thymidine incorporation with later additions
of colchicine was also correlated with an increasing number of cells responding
to lectin stimulation. The fact that the kinetics of inhibition of stimulation by
αMM and by colchicine are similar suggests that the inhibitory effect of the

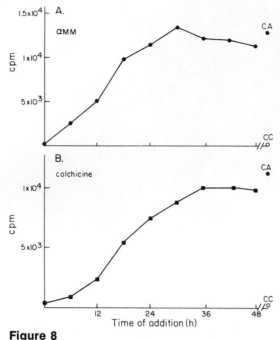

Figure 8

Effect of (*A*) αMM (0.1 M) and (*B*) colchicine (10^{-6} M) added at different times on the incorporation of [^{3}H]thymidine in human lymphocytes stimulated by Con A (20 μg/ml). At 48 hr the cultures were pulsed 2 hr with 6 μCi [^{3}H]thymidine. The points labeled CC and CA represent the incorporation of [^{3}H]thymidine by cultures containing cells alone and by those containing cells plus Con A, respectively.

drug might be temporally correlated with cellular commitment. This provides additional support for the notion that the SMA, or its components, plays a role in the regulation of signals for proliferation.

The SMA As a Signal Regulator

The experiments relating the modulation of cell surface receptor mobility with cell shape and motion and those relating modulation and microtubular assembly to the events in mitogenesis suggest the hypothesis that the SMA is a general signal regulator. A macromolecular assembly that connects components at the cell surface with cytoplasmic structures has exactly the properties required to coordinate the fundamental functions of cell motion, cell division and cell–cell interaction. The receptor portion of the assembly would provide the specificity for response to external signals, the microfilamentous arrays would provide a connection between motion and receptor motion, and the microtubules would provide a reversibly reactive anchorage as well as a component for interaction with small molecules such as activators of nuclear metabolism. The propagated nature of modulation would permit responses to signals induced at a particular portion of the cell surface, and the reversible properties of the system would allow the cell to move from one to another function depending upon external or internal stimuli. These properties are just those required for cell–cell interaction during development, a process in which cell surface receptor states, movement and mitosis are all intimately connected. At present there is no clear-cut evidence to indicate that modulation plays a role in such interactions, and much work must still be done on this complex problem before detailed hypotheses can be formulated. In particular, a search for natural (i.e., tissue-specific) modulating substances might be particularly profitable.

This search and the detailed examination of the structural interactions proposed to occur among the components of the SMA remain as major tasks before one can conclude that phenotypic regulation of cellular function is carried out by these means. Nonetheless, the evidence accumulated so far makes this an extremely attractive hypothesis.

Acknowledgments

This work was supported by U.S. Public Health Service Grants AM 04256, AI 09273 and AI 11378 from the National Institutes of Health and by Grant GB 37556 from the National Science Foundation.

REFERENCES

Andersson, J., G. M. Edelman, G. Möller and O. Sjöberg. 1972. Activation of B lymphocytes by locally concentrated concanavalin A. *Eur. J. Immunol.* **2**:233.

Bretscher, M. 1971. Major human erythrocyte glycoprotein spans the cell membrane. *Nature New Biol.* **231**:229.

Carson, D. A. and H. Metzger. 1974. Interaction of IgE with rat basophilic leukemia cells. IV. Antibody-induced redistribution of IgE receptors. *J. Immunol.* **113**:1271.

de Petris, S. 1974. Inhibition and reversal of capping by cytochalasin B, vinblastine and colchicine. *Nature* **250**:54.

Edelman, G. M., I. Yahara and J. L. Wang. 1973. Receptor mobility and receptor-cytoplasmic interactions in lymphocytes. *Proc. Nat. Acad. Sci.* **70**:1442.

Edidin, M. and A. Weiss. 1972. Antigen cap formation in cultured fibroblasts: A reflection of membrane fluidity and of cell motility. *Proc. Nat. Acad. Sci.* **69**:2456.

Frye, L. D. and M. Edidin. 1970. The rapid intermixing of cell surface antigens after formation of mouse-human heterokaryons. *J. Cell Sci.* **7**:319.

Goldberg, N. D., M. K. Haddox, R. Estensen, J. G. White, C. Lopez and J. W. Hadden. 1974. Evidence of a dualism between cyclic GMP and cyclic AMP in the regulation of cell proliferation and other cellular processes. In *Cyclic AMP, cell growth, and the immune response* (ed. W. Braun, L. Lichtenstein and C. Parker), p. 247. Academic Press, New York.

Greaves, M. F. and S. Bauminger. 1972. Activation of T and B lymphocytes by insoluble phytomitogens. *Nature New Biol.* **235**:67.

Gunther, G. R., J. L. Wang and G. M. Edelman. 1974. The kinetics of cellular commitment during stimulation of lymphocytes by lectins. *J. Cell Biol.* **62**:366.

Gunther, G. R., J. L. Wang, I. Yahara, B. A. Cunningham and G. M. Edelman. 1973. Concanavalin A derivatives with altered biological activities. *Proc. Nat. Acad. Sci.* **70**:1012.

Gwynn, I., R. B. Kemp, B. Jones and U. Groeschel-Stewart. 1974. Ultrastructural evidence for myosin of the smooth muscle type at the surface of trypsin-dissociated embryonic chick cells. *J. Cell Sci.* **15**:279.

Lazarides, E. 1975. Immunofluorescence studies on the structure of actin filaments in tissue culture cells. *J. Histochem. Cytochem.* **23**:507.

Lazarides, E. and K. Weber. 1974. Actin antibody: The specific visualization of actin filaments in non-muscle cells. *Proc. Nat. Acad. Sci.* **71**:2268.

Loor, F. 1974. Binding and redistribution of lectins on lymphocyte membrane. *Eur. J. Immunol.* **4**:210.

Mastro, A. M. and G. C. Mueller. 1974. Synergistic action of phorbol esters in mitogen-activated bovine lymphocytes. *Exp. Cell Res.* **88**:40.

Medrano, E., R. Piras and J. Mordoh. 1974. Effect of colchicine, vinblastine and cytochalasin B on human lymphocyte transformation by phytohemagglutinin. *Exp. Cell. Res.* **86**:295.

Oliver, J. M., T. E. Ukena and R. D. Berlin. 1974. Effects of phagocytosis and colchicine on the distribution of lectin-binding sites on cell surfaces. *Proc. Nat. Acad. Sci.* **71**:394.

Oliver, J. M., R. B. Zurier and R. D. Berlin. 1975. Concanavalin A cap formation on polymorphonuclear leucocytes of normal and beige (Chediak-Higashi) mice. *Nature* **253**:471.

Rutishauser, U., I. Yahara and G. M. Edelman. 1974. Morphology, motility and surface behavior of lymphocytes bound to nylon fibers. *Proc. Nat. Acad. Sci.* **71**:1149.

Ryan, G. B., J. Z. Borysenko and M. J. Karnovsky. 1974. Factors affecting the redistribution of surface-bound concanavalin A on human polymorphonuclear leukocytes. *J. Cell Biol.* **62**:351.

Sela, B., J. L. Wang and G. M. Edelman. 1975. Isolation of lectins of different specificities on a single affinity absorbent. *J. Biol. Chem.* **250**:7535.

Taylor, R. B., P. H. Duffus, M. C. Raff and S. de Petris. 1971. Redistribution and pinocytosis of lymphocyte surface immunoglobulin molecules induced by anti-immunoglobulin antibody. *Nature New Biol.* **233**:225.

Tilney, L. G. and K. R. Porter. 1967. Studies on the microtubules in Heliozoa. II. The effect of low temperature on the formation and maintenance of axopodia. *J. Cell Biol.* **34**:327.

Ukena, T. E. and R. D. Berlin. 1972. Effect of colchicine and vinblastine on the topographical separation of membrane functions. *J. Exp. Med.* **136**:1.

Ukena, T. E., J. Z. Borysenko, M. J. Karnovsky and R. D. Berlin. 1974. Effects of colchicine, cytochalasin B, and 2-deoxyglucose on the topographical organization of surface-bound concanavalin A in normal and transformed fibroblasts. *J. Cell Biol.* **61**:70.

Unanue, E. R. and M. J. Karnovsky. 1974. Ligand-induced movement of lymphocyte membrane macromolecules. V. Capping, cell movement, and microtubular function in normal and lectin-treated lymphocytes. *J. Exp. Med.* **140**:1207.

Unanue, E. R., W. D. Perkins and M. J. Karnovsky. 1972. Ligand-induced movement of lymphocyte membrane macromolecules. I. Analysis by immunofluorescence and ultrastructural radioautography. *J. Exp. Med.* **136**:885.

Wang, J. L., G. R. Gunther and G. M. Edelman. 1975. Inhibition by colchicine of the mitogenic stimulation of lymphocytes prior to the S phase. *J. Cell Biol.* **66**:128.

Wang, J. L., D. A. McClain and G. M. Edelman. 1975. Modulation of lymphocyte mitogenesis. *Proc. Nat. Acad. Sci.* **72**:1917.

Weisenberg, R. C., G. G. Borisy and E. W. Taylor. 1968. The colchicine-binding protein of mammalian brain and its relation to microtubules. *Biochemistry* **7**:4466.

Wessells, N. K., B. S. Spooner, J. F. Ash, M. O. Bradley, M. A. Luduena, E. L. Taylor, J. T. Wrenn and K. M. Yamada. 1971. Microfilaments in cellular and developmental processes. *Science* **171**:135.

Willingham, M. C., R. E. Ostlund and I. Pastan. 1974. Myosin is a component of the cell surface of cultured cells. *Proc. Nat. Acad. Sci.* **71**:4144.

Yahara, I. and G. M. Edelman. 1972. Restriction of the mobility of lymphocyte immunoglobulin receptors by concanavalin A. *Proc. Nat. Acad. Sci.* **69**:608.

―――. 1973a. The effects of concanavalin A on the mobility of lymphocyte surface receptors. *Exp. Cell Res.* **81**:143.

―――. 1973b. Modulation of lymphocyte receptor redistribution by concanavalin A, anti-mitotic agents and alterations of pH. *Nature* **246**:152.

―――. 1975a. Modulation of lymphocyte receptor mobility by locally bound concanavalin A. *Proc. Nat. Acad. Sci.* **72**:1579.

―――. 1975b. Electron microscopic analysis of the modulation of lymphocyte receptor mobility. *Exp. Cell. Res.* **91**:125.

Filaments, Microtubules and Colchicine Receptors in Capped Ovarian Granulosa Cells

John I. Clark and David F. Albertini*

Department of Anatomy and
Laboratory of Human Reproduction and Reproductive Biology
Harvard Medical School, Boston, Massachusetts 02115

The study of concanavalin A (Con A)-induced capping in ovarian granulosa cells from the mammalian ovary demonstrates that filaments, microtubules and fluorescein-labeled colchicine (FTC-CLC) receptors are clearly associated with capped Con A binding sites. Although the morphological observation of filaments and microtubules does not define their role in the capping event, filaments and microtubules appear to be involved in the formation of the cap (Ryan, Borysenko and Karnovsky 1974; Unanue and Karnovsky 1974; de Petris 1975; Yahara and Edelman 1975a). Our results have encouraged us to compare capping with phagocytosis, cytokinesis and cell motility, processes in which the involvement of surface and cytoplasmic elements may have a common basis.

Con A-induced capping on ovarian granulosa cells involves several discrete events occurring on the cell membrane: (1) the interaction of the tetravalent mitogen Con A with surface receptors; (2) cross-linking of lectin receptors to form clusters in a temperature-dependent but energy-independent step; (3) an energy-requiring directed translocation of clusters to central parts of the cell surface where patches and caps form; and (4) the interiorization of capped receptors. This paper describes the morphology of the capped surface receptors and the coincident cellular disposition of submembraneous filaments, microtubules and FTC-CLC receptors. Formation of the Con A-induced cap and subsequent phagocytosis will be described in detail elsewhere.

RESULTS

The distribution of Con A receptors was studied using hemocyanin- and horseradish peroxidase-labeled Con A (Albertini and Clark 1975). Following Con A labeling at 0°C, cells were warmed to 37°C, and 80% of the Con A-treated cells formed caps after incubation at 37°C for 35–45 minutes. The

*Present address: Department of Physiology, School of Medicine, University of Connecticut Health Center, Farmington, Connecticut 06032.

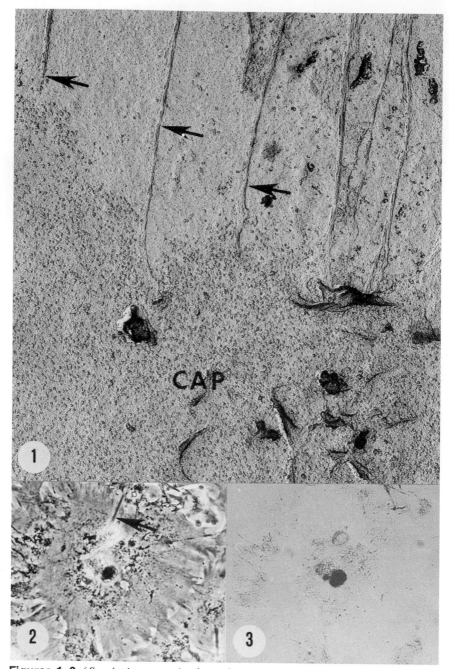

Figures 1–3 (*See facing page for legend*)

caps appeared as circular (7–12 μm in diameter) aggregates of Con A binding sites situated on the cell surface, and Con A receptors were completely
cleared from peripheral areas of the cell (Figs. 1 and 3). Uncapped cells
showed a dispersed distribution of hemocyanin- or horseradish peroxidase-
labeled Con A binding sites over the cell surface. Microspikes or retractile

fibers extended from the margins of the cap (Figs. 1 and 2), which was localized to a pronounced protrusion on the dorsal cell surface (Fig. 2).

Transmission electron microscopy showed both microfilaments and microtubules in the cytoplasm underlying the capped membrane receptors (Fig. 4). Thin filaments (6 nm) were organized into a dense ring appearing directly beneath the margin of the capped Con A receptors (Fig. 5). The band correlated with a cytoplasmic dense band seen with light microscopy to circumscribe the cell nucleus and confine cellular inclusions (Fig. 2). Intermediate filaments were found beneath the cap apparently associated with the filamentous dense band. Dense bodies were often observed embedded in the meshwork of thin filaments (Fig. 5). Untreated cells contained submembraneous filaments, but the filamentous ring and complex network associated with the capped Con A binding sites were never observed. Cytochalasin B-treated cells displayed many peripherally localized clusters and patches of Con A binding sites but rarely formed the large circular aggregates typical of normal caps.

In oblique sections through the cap, some microtubules were observed in close association with cell membranes, whereas others were oriented normal to the surface of the capped portion of the plasma membrane (Fig. 4) (Albertini and Clark 1975). Microtubules and filaments in capped cells were never observed in direct attachment to the cell membrane.

Colchicine and vinblastine had no apparent effect on the aggregation of the Con A binding sites. In the presence of colchicine and vinblastine sulfate, the caps were often located at the periphery of the cells, and cytoplasmic microtubules were rarely observed. Uncapped, vinblastine-treated cells contained modest numbers of small paracrystals that were not confined to any particular part of the cytoplasm. Vinblastine-induced paracrystals in capped cells were large and localized beneath the capped Con A receptors (Fig. 6).

FTC-CLC (Fig. 7) was used to observe the distribution of colchicine receptors in the Con A-treated cells. FTC-CLC fluorescence was concentrated in the capped region of the cells, and this concentration of FTC-CLC binding was absent in uncapped cells. Capped and uncapped cells incubated with FTC-CLC showed binding in association with the nuclear envelope, a nuclear inclusion and a pattern of diffuse, granular cytoplasmic fluorescence (Fig. 9). Metaphase cells were brilliantly fluorescent following treatment with FTC-CLC (Fig. 8). No fluorescence was observed in capped or uncapped cells treated with fluorescein alone, and all FTC-CLC fluorescence could be competed out with unlabeled colchicine.

Figures 1–3

(*Fig. 1*) Shadow-cast replica of a cell labeled with Con A and hemocyanin at 4°C and warmed to 37°C for 30 min prior to fixation. The aggregated hemocyanin-labeled Con A receptors appear as a large, granular cap on the cell surface. Prominent retractile fibers emanate from the periphery of the cap (arrows). 10,800×.

(*Fig. 2*) Con A-induced cap labeled with horseradish peroxidase and photographed using phase optics, demonstrating a large, centrally located cap and microspikes (arrow). 450×.

(*Fig. 3*) Corresponding light micrograph of a peroxidase-labeled Con A cap illustrating the localized aggregation of binding sites. 450×.

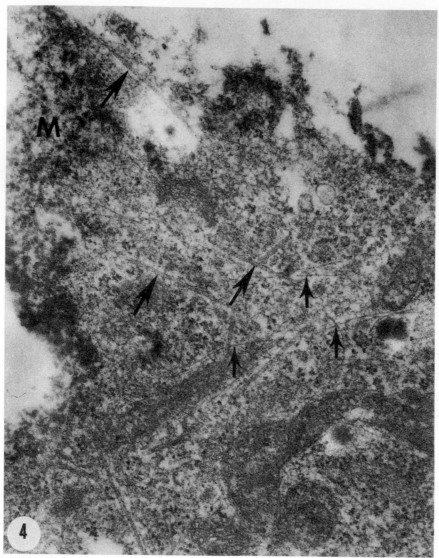

Figure 4
Transmission electron micrograph from a grazing section through the cap on a granulosa cell 30 min after treatment with Con A and horseradish peroxidase at 37°C. Note the dark peroxidase reaction product situated on the capped plasma membrane (M) and the closely associated cytoplasmic microtubules (large arrows) and intermediate filaments (small arrows). 48,420×.

DISCUSSION

The capping event induced by treatment of ovarian granulosa cells with Con A appears to be a special case of active membrane movement invoking the participation of cytoplasmic microtubules and filaments. The filamentous and microtubular structures observed beneath the capped Con A binding sites have

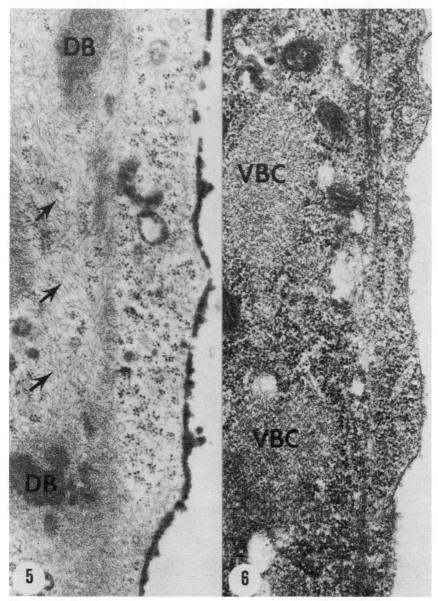

Figures 5–6

(*Fig. 5*) A portion of a Con A-peroxidase cap, showing the peroxidase-labeled capped membrane on the right and dense bodies (DB) embedded in a layer of thin filaments (6 nm) beneath the cap. Thicker filaments (arrows) are located below the band of thin filaments. 45,900×.

(*Fig. 6*) A vinblastine-treated capped cell, showing characteristic paracrystals (VBC) beneath the capped Con A binding sites labeled with hemocyanin. 34,920×.

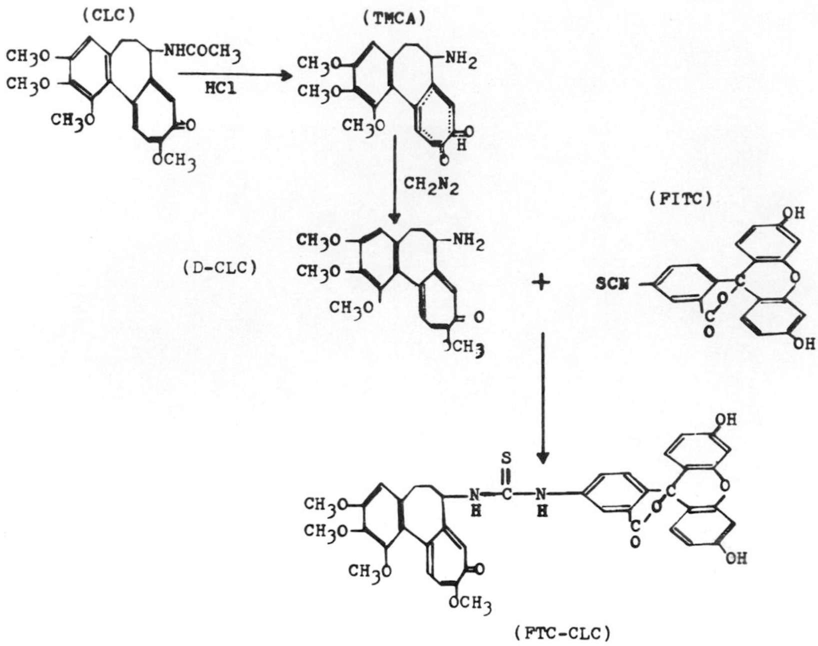

Figure 7
The synthesis of FTC-CLC follows the procedure of Wilson and Friedkin (1966) to give deacetyl colchicine (D-CLC), which is reacted with fluorescein isothiocyanate (FITC), yielding the product fluorescein thiocarbamyl colchicine, FTC-CLC. The compounds of each step are abbreviated: colchicine (CLC); trimethyl colchicinnic acid (TMCA).

correlates in mitotic cells (Schroeder 1973), macrophages (Reaven and Axline 1973) and migratory glial cells (Spooner, Yamada and Wessels 1971). For example, the filamentous band observed in association with the Con A-induced cap may have the contractile properties of the filamentous ring associated with the directed deformation of the plasma membrane occurring in cytokinesis (Schroeder 1972). Six-nanometer filaments comprise both structures, and each ring defines a plane that partitions the cell into two cytoplasmic compartments. In the mitotic cell, the compartments are of apparently equal composition, whereas the "cap compartment" is never cleanly separated as the daughter cell is during cytokinesis, suggesting that Con A treatment of ovarian granulosa cells may activate only part of the mitogenic response induced in lymphocytes (Wang, Gunther and Edelman 1975).

In addition to the formation of microfilaments, capping induced a local accumulation of microtubules in untreated cells and vinblastine paracrystals in vinblastine-treated cells. These tubulin-containing structures may appear in the cap by being translated with the surface Con A receptors during capping or by assembly subsequent to the migration of membrane-associated microtuble organizing centers. The absence of basal bodies, centrioles and other structures associated with microtubule organization (Pickett-Heaps 1969) might imply that microtubules or vinblastine crystals are not assembled in association with the cap but are moved as intact structures into the cap by

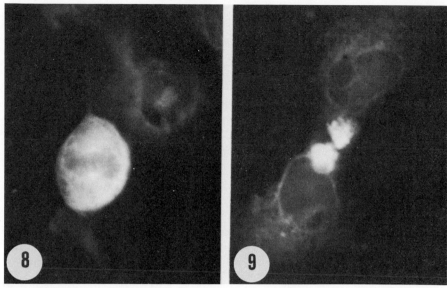

Figures 8–9

(*Fig. 8*) Fluorescence micrograph of cultured granulosa cells treated with FTC-CLC, showing the intense fluorescence associated with a metaphase cell. An adjacent interphase cell shows considerably less fluorescence. The cells were fixed at $-10°C$ in acetone and stained 30 min using 5×10^{-6} M FTC-CLC in 0.14 M phosphate buffer pH 7.2. 540×.

(*Fig. 9*) Two Con A capped granulosa cells treated with FTC-CLC demonstrate the intense fluorescence associated with the capped portions of each cell. 600×.

the same forces responsible for aggregation of the Con A binding sites. Alternatively, the localization of microtubules and tubulin paracrystals beneath the cap may suggest that a pool of tubulin subunits is concentrated in the cap during the migration of Con A binding sites. In the untreated cells, the subunits may assemble into cap-associated microtubules instead of vinblastine paracrystals. The possibility that polymerizable tubulin is preferentially associated with capped Con A binding sites suggests that membrane associated microtubule organizing centers can cap with the Con A receptors. Membrane-bound organizing sites for tubulin are implied in the demonstration of colchicine-sensitive tubulin complexes on polymorphonuclear leukocyte membranes (Becker, Oliver and Berlin 1975), [^{14}C]colchicine-binding to membrane fractions of homogenized liver (Stadler and Franke 1974) and tubulin associated with particulate components of isolated nerve endings (Shelanski et al. 1972).

Another consequence of capping was the striking localization of FTC-CLC receptors in the capped area of the cell. It is not clear whether FTC-CLC binds colchicine receptors on the capped membrane or binds tubulin associated with microtubule organizing sites of the capped membrane. Our interpretation of the FTC-CLC localization considers that colchicine (and FTC-CLC) alters the equilibrium between assembled and disassembled microtubules so as to inhibit assembly while favoring disassembly (Inoué and Sato 1967). Apparently FTC-CLC labels cellular concentrations of polymerizable tubulin necessary for the formation of cap-associated microtubules. Similarly, the apparent increase in FTC-CLC binding to cells in metaphase was attributed to the increased concentrations of polymerizable tubulin expected for the formation of mitotic apparatus microtubules (Stephens 1973). Although remarkable, the existence of local concentrations of FTC-CLC binding tubulin in association with the formation of intact microtubules agrees with in vivo and in vitro studies showing that microtubule polymerization is concentration-dependent (Gaskin, Cantor and Shelanski 1974; Inoué, Borisy and Kiehart 1974). Concentrated FTC-CLC receptors, localized vinblastine-induced paracrystals and numerous microtubules beneath capped Con A binding sites appear to support the provocative suggestion that membrane elements as well as cytoplasmic organizing centers may influence the equilibrium between assembled and disassembled microtubules (Yahara and Edelman 1975b; Albertini and Clark 1975).

Analogous interactions between membranes and microtubules may be responsible for directing membrane displacements during mitosis (Rappaport 1973), flagellar growth (Rosenbaum and Child 1967) and cell motility (Vasiliev et al. 1970). If the mobility of surface components is modulated by microtubules, as suggested in a number of models, conclusive evidence for direct linkages from microtubules to or through the membrane await demonstration. This will be difficult if the mechanism of control is not by direct linkages between microtubules and membranes but instead involves calcium and magnesium fluxes between these components, one of the earliest functions ascribed to microtubules (Slautterback 1963). That the capped Con A receptors may include or combine with tubulin organizing elements within the membrane to induce the local concentration of tubulin promoting microtubule assembly is equally interesting as speculation. Although the specific nature of the membrane-mediated appearance of microtubules and microfilaments remains unclear, we feel that the concomitant reorganization of membranes and cytoplasmic components elicited by capping of ovarian granulosa cells is related to other morphogenetic events where surface alterations ultimately affect the entire cell structure and function.

Acknowledgments

This research was supported by Training Grant 5 T01 GM00406 from the National Institutes of Health, United States Public Health Service, and a postdoctoral fellowship from the Pharmaceutical Manufacturers Association Foundation to J. I. C.

REFERENCES

Albertini, D. F. and J. I. Clark. 1975. Membrane-microtubule interactions: Concanavalin A capping induced redistribution of cytoplasmic microtubules and colchicine binding proteins. *Proc. Nat. Acad. Sci.* **72**:4976.

Becker, J. S., J. M. Oliver and R. D. Berlin. 1975. Fluorescence techniques for following interactions of microtubule subunits and membranes. *Nature* **254**:152.

de Petris, S. 1975. Concanavalin A receptors, immunoglobulins, and O antigen of the lymphocyte surface. *J. Cell Biol.* **65**:123.

Gaskin, F., C. R. Cantor and M. L. Shelanski. 1974. Turbidimetric studies of the *in vitro* assembly and disassembly of porcine neurotubules. *J. Mol. Biol.* **89**:737.

Inoué, S. and H. Sato. 1967. Cell motility by labile association of molecules. *J. Gen. Physiol.* (Suppl.) **50**:259.

Inoué, S., G. Borisy and D. P. Kiehart. 1974. Growth and liability of Chaetopterus oocyte mitotic spindles isolated in the presence of porcine brain tubulin. *J. Cell Biol.* **62**:175.

Pickett-Heaps, J. D. 1969. The evolution of the mitotic apparatus: An attempt at comparative ultrastructural cytology in dividing plant cells. *Cytobios* **1**:257.

Rappaport, R. 1973. Cleavage furrow establishment—A preliminary to cylindrical shape change. *Amer. Zool.* **13**:941.

Reaven, E. P. and S. G. Axline. 1973. Subplasmalemmal microfilaments and microtubules in resting and phagocytizing cultivated macrophages. *J. Cell Biol.* **59**:12.

Rosenbaum, J. L. and F. M. Child. 1967. Flagellar regeneration in protozoan flagellates. *J. Cell Biol.* **34**:345.

Ryan, G. B., J. Z. Borysenko and M. J. Karnovsky. 1974. Factors affecting the redistribution of surface bound concanavalin A on human polymorphonuclear leukocytes. *J. Cell Biol.* **62**:351.

Schroeder, T. E. 1972. The contractile ring. *J. Cell Biol.* **53**:419.

———. 1973. Cell constriction: Contractile role of microfilaments in division and development. *Amer. Zool.* **13**:949.

Shelanski, M. L., H. Feit, R. W. Berry and M. P. Daniels. 1972. Some biochemical aspects of neurotubule and neurofilament proteins. In *Functional and structural proteins of the nervous system* (ed. A. N. Davison, P. Mandel and I. G. Morgan), p. 55. Plenum Press, New York.

Slautterback, D. B. 1963. Cytoplasmic microtubules. *J. Cell Biol.* **18**:367.

Spooner, B. S., K. M. Yamada and N. K. Wessels. 1971. Microfilaments and cell locomotion. *J. Cell Biol.* **49**:595.

Stadler, J. and W. W. Franke. 1974. Characterization of the colchicine binding of membrane fractions from rat and mouse liver. *J. Cell Biol.* **60**:297.

Stephens, R. E. 1973. A thermodynamic analysis of mitotic spindle equilibrium at active metaphase. *J. Cell Biol.* **57**:133.

Unanue, E. R. and M. J. Karnovsky. 1974. Ligand induced movement of lymphocyte membrane macromolecules. V. Capping, cell movement and microtubular function in normal and lectin treated lymphocytes. *J. Exp. Med.* **140**:1207.

Vasiliev, J. M., I. M. Gelfand, L. V. Domnina, O. Y. Ivanova, S. G. Komm and L. V. Olshevskaja. 1970. Effect of colcemid on the locomotory behavior of fibroblasts. *J. Embryol. Exp. Morphol.* **24**:625.

Wang, J. L., G. R. Gunther and G. M. Edelman. 1975. Inhibition by colchicine of the mitogenic stimulation of lymphocytes prior to the S phase. *J. Cell Biol.* **66**:128.

Wilson, L. and M. Friedkin. 1966. Biochemical events of mitosis. I. Synthesis and properties of colchicine labeled with tritium in its acetyl moiety. *Biochemistry* **5**:2463.

Yahara, I. and G. M. Edelman. 1975a. Electron microscopic analysis of the modulation of lymphocyte receptor mobility. *Exp. Cell Res.* **91**:125.

———. 1975b. Modulation of lymphocyte receptor mobility by concanavalin A and colchicine. *Ann. N.Y. Acad. Sci.* **253**:455.

Section 4

THE USE OF ANTIBODIES
IN THE LOCALIZATION
OF CONTRACTILE PROTEINS
AND MICROTUBULES
IN MUSCLE AND NON-MUSCLE CELLS

Introductory Remarks: Use of Antibodies for the Intracellular Localization of Proteins Involved in Cell Motility

Gerald Goldstein

Department of Medicine, University of Virginia
Charlottesville, Virginia 22903

This section focuses almost exclusively upon the proper utilization and interpretation of immunofluorescent techniques. With the exceptions of the opening paper by Pepe and that by Schollmeyer on immunoelectron microscopy techniques, papers in this section attempt to examine, in isolation, individual steps in the utilization of fluorescent antibodies. These introductory remarks will focus on some of the items of the immunofluorescent sequence as listed below.

Purification of antigen
Immunization of animals
Characterization of antisera
Purification of antibodies
Fluorescent labeling
Fixation of cells, tissues
Fluorescence microscopy and filters
Direct and Indirect immunofluorescence
Interpretation of staining

The extraordinarily rich variety of structures included under the term "antigen" precludes any technical discussion of the techniques for purifying antigen. In any complex biological preparation, the purification of the compound of interest, i.e., the "antigen," represents a loss of antigen at each step but is accompanied by a greater loss of the other compounds in the original preparation. Viewed from the perspective of immunization techniques, the losses entailed in obtaining a high degree of purity may not be justified.

The methods for obtaining large amounts of antibody from immunized animals are excellently reviewed in terms of principles and details by Chase (1967). For immunofluorescence it is desirable to have high-titered antibodies. A commonly used technique for raising such antisera is to incorporate "the antigen" in some form of antigen. Two effects of the adjuvant are cited. To some degree, the impact of the adjuvant, by magnifying the immunogenicity of all components in the antigen preparation, negates the effects of antigen purification. Therefore at some point in the purification of antigen for immunization it may be possible to eliminate some later steps.

A second feature of adjuvants is that they may contain their own antigen components. The Mycobacteria widely used in Freund's adjuvant contains many antigens. At least 11 antigen-antibody systems have been defined in diagnostic tuberculins prepared from Mycobacteria (Chase and Kawata 1975). Not all animals produced antibodies to all or many of the components.

Formation of precipitin bands in agar gel diffusion is the common method used to characterize the antibody response. The use of detergents in preparing antigens does not interfere with precipitin band formation. The presence of more than one precipitin band is indicative of the presence of antibody to more than one antigenic determinant. However, the converse is not true; the detection of only one precipitin band does not indicate that the antisera contains antibodies to only a single antigen. With well-defined antigen-antibody systems, agar gel diffusion can be used as a semiquantitative measure of antibody concentration (Beutner et al. 1970). The formation of an antigen-antibody precipitin band requires an interaction between molecules each of which has more than one reactive site. The localization of antigen by immunofluorescence requires only the primary interaction between an antigenic determinant and the antibody combining site. Thus nonprecipitating antigen-antibody systems may be successfully visualized by immunofluorescence.

The following papers do not discuss the techniques of purification of antibodies and fluorescent labeling (the reader is directed to the earlier work of Cebra and Goldstein [1965] and Herbert, Pittman and Cherry [1971]), and attention is only briefly focused on the tissue binding property of the Fc portion of the antibody molecule. Elimination of this region by enzymatic splitting may yield a fluorescent reagent with less nonimmunologic interactions.

The effect of fixatives upon the structure of antigenic components, however, is extensively discussed. No comprehensive fixation procedure is applicable to all antigens. Feltkamp-Vroom (1975) and Arnold, Kalden and Von Mayersbach (1975) discuss fixation procedures for immunofluorescence. It is considered desirable to prepare tissue for both fluorescence microscopy and electron microscopy by the same fixation procedure. In this way, the actual structural integrity can be determined by electron microscopy and the immunologic studies done by fluorescence microscopy. (For an excellent review of fluorescence microscopy, see Ploem 1973.)

The major advantages of the direct and indirect immunofluorescent techniques are listed below. Where precipitating antibodies are present it is usually possible to prepare good direct fluorescent antibody reagents. There are many illustrations of the use of contrastingly labeled antibodies (Goldstein 1971; Hijmans, Schuit and Hulsing-Hesselink 1971; Brandtzaeg 1975).

Advantages of Indirect Immunofluorescence

1. Single fluorescent reagent useful with different antibodies
2. Enhanced sensitivity

Advantages of Direct Immunofluorescence

1. Permits simultaneous visualization of two antigens with contrasting colored antibodies
2. Permits use of inhibition tests for evaluating specificity of staining

Questions about the interpretation of staining arise throughout this section. In using indirect immunofluorescence it is difficult to establish the immunologic basis of the staining by directly inhibiting the fluorescence with unlabeled antibody to the primary antigen. Reliance upon the specificity of staining is achieved by two features. Using the same fluorescent antibody, antibody to tubulin is located at different sites than antibody to actin (Weber, this volume). Furthermore, absorption of antibody by insolubilized antigen eliminates the fluorescent staining, whereas absorption with unrelated antigens does not.

Where the amounts of antigen to be localized are abundant and known to be present, rigorous proof for the localization of antigen is not required. If fluorescence is seen at places where antigen is not thought to be present, the fluorescence can be eliminated by intellectual absorption.

Direct immunofluorescence can be tested for specificity by inhibition with unlabeled antibody. All immunologic staining should be completely inhibited. If all staining is inhibited, it only establishes that the reactions are on an immunologic basis. It does not establish the identity of the antigen-antibody system. Purification of specific antibody, removal of Fc fragments, and proper labeling of antibody increase the confidence with which one can interpret immunohistochemical localization of trace amounts of antigen.

REFERENCES

Arnold, W., J. R. Kalden and H. Von Mayersbach. 1975. Influence of different histologic preparation methods on preservation of tissue antigens in the immunofluorescent antibody technique. *Ann. N.Y. Acad. Sci.* **254**:27.

Beutner, E. H., G. Wick, M. Sepuldeva and K. Morrison. 1970. A reverse immunodiffusion assay for antibody protein concentration in antisera or conjugates to human IgG. In *Standardization in immunofluorescence* (ed. E. J. Holborow), p. 165. Blackwell Scientific Publications, Oxford.

Brandtzaeg, P. 1975. Rhodamine conjugates: Specific and nonspecific binding properties in immunohistochemistry. *Ann. N.Y. Acad. Sci.* **254**:35.

Cebra J. J. and G. Goldstein. 1965. Chromatographic purification of tetramethylrhodamine immunoglobulin conjugates and their use in the cellular localization of rabbit gamma globulin polypeptide chains. *J. Immunol.* **95**:230.

Chase, M. W. 1967. Production of antiserum. In *Methods in immunology and immunochemistry* (ed. C. A. Williams and M. W. Chase), vol. 1, p. 197. Academic Press, New York.

Chase, M. W., and H. Kawata. 1975. Multiple mycobacterial antigens in diagnostic tuberculins. *Devel. Biol. Standard.* **29**:308.

Feltkamp-Vroom, T. 1975. Preparation of tissues and cells for immunohistochemical processing. *Ann. N.Y. Acad. Sci.* **254**:21.

Goldstein, G. 1971. Immunofluorescent detection of human antibodies reactive with tumor. *Ann. N.Y. Acad. Sci.* **177**:279.

Herbert, G. A., B. Pittman and W. B. Cherry. 1971. The definition and application of evaluation techniques as a guide for the improvement of fluorescent antibody reagents. *Ann. N.Y. Acad. Sci.* **177**:54.

Hijmans, W., H. R. E. Schuit and E. Hulsing-Hesselink. 1971. An immunofluorescence study on the intracellular immunoglobulins in human bone marrow cells. *Ann. N.Y. Acad. Sci.* **177**:290.

Ploem, J. D. 1973. Immunofluorescence microscopy comparisons of conventional systems with interference filters and epi-illumination. In *Immunopathology of the skin* (ed. E. H. Beutner et al.), p. 248. Dowden, Hutchinson and Ross, Stroudsburg, Pennsylvania.

Detectability of Antibody in Fluorescence and Electron Microscopy

Frank A. Pepe

Department of Anatomy, Medical School
University of Pennsylvania, Philadelphia, Pennsylvania 19174

One of the main problems we face in the use of antibody techniques is establishing the specificity of the antibody. In general, if immunodiffusion of the antibody against a crude preparation of the antigen gives a single precipitin line, then we are satisfied that the antibody is specific and that the localization observed therefore represents specific localization of the antigen. This presentation has three main purposes: (1) to show that immunodiffusion is not always an adequate criterion for antibody specificity; (2) to show that it is possible to observe localization of unwanted antibodies in fluorescent and electron microscopy even though these antibodies may not be readily detectable in immunodiffusion; and (3) to describe how it is possible to improve the specificity of localization of the antibody.

These studies have been made with skeletal muscle. However, the implications of the findings with respect to detectability and specificity of antibody localization are generally applicable.

Immunodiffusion

The antibody preparations used in this work were obtained using either myosin (purified by repeated precipitation) or C-protein as an immunogen in either rabbits or goats (Pepe 1967a,b, 1973, 1975; Pepe and Drucker 1975). More specific antibody was isolated from these antibody preparations by using column-purified myosin (Richards et al. 1967) or C-protein (Offer, Moos and Starr 1973) coupled to PAB (p-aminobenzyl)-cellulose to bind the specific antibody; the specific antibody was then dissociated by reducing the pH and concentrated by precipitation with ammonium sulfate (Pepe 1973; Pepe and Drucker 1975).

Contaminating antibodies were easily detectable in the anti-myosin obtained when myosin purified by repeated precipitation was used as the immunogen (anti-precipitated myosin) (Pepe 1973; Pepe and Drucker 1975; G. Offer, unpubl.). However, these were only detectable if crude myosin (i.e., the first

337

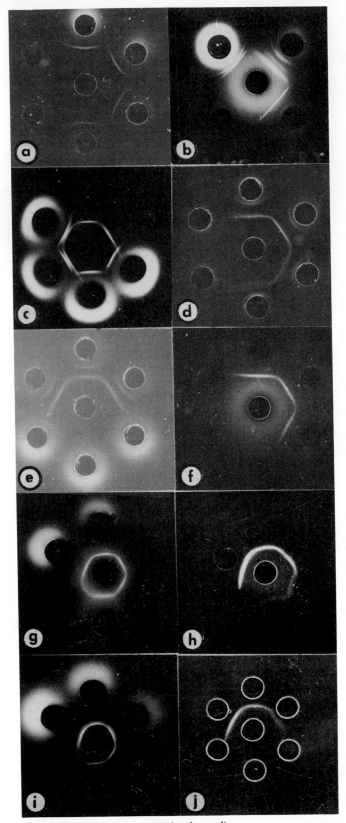

Figure 1 (*See facing page for legend*)

precipitate of the crude myosin extract, Fig. 1b) or the impurities (Fig. 1d) isolated from myosin preparations by column chromatography (Richards et al. 1967; Pepe 1973) were diffused against the antibody. When the precipitated myosin used as the immunogen was diffused against the antibody, a single precipitin line was observed (Fig. 1a), with no indication of the presence of antibody to impurities (Pepe 1973). Therefore although the amount of impurities present in the immunogen was sufficient to produce large amounts of antibody, it was not sufficient to be detectable in the immunogen by immunodiffusion.

The major antibody contaminant present in anti-precipitated myosin has been identified as anti-C-protein (Pepe 1973; Offer 1973 and unpubl.; Pepe and Drucker 1975). This can be seen clearly in Figure 1c, where two precipitin lines are observed with crude myosin, one of the lines being continuous with that produced with C-protein and the other with that produced with column-purified myosin. On serial dilution of crude myosin against the antibody (Fig. 1b), another unidentified antibody is detectable. Both of these unwanted antibodies are detectable when the impurities isolated from myosin by column chromatography are diffused against the antibody (Fig. 1d). In Figure 1e the antibody was absorbed with column-purified myosin prior to use in immunodiffusion. No antibody specific for myosin is detectable by

Figure 1

Immunodiffusion. (*a*) Anti-precipitated chicken myosin in the center well (29 mg/ml IgG). Serial twofold dilution of reprecipitated chicken myosin (immunogen) starting at 5 mg/ml in top well. (*b*) Anti-precipitated chicken myosin in the center well (20 mg/ml IgG). Serial fourfold dilution of crude (precipitate from the initial extract) chicken myosin starting at 4 mg/ml. (*c*) Anti-precipitated chicken myosin in the center well (15 mg/ml IgG). Two wells at top right contain chicken C-protein at 1 mg/ml. Two wells at bottom right contain column-purified chicken myosin at 5 mg/ml. Two wells at bottom left contain crude chicken myosin at 6 mg/ml. (*d*) Anti-precipitated chicken myosin in the center well (29 mg/ml IgG). The outer wells contain the impurities isolated from chicken myosin on column purification: the first two wells at a concentration of 0.5 mg/ml, the next two at 0.25 mg/ml, and the last two at 0.125 mg/ml. (*e*) Anti-precipitated chicken myosin after absorbing with column-purified chicken myosin insolubilized on PAB cellulose (20 mg/ml IgG). Top three wells contain impurities (as in *d*) at a concentration of 0.5 mg/ml. The bottom three wells contain column-purified myosin at 5 mg/ml. (*f*) Anti-precipitated chicken myosin in the center well (29 mg/ml IgG). Serial twofold dilution of C-protein in the outer wells beginning at 1 mg/ml. (*g*) Anti-chicken C-protein in the center well (30 mg/ml IgG). Serial twofold dilution of impurities (as in *d*) starting with a concentration of 4.8 mg/ml. (*h*) Anti-chicken C-protein in the center well (33 mg/ml IgG). Serial twofold dilution of chicken C-protein starting at 1 mg/ml. (*i*) Specific anti-C isolated from anti-chicken C-protein using C-protein isolubilized on PAB-cellulose in the center well (0.5 mg/ml). Serial twofold dilution of impurities (as in *d*) starting with a concentration of 4.8 mg/ml. (*j*) Specific anti-C (as in *i*) in the center well (0.6 mg/ml). Serial twofold dilution of chicken C-protein starting with a concentration of 1 mg/ml.

immunodiffusion, but the two lines with impurities are clearly visible. This verifies the observation in Figure 1 b and d that there are at least two different antibodies presented in addition to antibody specific for myosin (Pepe 1973).

These results with anti-precipitated myosin show that antibodies to impurities can be detected in the anti-precipitated myosin by immunodiffusion against either a sufficiently crude myosin preparation or the impurities isolated from myosin by column chromatography. Antibody to impurities present in the anti-C were not as easily detectable.

C-protein was prepared from rabbit and chicken muscle as described by Offer, Moos and Starr (1973) with a slight modification in the procedure used with chicken muscle (Pepe and Drucker 1975). On immunodiffusion of this C-protein against anti-precipitated myosin, only a single precipitin line is observed (Fig. 1c,f), indicating that the C-protein is not significantly contaminated with the other non-myosin component detectable in crude myosin preparations. Antibody prepared using this C-protein as the immunogen gave a single precipitin line on diffusion against either the immunogen (Fig. 1h) or the impurities isolated on column purification of myosin (Fig. 1g). The impurities isolated from myosin constitute the crudest available C-protein preparation. Therefore since only a single precipitin line is observed with the anti-C even against the crudest available C-protein preparation and since only a single precipitin line is observed when C-protein is diffused against anti-precipitated myosin (Fig. 1c,f), it appears from immunodiffusion that the antibody is specific for C-protein and the immunogen is not significantly contaminated with impurities.

In addition to using the IgG fraction of the antiserum, specific anti-C was isolated using C-protein coupled to PAB-cellulose (Pepe 1973; Pepe and Drucker 1975). In this way, specific anti-C was isolated from anti-precipitated myosin as well as from anti-C-protein. Both of these isolated antibodies behaved similarly to the anti-C-protein in that a single line was observed on immunodiffusion against either C-protein or the impurities isolated from myosin preparations (Fig. 1i,j). Therefore no antibody other than that specific for C-protein is detectable in either the anti-C-protein or the isolated, specific anti-C preparations by these immunodiffusion studies.

Detectability of Antibody in Fluorescence and Electron Microscopy

The differences in the detectability of antibody in fluorescence and electron microscopy have been discussed previously (Pepe 1968, 1975; Pepe and Drucker 1975). These can be summarized as follows: (1) Fluorescein-tagged antibody can be identified in fluorescent microscopy by its fluorescence on irradiation with ultraviolet light (Nairn 1964; Goldman 1968). (2) In electron microscopy, antibody tagged with a marker such as ferritin (Singer and Schick 1961; Ohtsuki et al. 1967; Samosudova et al. 1968) or horseradish peroxidase (Nakane and Pierce 1966; Nakane and Kawaoi 1974) can be identified by visualizing the marker or the products of the enzymatic reaction. The difficulty encountered with tagged antibodies is that in some cases the size of the tag makes the tagged antibody sterically inaccessible to the antigenic sites (Pepe 1968). The localization of untagged antibody in electron microscopy (Pepe, Finck and Holtzer 1961; Pepe and Huxley, 1964; Pepe 1966a,b, 1967a,b, 1968) can be identified by visualizing an increase in protein density in a resolvable area or the alteration of a structure due to the adherence of antibody protein. In electron microscopy, the use of a tag provides a more direct method of antibody localization than the use of untagged antibodies. So far, only ferritin-tagged antibody has been used to study muscle (Ohtsuki et al. 1967; Samosudova et al. 1968), and in these cases, the ferritin-tagged antibody did not enter the region of overlap of myosin and actin filaments, presumably because the tag increased the size of the antibody sufficiently to prevent entrance in the space between the filaments (Pepe 1968).

Some dramatic examples of the difference in detectability of untagged antibody in electron microscopy and fluorescein-tagged antibody in fluorescence microscopy are shown in Figure 2. The localization of anti-precipitated myosin in electron microscopy and fluorescence microscopy is shown in Figure 2 a and b, respectively, and the comparable localization of the specific anti-myosin isolated from it is shown in Figure 2 c and d, respectively. Although the fluorescent localization patterns are essentially the same (Fig. 2b,d), the localization observed in electron microscopy is quite different for anti-precipitated myosin (Fig. 2a) than for the specific anti-myosin isolated from it (Fig. 2c). The difference observed in electron microscopy is due to the presence of antibody to C-protein in the anti-precipitated myosin, giving the stripes in each half of the A-band (Fig. 2a; Pepe 1967a, 1973; Offer 1973; Pepe and Drucker 1975; Craig and Offer 1975). Antibody specific for C-protein localizes heavily in a periodic fashion, thus introducing large amounts of antibody protein into a small area resolvable in electron microscopy. Antibody specific for myosin is introduced more diffusely in the A-band and therefore, although a considerable amount of antibody has been introduced as detected by fluorescence microscopy (Fig. 2d), it is not easily visible in electron microscopy. Specific anti-chicken myosin (Fig. 2c) and specific anti-rabbit myosin (Fig. 2e) isolated from the corresponding anti-precipitated myosin showed some difference in that a single stripe was observed in each half of the A-band with one but not with the other. The reason for this difference is unclear (Pepe 1973, 1975).

Since detectability of untagged antibody in electron microscopy is dependent on how the antibody is distributed, i.e., diffusely or periodically, it is possible that antibody which is barely detectable in fluorescence because the total amount introduced is low may be easily detectable in electron microscopy if that small amount is localized heavily in periodic areas resolvable in electron microscopy (Pepe 1967a,b; Pepe and Drucker 1975). It is also possible that diffusely distributed antibody can be observed in electron microscopy if sufficiently large amounts are introduced and if this changes the normal structure. In Figure 2e is part of a rabbit psoas fiber that has been immersed in antibody specific for rabbit myosin. The sarcomeres are all about the same length. Note that the H-zone is clearly visible in the more lightly stained sarcomeres deep within the fiber than in those close to the fiber surface which presumably were most heavily stained. The H-zone has been filled in with antibody in the more heavily stained peripheral fibrils.

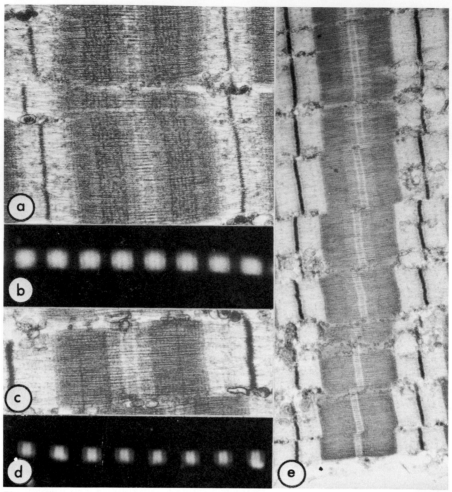

Figure 2

Antibody localization observed in fluorescence and electron microscopy. (*a*) Localization of anti-precipitated chicken myosin observed in electron microscopy. Seven stripes are observed in each half of the A-band. (*b*) Localization of anti-precipitated chicken myosin observed in fluorescence microscopy. Two bright fluorescent bands are observed in the middle of the A-band and one less bright band at each edge of the A-band. (*c*) Localization of specific anti-chicken myosin isolated from anti-precipitated chicken myosin observed in electron microscopy. The seven stripes are no longer observed in each half of the A-band. (*d*) Localization of specific anti-chicken myosin (as in *c*) observed in fluorescence microscopy. The pattern of localization is the same as that in *b*. (*e*) Localization of specific anti-rabbit myosin isolated from anti-precipitated rabbit myosin. Fibrils at bottom are at the periphery of a fiber and those at top are deep within the fiber. The H-zone (zone of no overlap of thin and thick filaments in the middle of the A-band) is clearly observed in the less heavily stained deeper fibrils and is filled in by antibody in the more heavily stained peripheral fibers. The region of the filaments devoid of myosin cross bridges (pseudo-H-zone) is not filled in even in the heavily stained peripheral fibers.

Detection of Unwanted (Contaminating) Antibodies by Fluorescence and Electron Microscopy

As is clear from Figure 1c, the presence of antibody specific for C-protein in anti-precipitated myosin is clearly identifiable in immunodiffusion. This C-protein can be related to the stripes observed in Figure 2a in electron microscopy. The other contaminating antibody present in anti-precipitated myosin (Fig. 1b,d,e) has not been identified and has not been related to any part of the A-band patterns observed in fluorescence or electron microscopy. In contrast to these findings, contaminating antibody present in anti-C was not clearly detectable in immunodiffusion but was clearly detectable in fluorescence microscopy (Fig. 3d) and electron microscopy (Fig. 3a), some contaminating antibody still being detectable in electron microscopy when the more specific isolated anti-C was used (Fig. 3b,c,g).

Figure 3d shows the pattern observed with anti-C-protein in fluorescence microscopy. The antibody localizes lightly on the Z-band and heavily in a wide band in the middle of the A-band. In some sarcomeres it is possible to see that there is a somewhat brighter narrow band in the middle of the broad band in the A-band; this corresponds to the M-band. Correspondingly, in electron microscopy (Fig. 3a) we see that the M-band is dense (due to the presence of antibody). We can also see the seven dense stripes in each half of the A-band that we have identified as due to specific localization of C-protein (Pepe 1973; Offer 1973; Pepe and Drucker 1975; Craig and Offer 1975). No evidence of antibody localization to the Z-band is observed in electron microscopy primarily because (a) the amount of antibody binding to the Z-band is small, as indicated by the fluorescent localization (Fig. 3d), and (b) the normal density of the Z-band makes it difficult to observe the introduction of relatively small amounts of antibody.

The more specific isolated anti-C-protein shows no evidence of M- or Z-band localization in fluorescence microscopy (Fig. 3e,f). In electron microscopy some M-band localization is observed in the heavily stained peripheral fibrils of a fiber but not in the more lightly stained deeper fibrils (Fig. 3g). Therefore although the anti-C-protein and the more highly specific isolated anti-C-protein are not clearly distinguishable by immunodiffusion in terms of the presence of contaminating antibodies, they are clearly distinguishable by observation of antibody localization.

In addition to antibodies to M- and Z-band proteins, the presence of another two specific antibodies can be deduced from observations of antibody localization (Pepe and Drucker 1975; Craig and Offer 1975). In Figure 3a,b,c it is clear that there are seven stripes in each half of the A-band which are of about equal density and that there are two other stripes on the side closest to the M-band. These other stripes are variable in that (a) they may be entirely missing (Pepe 1967a, 1973), (b) they may both be less dense than the other seven stripes (Fig. 3a,b; Pepe and Drucker 1975; Craig and Offer 1975), or (c) one of them may be less dense than the other or may be missing (Fig. 3c; Pepe and Drucker 1975). The two more specific isolated anti-C-protein preparations used in Figure 3 b and c were isolated from anti-C-protein and from anti-precipitated myosin, respectively, using the same

C-protein preparation coupled to PAB-cellulose; thus the difference between them must be due to a difference in the antibody present in the two preparations. These observations mean that the two variable stripes are due to two different antibodies present in addition to antibody to C-protein. This is strengthened by comparing the observation from Figure 3c that the eighth stripe is less dense than the ninth, or is missing, with the observation made

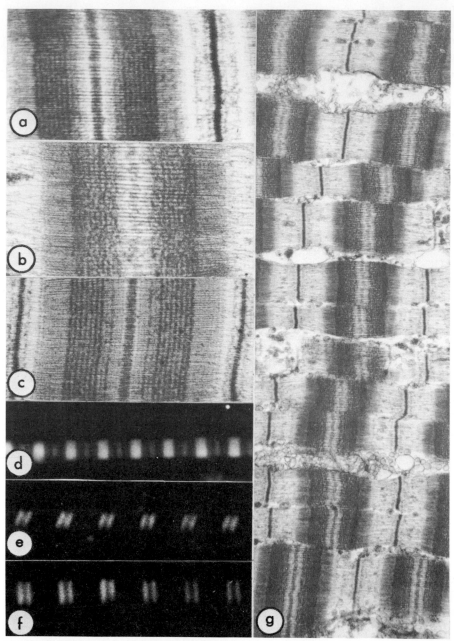

Figure 3 (*See facing page for legend*)

by Craig and Offer (1975) that only the ninth stripe is observed after absorbing anti-C-protein with C-protein. Taken together, these observations provide strong evidence that the two stripes (i.e., eighth and ninth stripes) represent localization of two different antigens, neither of which is C-protein.

CONCLUSIONS

From these findings we might conclude that only under special circumstances is it possible to identify the presence of small amounts of unwanted antibodies, which could, however, contribute substantially to the staining pattern. Because of the high sensitivity of antibody localization, we could be observing localization of the small amount of unwanted antibody rather than the major antibody component that we are interested in. Although it is wise to be cautious about interpreting antibody localization studies, it is still one of the best techniques, and in some cases the only technique, available for the in situ localization of proteins. With proper controls and healthy skepticism it can provide information not available in any other way. From the results observed with anti-C it is obvious that the isolation of specific antibody from the IgG can lead to improvement in the specificity of localization even under conditions where improvement may not seem necessary. The isolation of specific antibody by binding it to insolubilized, highly purified antigen should be used wherever possible.

Acknowledgment

This work was supported by U.S. Public Health Service Grant HL 15835 to the Pennsylvania Muscle Institute.

Figure 3

Detectability of contaminating antibodies not detectable in immunodiffusion. (a) Localization of anti chicken C-protein observed in electron microscopy. Seven stripes are observed in each half of the A-band. The M-band is also stained. (b) Localization of specific anti-C-protein isolated from the anti chicken C-protein in a. The M-band is not stained. Seven dense stripes plus two less dense stripes (added to the side close to the M-band) are observed in each half of the A-band. (c) Localization of specific anti-C-protein isolated from anti-precipitated chicken myosin. Seven dense stripes plus one light stripe and one dense stripe are observed in each half of the A-band. (d) Localization of anti-chicken C-protein (same as a) observed in fluorescence microscopy. A broad, bright band in the middle of the A-band and lightly fluorescent Z-bands are observed. (e, f) Localization of the specific anti-C-proteins (same as b and c, respectively) observed in fluorescence microscopy. Two bands are observed in the A-band. No M-band or Z-band staining is observed. (g) Localization of specific anti-C-protein (same as in b) observed in electron microscopy. Fibrils at bottom are located at the periphery of the fiber, whereas those at top are located deep within the fiber. In the more heavily stained fibrils at the periphery of the fiber, the M-band is stained, whereas no M-band staining is observed in the more lightly stained deeper fibrils.

REFERENCES

Craig, R. and G. Offer. 1975. The location of C-protein in rabbit skeletal muscle. *Proc. Roy. Soc. B.* **192**:451.

Goldman, M. 1968. *Fluorescent antibody methods.* Academic Press, New York.

Nairn, R. C. 1964. *Fluorescent protein tracing.* Williams and Wilkins, Baltimore, Maryland.

Nakane, P. K. and A. Kawaoi. 1974. Peroxidase-labeled antibody: A new method of conjugation. *J. Histochem. Cytochem.* **22**:1084.

Nakane, P. K. and G. B. Pierce. 1966. Enzyme-labeled antibodies: Preparation and application for localization of antigens. *J. Histochem. Cytochem.* **14**:929.

Offer, G. 1973. C-protein and the periodicity in the thick filaments of vertebrate skeletal muscle. *Cold Spring Harbor Symp. Quant. Biol.* **37**:87.

Offer, G., C. Moos and R. Starr. 1973. A new protein of the thick filament of vertebrate skeletal myofibrils. *J. Mol. Biol.* **74**:653.

Ohtsuki, I., T. Masaki, Y. Nonomura and S. Ebashi. 1967. Periodic distribution of troponin along the thin filament. *J. Biochem.* **61**:817.

Pepe, F. A. 1966a. Organization of myosin molecules in the thick filaments of striated muscle as revealed by antibody staining in electron microscopy. *Electron Microsc.* **2**:53.

———. 1966b. Some aspects of the structural organization of the myofibril as revealed by antibody-staining methods. *J. Cell Biol.* **28**:505.

———. 1967a. The myosin filament. I. Structural organization from antibody-staining in electron microscopy. *J. Mol. Biol.* **27**:203.

———. 1967b. The myosin filament. II. Interaction between myosin and actin filaments observed using antibody-staining in fluorescent and electron microscopy. *J. Mol. Biol.* **27**:227.

———. 1968. Analysis of antibody staining patterns obtained with striated myofibrils in fluorescence microscopy and electron microscopy. *Int. Rev. Cytol.* **24**:193.

———. 1973. The myosin filament: Immunochemical and ultrastructural approaches to molecular organization. *Cold Spring Harbor Symp. Quant. Biol.* **37**:97.

———. 1975. Structure of muscle filaments from immunohistochemical and ultrastructural studies. *J. Histochem. Cytochem.* **23**:543.

Pepe, F. A. and B. Drucker. 1975. The myosin filament. III. C-protein. *J. Mol. Biol.* **99**:609.

Pepe, F. A. and H. E. Huxley. 1964. Antibody staining of separated thin and thick filaments of striated muscle. In *Biochemistry of muscle contraction* (ed. J. Gergely), p. 320. Little, Brown and Company, Boston.

Pepe, F. A., H. Finck and H. Holtzer. 1961. The use of specific antibody in electron microscopy. III. Localization of antigens by the use of unmodified antibody. *J. Biophys. Biochem. Cytol.* **11**:533.

Richards, E. G., C. S. Chung, D. B. Menzel and M. S. Olcott. 1967. Chromatography of myosin on diethylaminoethyl-Sephadex A-50. *Biochemistry* **6**:528.

Samosudova, N. V., M. M. Ogievetskaya, M. B. Kalamkarova and G. M. Frank. 1968. Use of ferritin antibodies for the electron microscopic study of myosin. *Biofizika* **13**:877.

Singer, S. J. and A. F. Schick. 1961. The properties of specific stains for electron microscopy prepared by the conjugation of antibody molecules with ferritin. *J. Biophys. Biochem. Cytol.* **9**:519.

Aspects of the Structural Organization of Actin Filaments in Tissue Culture Cells

Elias Lazarides*

Cold Spring Harbor Laboratory
Cold Spring Harbor, New York 11724

One of the characteristics of the epithelial and fibroblastic cell types grown in tissue culture is their ability to adhere to and flatten out on plastic or glass substrata and assume lengths of 100 μm or more. When such cells are examined with phase-contrast optics, they exhibit a number of phase-contrast-dense, straight fibers that usually run along the length of the cell in parallel arrays (Buckley and Porter 1967; Goldman 1971; Goldman and Knipe 1973; Wessells, Spooner and Luduena 1973). By a number of electron microscopic, immunofluorescent and biochemical criteria, these fibers have been shown to correspond to the microfilament bundles that are usually found beneath the plasma membrane of these cells and to have as one of their major constituents actin (Ishikawa, Bischoff and Holtzer 1969; Goldman and Knipe 1973; Goldman et al. 1975; Goldman 1975; Sanger 1975). In non-muscle cells, these actin filaments have been implicated in a number of cellular functions, including cytokinesis, endocytosis and exocytosis, cell adhesion to a substratum, cell locomotion, membrane ruffling, maintenance of cell shape and cell division (Buckley and Porter 1967; Schroeder 1970; Orr, Hall and Allison 1972; Goldman and Knipe 1973; Wessells et al. 1973; Sanger 1975).

The elucidation of the mechanism(s) by which actin filaments mediate specific cellular functions in non-muscle tissue-culture cells was hampered due to a lack of knowledge both of the cellular structure of these filaments and of the involvement of other structural proteins in the regulation of their assembly and function. However with the recent demonstration that antibodies to actin can be used in indirect immunofluorescence to study the distribution of this protein in both muscle and non-muscle cells (Lazarides and Weber 1974; Lazarides 1975b), the possibility opened up that antibodies to other muscle structural proteins could be used to study their distribution and possible function in non-muscle cells. Immunofluorescence studies with antibodies to the muscle structural proteins tropomyosin and α-actinin, as well as myosin, on

* Present Address: Division of Biology, California Institute of Technology, Pasadena, California 91125.

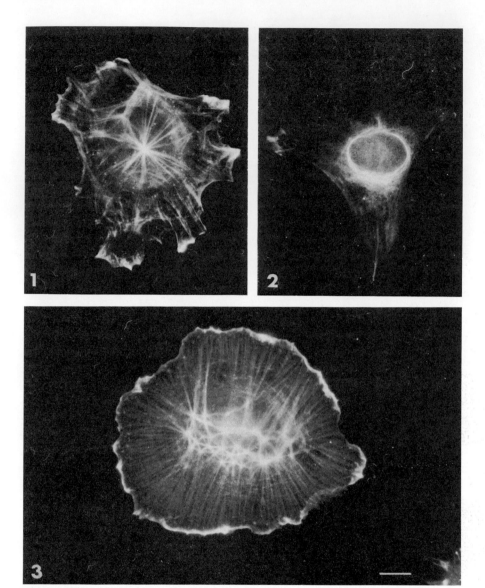

Figures 1–3

(*Fig. 1*) Indirect immunofluorescence on a culture of spreading rat embryo cells using the actin antibody. Details of the indirect immunofluorescence technique and description of the specificity of the antibodies used in this work have been presented previously (Lazarides 1975a,b, 1976). Cells were removed at confluency with 0.05% trypsin, 0.5 mM EDTA in phosphate-buffered saline and plated on glass cover slips. The cells were fixed for indirect immunofluorescence approximately 3.5 hr after plating.

(*Fig. 2*) Indirect immunofluorescence on the same culture of rat embryo cells as that used in Fig. 1, using the tropomyosin antibody. Cells were fixed approximately 3.5 hr after plating.

(*Fig. 3*) Indirect immunofluorescence on the same population of spreading rat embryo cells as that used in Fig. 1. Staining was done using the actin antibody. The cells were fixed for indirect immunofluorescence approximately 5.5 hr after plating. Bar = 10 μm for all three figures.

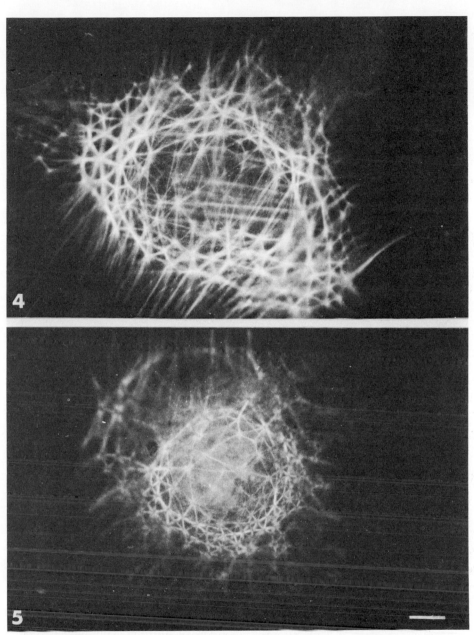

Figures 4–5

Indirect immunofluorescence on the same population of spreading rat embryo cells as that used in Fig. 1, approximately 8 hr after plating, using the actin antibody (Fig. 4) and the tropomyosin antibody (Fig. 5). Note that in Fig. 5, the vertices of the network do not react with the tropomyosin antibody. Bar = 40 μm.

non-muscle cells grown in tissue culture have shown a close association between these three proteins and actin or actin filament bundles in fully spread out cells (Weber and Groeschel-Stuart 1974; Lazarides 1975a; Lazarides and Burridge 1975; Wang, Ash and Singer 1975; Pollard et al., this volume).

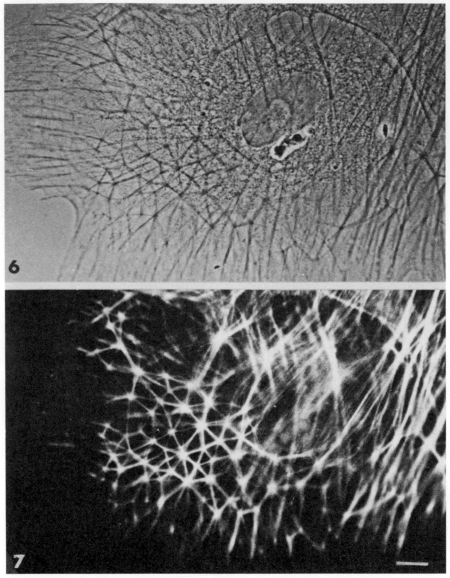

Figures 6–7
Indirect immunofluorescence on a population of rat embryo cells, 8.5 hr after plating, using the actin antibody. Cells were reacted with the actin antibody and photographed with phase-contrast optics (Fig. 6) and epifluorescence optics (Fig. 7). Bar = 10 μm.

When such fully spread out cells go into division or are exposed to proteolytic enzymes such as trypsin or to divalent cation chelating agents like EDTA or EGTA, they lose their adhesive properties and round up, assuming a spherical shape. Concomitant with this loss of shape, the actin filament bundles disaggregate rapidly and assume, presumably, a random cellular distribution. When the cells are allowed to flatten out, they slowly spread onto the substratum, with a slow reorganization of their straight actin filament bundles. This process of the reorganization of these filament bundles allows the study of the mechanism(s) by which these filaments assemble as well as the study of the way in which specific cellular structural proteins might interact to bring about the highly organized filament bundles that are seen in fully spread out cells.

In order to study the role that tropomyosin and α-actinin might play in the assembly of actin filament bundles I have examined their localization during the formation of these filament bundles in the early stages of spreading onto a glass substratum of rat embryo cells.

Structural Study of the Actin Filament Bundles in Fully Spread Cells

There is increasing evidence that besides actin and myosin, two more muscle structural proteins, α-actinin and tropomyosin, are present in the cytoplasm of non-muscle cells (Cohen and Cohen 1972; Fine et al. 1973; Schollmeyer et al. 1974; Lazarides and Burridge 1975). The non-muscle forms of α-actinin and tropomyosin appear to be antigenically related to their muscle counterparts, and antibodies raised against the muscle forms of α-actinin and tropomyosin can be used, as is the case with actin, to study the localization and distribution of these two proteins in non-muscle cells.

Using the phase-contrast fibers of the fully spread out cells as a frame of reference, it can be shown by indirect immunofluorescence that all three proteins, actin, α-actinin and tropomyosin, are closely associated within these filament bundles (Lazarides 1975a,b; Lazarides and Burridge 1975; see also Figs. 14–17). However, the distribution of these three proteins within the filament bundles appears to be distinct. The actin antibody invariably reveals a continuous fluorescence along these phase-contrast fibers, whereas the tropomyosin and α-actinin antibodies show a periodic fluorescence. In the case of tropomyosin, this fluorescent periodicity is adequately resolved only at higher magnifications (100 × oil immersion) (Lazarides 1975a,b), although it is usually obscured along the thicker filament bundles. Such bundles therefore exhibit a continuous fluorescence when reacted with the tropomyosin antibody (Fig. 15). The periodicity seen with the α-actinin antibody is adequately resolved even at lower magnifications. The lengths of the fluorescent and nonfluorescent segments seen with the α-actinin antibody seem to be distinct from those seen with the tropomyosin antibody (Lazarides 1975b; Lazarides and Burridge 1975). However even in the case of the α-actinin antibody, the periodicity is only poorly, or not at all, resolved along the thicker filament bundles (see Fig. 17). Measurements of well-resolved fluorescent and nonfluorescent segments seen with each antibody indicate that the localization of these two antigens is complementary, i.e., the lengths of the fluorescent segments seen with the tropomyosin antibody are quite similar to those of the nonfluorescent segments seen with the α-actinin antibody. Similarly, the lengths of the fluorescent segments seen with the α-actinin antibody are very similar to those of the nonfluorescent segments seen with the tropomyosin antibody. Furthermore, reaction of the cells with both antibodies sequentially reveals no fluorescent periodicities (Lazarides and Burridge 1975). Such observations indicate that tropomyosin and α-actinin might alternate their localization along the actin filaments. The observation that, in general, the length of the striations is variable, or occasionally poorly resolved,

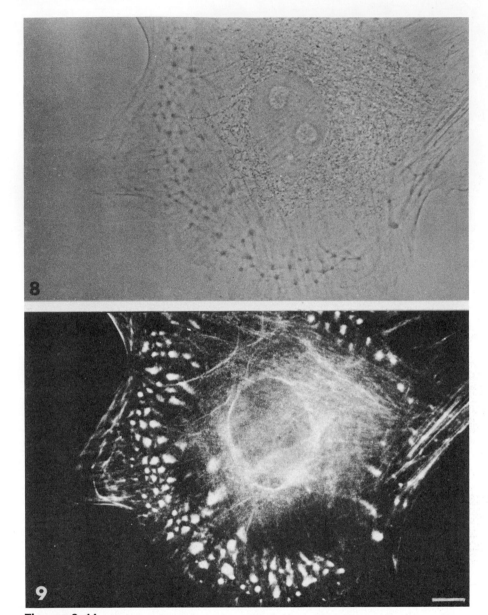

Figures 8–11

Same procedure as for Figs. 6 and 7, except that the α-actinin antibody was used. The cells were photographed with phase-contrast optics (Figs. 8 and 10) and epifluorescence optics (Figs. 9 and 11). Note that only the vertices of the network stain with the α-actinin antibody. Also note that α-actinin is localized at the tip of filament bundles that extend into cellular lamellar projections.

might further indicate that the individual actin filaments within a filament bundle are able to show lateral freedom of motion and slide past each other. This conclusion might be valid if this technique labels α-actinin and tropomyosin at consistent locations on the actin filaments. This has been shown to be

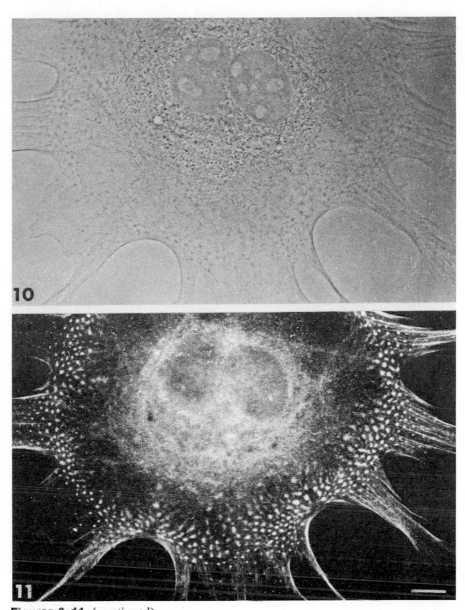

Figures 8–11 (*continued*)

the case for striated muscle (Pepe 1966; Lazarides 1975; Lazarides and Burridge 1975). The presence of myosin within these bundles (Pollard et al., this volume) also leads us to believe that at least some of the actin filaments within a filament bundle are capable of sliding.

Reorganization of Actin Filament Bundles in Spreading Cells

When a tissue culture cell is exposed to trypsin, it loses its adhesiveness to the substratum and rounds up. Electron microscopic examination of such cells

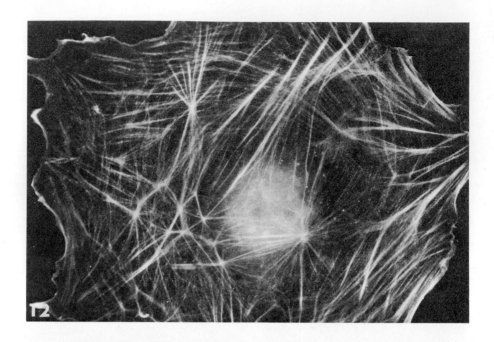

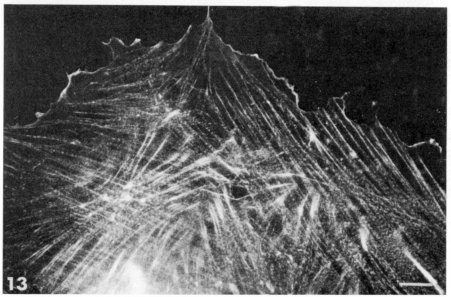

Figures 12–13

Indirect immunofluorescence on a population of spreading rat embryo cells, approximately 9.5 hr after plating, using the actin antibody (Fig. 12) and the α-actinin antibody (Fig. 13). Note that the network is being transformed into the straight filament bundles normally seen in fully spread out cells. A few vertices of the network are still visible, which seem to be transformed into fluorescent-dense areas of convergence of filaments, giving the impression of "focal points." By this time, the α-actinin antibody reveals a clearly resolved fluorescent periodicity along the straight filament bundles. Bar = 10 μm.

354

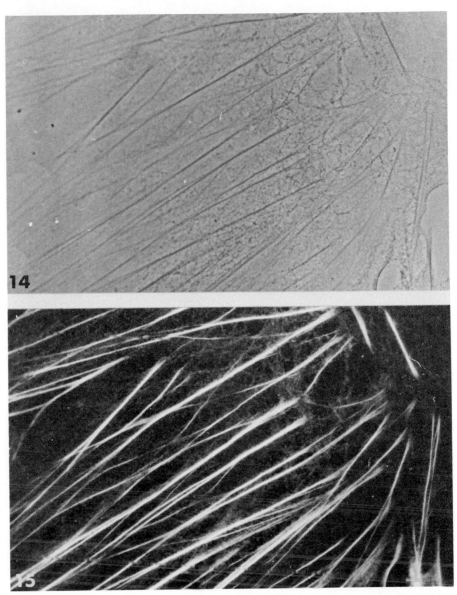

Figures 14–15
Indirect immunofluorescence on a population of fully spread out rat embryo cells 24 hr after plating. The cells were reacted with the tropomyosin antibody and photographed with phase-contrast (Fig. 14) and epifluorescence optics (Fig. 15). Similar images are obtained when the cells are reacted with the actin antibody. Bar = 10 μm.

indicates that the highly organized filament bundles have become disorganized, and that occasional individual actin filaments are visible scattered throughout the cytoplasm (Goldman and Knipe 1973). Thus exposure of the cell to trypsin mimics morphologically the rounding up and the disorganization of actin filaments that takes place during mitosis.

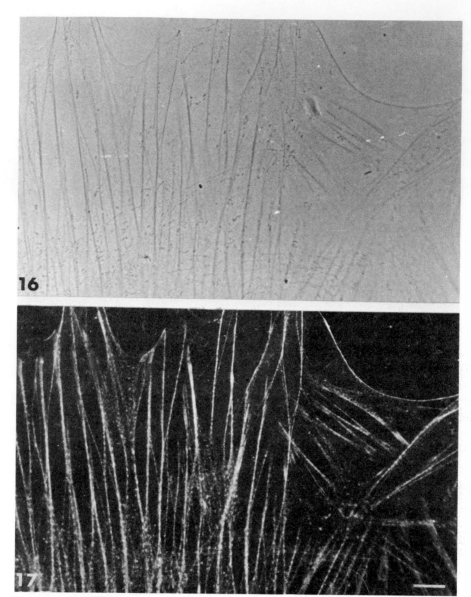

Figures 16–17
Same procedure as Figs. 14 and 15, except that the α-actinin antibody was used. Cells were reacted with α-actinin antibody and photographed with phase-contrast (Fig. 16) and epifluorescence (Fig. 17) optics. Note the fine fluorescent striations that become apparent with this antibody. Bar = 10 μm.

When the cells are allowed to reestablish contact with the substratum, they begin to flatten out slowly, with a concomitant slow reorganization of their actin filament bundles.

Within the first 2 hours after plating, the cells are still quite round, and all three antibodies, actin, α-actinin and tropomyosin, exhibit a diffuse cytoplasmic fluorescence. Thus, using immunofluorescence, no distinct filament organization is visible during this time (Lazarides 1975b).

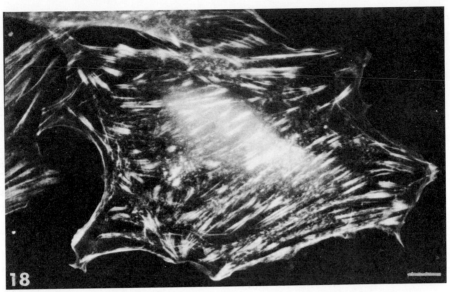

Figure 18
Indirect immunofluorescence on a rat embryo cell, 9 hr after plating, using the
α-actinin antibody. Note the "patchy" localization of α-actin. Bar = 10 μm.

Within the next 4 hours, approximately 40% of the cells in a heterogeneous
population of spreading primary rat embryo cells develop a highly regular
actin filament network (Fig. 4). Figures 6 and 7 show that the networks
seen with the actin antibody correspond to identical patterns in the fixed and
antibody-treated cells in phase-contrast optics. Using the networks seen in
phase-contrast optics as a frame of reference it can be shown by indirect immunofluorescence that the vertices of the actin network contain α-actinin but
no tropomyosin (Figs. 8–11). The connecting fibers between the vertices contain tropomyosin but no α-actinin (Fig. 5). Fibers that attach to the vertices
of the network at only one end can also be seen. The other ends of such fibers
extend all the way to the edge of the spreading membrane. These latter fibers
contain all three molecules, actin, α-actinin and tropomyosin, distributed in
a manner indistinguishable from that seen within the actin filament bundles
of the fully spread out cells, as discussed above (Figs. 3–5, 11). In this
respect, these latter fibers are morphologically indistinguishable from the
straight actin filament bundles seen in fully spread out cells. Thus the vertices of the network may act as nucleation sites as well as cytoplasmic organization centers for the newly forming actin filament bundles. As revealed by
the actin antibody, the network appears to form from a common convergence
point localized above the nucleus (Fig. 1). Within the next 2 hours it is
augmented, and numerous filament bundles radiate from its vertices to the
edges of the cell (Fig. 3). During these intitial stages in the formation of
the network, α-actinin is localized at its vertices (Lazarides 1976). However,
in contrast to the α-actinin antibody, the tropomyosin antibody reveals a perinuclear filamentous localization of this antigen but no association with the
networks themselves (Fig. 2). The tight association of tropomyosin with the
filaments that connect the vertices of the network occurs approximately 2

hours after the association of α-actinin with the vertices of the network. This observation indicates that α-actinin precedes tropomyosin in their respective associations with the newly formed actin filament network. It suggests further that α-actinin, by binding to the vertices of the network, restricts, and may therefore determine, the binding of tropomyosin to the filaments that connect the vertices of the network. The subsequent binding of tropomyosin solely to the actin filament bundles that connect the vertices of the network, and not to the vertices themselves, clearly demonstrates the complementary interaction of these two molecules within the actin filament bundles.

The network itself is only transiently formed. At the last stages of spreading it is replaced by the straight "stress fibers" normally seen in fully spread out cells. The process by which the network is transformed into the stress fibers is difficult to follow, primarily because by this time (approximately 9 hr after plating) the spreading of the cells has become asynchronous and a number of them have begun to locomote. However, a few characteristic patterns have been observed: the pattern associated with the actin antibody is shown in Figure 12 and that associated with the α-actinin antibody, in Figures 13 and 18. Newly formed stress fibers exhibit areas of convergence where the fluorescence is more intense, which gives the appearance of filamentous focal points. These focal points may represent an intermediate stage in the conversion of the network actin filament bundles to the straight stress fibers.

CONCLUSIONS

The population of network-forming cells provides an example of the characteristic interactions of specific cytoplasmic structural proteins with actin in the formation of the actin filament bundles during cell spreading onto a glass substratum.

The localization of the vertices of the network is at present still undetermined. However, they might possibly be attached either to the cell membrane or to any of the membraneous organelles within the cell. These vertices may be nucleation and cytoplasmic organization sites for the aggregation of actin filament towards the spreading membrane or outgrowing lamellar projections. Occasionally the fully developed network covers the whole volume of the spreading cell and may assume the shape of a dome that encompasses the nuclear region (see Figs. 3–5; see also Lazarides 1976). Thus the network may act as a filamentous structural support to the spreading cell, and the filament bundles that extend from the vertices of the network to the edges of the cell may exert a supporting effect on the spreading membrane and subject the whole cell to structural "stress." For this reason, the actin filament bundles have been correctly termed "stress fibers" (Buckley and Porter 1967). It should be borne in mind that these stress fibers are dynamic structures and can easily disaggregate despite their high degree of molecular organization. The specific cytoplasmic factors that may regulate this reversible disaggregation of actin filament bundles, either when the cell goes into mitosis or when exposed to proteolytic enzymes, are unknown. However, they provide another challenging aspect of investigation in attempting to understand the organization and assembly of actin filament bundles in tissue culture cells.

Acknowledgments

I am grateful to Dr. J. D. Watson for his advice and support throughout this work, to Dr. R. Pollack for the generous use of his laboratory facilities and for providing me with the cells used in this work, to Dr. G. Albrecht-Buehler for his valuable advice, and to Dr. F. Miller at the State University of New York at Stony Brook for his advice and help in the preparation of the antibodies.

The investigation was supported by the National Institutes of Health, Research Grant Ca-13106 from the National Cancer Institute.

REFERENCES

Buckley, I. and K. Porter. 1967. Cytoplasmic fibrils in living cultured cells. A light and electron microscope study. *Protoplasma* **64**:349.

Cohen, I. and C. Cohen. 1972. A tropomyosin-like protein from human platelets. *J. Mol. Biol.* **68**:383.

Fine, R., A. Blitz, S. Hitchcock and B. Kaminer. 1973. Tropomyosin in brain and growing neurons. *Nature New Biol.* **245**:182.

Goldman, R. 1971. The role of three cytoplasmic fibers in BHK-21 cell motility. I. Microtubules and the effects of colchicine. *J. Cell Biol.* **51**:752.

————. The use of heavy meromyosin binding as an ultrastructural cytochemical method for localizing and determining the possible function of actin-like microfilaments in nonmuscle cells. *J. Histochem. Cytochem.* **23**:529.

Goldman, R. and D. Knipe. 1973. Functions of cytoplasmic fibers in non-muscle cell motility. *Cold Spring Harbor Symp. Quant. Biol.* **37**:523.

Goldman, R., E. Lazarides, R. Pollack and K. Weber. 1975. The distribution of actin in nonmuscle cells: The use of actin antibody in the localization of actin within the microfilament bundles of mouse 3T3 cells. *Exp. Cell Res.* **90**:333.

Ishikawa, H., R. Bischoff and H. Holtzer. 1969. Formation of arrowhead complexes with heavy meromyosin in a variety of cell types. *J. Cell Biol.* **43**:312.

Lazarides, E. 1975a. Tropomyosin antibody: The specific localization of tropomyosin in non-muscle cells. *J. Cell Biol.* **65**:549.

————. 1975b. Immunofluorescence studies on the structure of actin filaments in tissue culture cells. *J. Histochem. Cytochem.* **23**:507.

————. 1976. Actin, alpha-actinin and tropomyosin interaction in the structural organization of actin filaments in nonmuscle cells. *J. Cell Biol.* **68**:202.

Lazarides, E. and K. Burridge. 1975. Alpha-actinin: Immunofluorescent localization of a muscle structural protein in nonmuscle cells. *Cell* **6**:289.

Lazarides, E. and K. Weber. 1974. Actin antibody: The specific visualization of actin filaments in nonmuscle cells. *Proc. Nat. Acad. Sci.* **71**:2268.

Orr, T., D. Hall and A. Allison. 1972. Role of contractile microfilaments in the release of histamine from mast cells. *Nature* **236**:350.

Pepe, F. 1966. Some aspects of the structural organization of the myofibril as revealed by antibody staining methods. *J. Cell Biol.* **28**:505.

Sanger, J. 1975. Changing patterns of actin localization during cell division. *Proc. Nat. Acad. Sci.* **72**:1913.

Schollmeyer, J., D. Goll, L. Tilney, M. Mooseker, R. Robson and M. Stromer. 1974. Localization of alpha-actinin in non-muscle material. *J. Cell Biol.* **63**:304a.

Schroeder, T. 1970. The contractile ring. I. Fine structure of dividing mammalian (HeLa) cells and the effects of cytochalasin B. *Z. Zellforsch. Mikrosk. Anat.* **109**:431.

Wang, K., F. Ash and S. Singer. 1975. Filamin: A new high-molecular-weight protein found in smooth muscle and nonmuscle cells. *Proc. Nat. Acad. Sci.* **72**:4483.

Weber, K. and U. Groeschel-Stuart. 1974. Antibody to myosin: The specific visualization of myosin-containing filaments in nonmuscle cells. *Proc. Nat. Acad. Sci.* **71**:4561.

Wessells, N., B. Spooner and M. Luduena. 1973. Surface movements, microfilaments and cell locomotion. In *Locomotion of tissue cells. Ciba Foundation Symposium,* vol. 14, p. 53. Associated Scientific Publishers, New York.

Localization of Contractile Proteins in Smooth Muscle Cells and in Normal and Transformed Fibroblasts

Judith E. Schollmeyer and L. T. Furcht

Department of Laboratory Medicine and Pathology
University of Minnesota, Minneapolis, Minnesota 55455

Darrel E. Goll, R. M. Robson and M. H. Stromer

Muscle Biology Group, Iowa State University
Ames, Iowa 50011

Although the composition and molecular architecture of myofilaments in striated muscle are relatively well established, considerable uncertainty exists concerning the composition and molecular structure of the contractile filaments and their associated structures in smooth muscle. It is now generally agreed that smooth muscle cells contain thick filaments composed largely of myosin (Panner and Honig 1967; Kelly and Rice 1968; Schoenberg et al. 1969; Nonomura 1968; Somlyo et al. 1973), but it is not known whether smooth muscle thick filaments contain accessory proteins like the C-protein found in thick filaments from skeletal muscle (Morimoto and Harrington 1973; Offer 1973; Offer, Moos and Starr 1973). The protein composition and molecular structure of thin filaments from smooth muscle also are not well understood. Thin filaments from skeletal muscle evidently are composed of a double-helical actin backbone (Hanson and Lowy 1963; Huxley 1963) with a strand of tropomyosin lying along each of the two grooves of the actin helix (Ebashi, Endow and Ohtsuki 1969; Spudich, Huxley and Finch 1972; Hanson et al. 1973; Parry and Squire 1973) and the three subunits of troponin arranged periodically at 38.5-nm intervals along the length of the filament (Ebashi, Endo and Ohtsuki 1969; Spudich, Huxley and Finch 1972; Hanson 1973; Ohtsuki 1975; Wakabayashi et al. 1975). Thin filaments from smooth muscle also contain actin and tropomyosin (Driska and Hartshorne 1975; Sobieszek and Small 1976) but in addition seem to contain proteins having subunit molecular weights of 110,000 and 130,000 daltons (Driska and Hartshorne 1975) when analyzed on SDS-polyacrylamide gels. Although Sobieszek and Bremel (1975) suggest that the 110,000-dalton component is smooth muscle α-actinin, the nature of the 130,000-dalton component is unknown (Driska and Hartshorne 1975; Sobieszek and Bremel 1975). Moreover, although smooth muscle thin filaments contain tropomyosin, smooth muscle tropomyosin is more soluble (Sobieszek and Bremel 1975) and is present in a higher proportion relative to myosin than it is in skeletal muscle

(Laszt and Hamoir 1961; Needham and Schoenberg 1967; Sobieszek and Bremel 1975). Murphy, Herlihy and Megerman (1974) state that the ratio of tropomyosin to actin is the same in smooth and skeletal muscle, but several other reports (Laszt and Hamoir 1961; Needham and Schoenberg 1967; Sobieszek and Bremel 1975) have indicated that the tropomyosin to actin ratio is higher in smooth muscle. Despite an intensive search, no Ca^{++}-binding protein with the molecular parameters of striated muscle troponin has been isolated from smooth muscle (Driska and Hartshorne 1974; Sobieszek and Bremel 1975). Consequently, it is not clear whether the precise ratios and positioning of the "regulatory" proteins relative to actin, such as ostensibly exist in skeletal muscle thin filaments (Greaser and Gergely 1971; Hitchcock, Huxley and Szent-Györgyi 1973), are also mandatory in smooth muscle thin filaments, and the molecular architecture of thin filaments from smooth muscle remains unclear.

In addition to the uncertainties concerning composition and structure of thick and thin filaments, smooth muscle cells contain densely staining structures, called dense bodies, that are not found in striated myofibrils. Because intracellular dense bodies and attachment plaques, which are densely staining regions subjacent to the plasma membrane, have been observed in association with both thin filaments (Somlyo et al. 1973) and intermediately sized filaments, 10 nm in diameter (Somlyo et al. 1971; Cooke and Chase 1971; Cooke and Fay 1972), it has been suggested (Bois 1973; Somlyo et al. 1973) that dense bodies are smooth muscle analogs of Z-disks found in striated muscle myofibrils. The protein composition of dense bodies is not known, however, and the possible relationship of dense bodies to striated myofibril Z-disks remains unclear. Rice and Brady (1973) reported that low ionic strength extraction removes amorphous material from dense bodies just as it does from Z-disks in skeletal muscle (Stromer, Hartshorne and Rice 1967, 1969). SDS-polyacrylamide gel electrophoresis of the protein solubilized from dense bodies by low ionic strength extraction, however, revealed only two proteins having subunit molecular weights of 16,000 and 18,000 daltons (Rice and Brady 1973). The residue remaining after low ionic strength extraction of dense bodies contained two high molecular weight components that migrated with molecular weights of 85,000 and 105,000 daltons on SDS-polyacrylamide gels (Rice and Brady 1973). SDS-polyacrylamide gel electrophoresis of low ionic strength extracts of striated myofibrils, on the other hand, reveals many proteins, with the major species having molecular weights of 100,000, 44,000 and 37,000 daltons (Stromer et al. 1976; J. Schollmeyer, in prep.). Biochemical analyses of low ionic strength extracts from striated myofibrils have shown that the 100,000-dalton component in these extracts is α-actinin (Robson et al. 1970), which is known to be one of the primary constituents of striated muscle Z-disks (Schollmeyer, Stromer and Goll 1972; Schollmeyer 1973, 1974a). No 16,000- or 18,000-dalton components have been found in low ionic strength extracts of striated myofibrils, and striated myofibrils contain no 85,000-dalton component (R. Allen, L. T. Tabatabai, D. E. Goll, R. M. Robson and M. H. Stromer, in prep.) such as Rice and Brady (1973) observed in the residue left after low ionic strength extraction of dense body preparations from smooth muscle. Recent studies on myofibrils prepared from smooth muscle by using Triton X-100 (Sobieszek and Bremel 1975; Sobieszek and Small 1976) show that these myofibrils possess large amounts of a 55,000-dalton component that is either absent or present in only very small amounts in striated myofibrils. Cooke (1975) has recently suggested that this 55,000-dalton component is a constituent of the intermediate, 10-nm filaments that are found in abundance in smooth muscle cells and that, together with thin filaments, ostensibly insert into dense bodies (Cooke and Chase 1971; Rice and Brady 1973; Somlyo et al. 1973; Cooke 1975; Ashton, Somlyo and Somlyo 1975).

Because of the uncertainties concerning the protein composition and molecular architecture of smooth muscle myofibrils and the conflicting results on the relationship of dense bodies and attachment plaques of smooth muscle to Z-disks in striated myofibrils (Small and Squire 1972; Somlyo et al. 1973), we have begun a detailed analysis of the composition and architecture of smooth muscle myofibrils by using SDS-polyacrylamide gel electrophoresis, selective extraction and localization of binding of antibodies made against specific proteins purified from smooth muscle cells. Contractile proteins similar to muscle actin and myosin have now been found in a wide variety of cells (Pollard and Weihing 1974), and smooth muscle cells exhibit many of the features observed in these non-muscle motile systems. Several recent reports (Pollack, Osborn and Weber 1975; Robbins et al. 1975; Wickus et al. 1975) have described a dramatic loss of filamentous actin structures in transformed cultured cells without any detectable loss of total actin. Because loss of actin filaments could be related to loss of attachment plaques for these actin filaments, we have also done a few preliminary studies localizing binding of antibodies to smooth muscle α-actinin in normal and transformed fibroblasts to determine whether α-actinin, which evidently binds actin filaments to Z-disks in striated myofibrils (Goll et al. 1972; Robson and Zeece 1973; Suzuki et al. 1973), was lost or dispersed upon viral transformation of cultured cells. Our results, showing that α-actinin is located in dense bodies and attachment plaques of smooth muscle cells and that viral transformation of cultured fibroblasts is accompanied by loss of α-actinin-containing dense areas in these cells, suggest that α-actinin may have a physiological function in anchoring and organizing actin filaments and in regulating the state of actin aggregation.

Localization of α-Actinin and Tropomyosin in Chicken Gizzard

All localization studies described in this paper were done by determining the site of binding of antibodies elicited against highly purified proteins to sections of chicken gizzard tissue or cultured fibroblasts. With the exception of a few preliminary studies done on localization of M-protein, these studies were done at the electron microscope level of resolution to obtain the maximum detail possible on the intracellular distribution of the proteins studied. Highly purified α-actinin (Fig. 1) for use as an antigen was prepared from chicken gizzard muscle as described by R. M. Robson et al. (in prep). Tropomyosin was isolated from an ethanol-ether powder of chicken gizzard

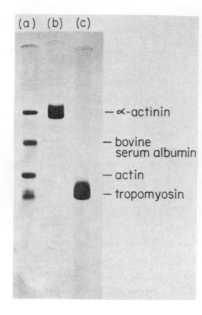

(a) (b) (c)

— α-actinin

— bovine
serum albumin

— actin

— tropomyosin

Figure 1
SDS (7.5%) polyacrylamide gels of (*b*) chicken gizzard α-actinin and (*c*) tropomyosin. Gel *a* shows the standard proteins, α-actinin, bovine serum albumin, actin and tropomyosin, electrophoresed under the same conditions as gels *b* and *c*. Gels *b* and *c* are each loaded with 33 μg of protein. Electrophoresis was done according to the method of Weber and Osborn (1969).

according to the method of Greaser and Gergely (1971) and was further purified by preparative slab electrophoresis in SDS. This electrophoretic purification was necessary in order to obtain highly purified tropomyosin (Fig. 1) free from a contaminating protein that resembled M-protein antigenically. With the exception of α-actinin, which was obtained in an antigenically homogeneous form by using column chromatography (R. M. Robson, T. Huiatt, M. G. Zeece, D. E. Goll and M. H. Stromer, in prep.), all proteins used as antigens in this study were purified by preparative electrophoresis in SDS. Details of this preparative electrophoresis procedure will be given elsewhere (J. Schollmeyer, D. Goll, R. Robson and M. Stromer, in prep.).

Antibodies to α-actinin or electrophoretically purified tropomyosin were produced by four weekly injections of the antigenic protein into rabbits. After suitable partial purification, the antibody fractions obtained all gave a single precipitin line in both double diffusion and immunoelectrophoresis when reacted against their respective antigens. Unconjugated antibodies to smooth muscle α-actinin or tropomyosin were used as the first reagent in the indirect method for immunocytochemistry at the electron microscope level. Sheep anti-rabbit IgG was coupled with horseradish peroxidase according to the procedure of Avrameas (1971) and was used as the second reagent in the indirect method. Details on antibody production and purification and on incubation of the antibodies with sections of chicken gizzard tissue will be given elsewhere (J. Schollmeyer, D. Goll, R. Robson, and M. Stromer, in prep.). Electron-dense reaction products were produced by incubating the peroxidase-labeled tissue sections with 0.05 M 3,3′-diaminobenzidene and 0.002% H_2O_2 (Nakane and Pierce 1967; J. Schollmeyer, D. Goll, R. Robson and M. Stromer, in prep.), and these reaction products were used to localize the site of binding of antibodies to specific proteins.

Because the horseradish procedure requires use of unstained or very lightly stained sections to facilitate detection of the small electron-dense reaction products, a postembedment, antibody-binding procedure was used in some instances to confirm identity of the structures that bound antibody. In this procedure, two semiadjacent sections are taken from tissue that had been fixed and embedded in the normal Epon-Araldite mixture (Anderson and Ellis 1965) used for electron microscopy. One section is double stained with uranyl acetate and lead citrate (Reynolds 1963) and examined in the electron microscope, whereas the semiadjacent section is subjected to a postembedment, antibody-binding and immunocytochemical staining procedure. Details of this postembedment, antibody-binding procedure will be given elsewhere (J. Schollmeyer, D. Goll, R. Robson and M. Stromer, in prep.).

Use of the horseradish peroxidase procedure shows that anti-α-actinin seems to bind to both the attachment plaques and the intracellular dense bodies of chicken gizzard (Fig. 2). The conclusion that anti-α-actinin binds to dense bodies is confirmed by examination of semiadjacent sections (Fig. 3), which show little or no reaction product when normal rabbit serum (Fig. 3b) is substituted for anti-α-actinin-containing serum in the immunoperoxidase procedure used in this study. Anti-tropomyosin, on the other hand,

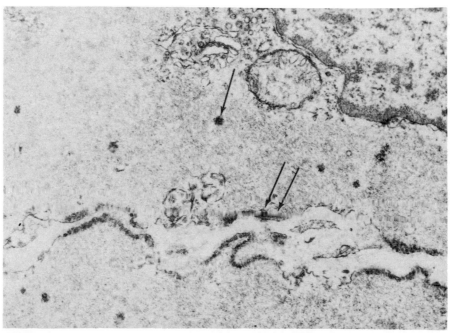

Figure 2
Chicken gizzard smooth muscle that has been incubated before embedding in Epon-Araldite with anti-α-actinin followed by sheep anti-rabbit IgG conjugated with peroxidase (indirect method). The reaction product is located both at the periphery of the cell (double arrow) and within the cytoplasm (single arrow). This section has been counterstained with lead citrate. Magnification, 32,000×.

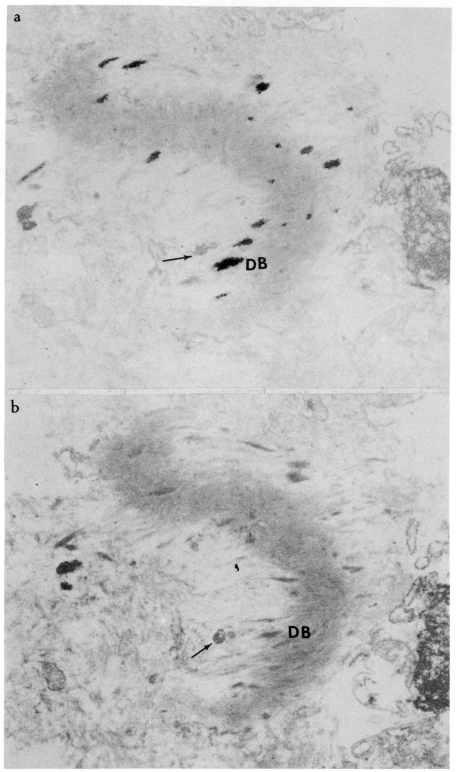

Figure 3 (*See facing page for legend*)

seems to bind both to intracellular dense bodies and to the filaments associated with these dense bodies (Fig. 4), although reaction product associated with the filaments is sparse in the particular example shown. The limitations imposed by the requirement of lightly stained sections for the peroxidase immunocytochemical procedure prevents separate identification of 6-nm and 10-nm filaments in the anti-tropomyosin-incubated samples. In several instances, filaments that show little evidence of the reaction product are seen in close association with putative dense bodies.

Composition of Intermediate Filaments in Chicken Gizzard

Numerous investigators (Cooke and Chase 1971; Cooke and Fay 1972; Fay and Cooke 1972; Rice and Brady 1973; Somlyo et al. 1973; Cooke 1975) have observed a sizable population of filaments 10 nm in diameter in various smooth muscles. These filaments, which have been called intermediate filaments, seem to be associated with dense bodies, both in sections of fixed intact tissue (Cooke and Fay 1972; Fay and Cooke 1972; Rice and Brady 1973; Somlyo et al. 1971, 1973; Ashton, Somlyo and Somlyo 1975) and in insoluble residues left after extraction of actin and myosin with 0.6 M KCl at a slightly alkaline pH (Cooke and Chase 1971; Rice and Brady 1973). Several recent reports have described conflicting results showing that intermediate filaments contain peptide chains with molecular weights of 85,000 and 105,000 (Rice and Brady 1973), 150,000 (Rice, Brady and Baldassore 1974) or 55,000 (Cooke 1975; Holtzer et al., this volume) daltons. Consequently, the protein composition and physiological role of these intermediate filaments remain unclear. Because both intermediate and thin actin-containing filaments are associated with dense bodies in smooth muscle (Somlyo et al. 1971, 1973; Ashton, Somlyo and Somlyo 1975) and because dense bodies have been suggested to be smooth muscle analogs of skeletal muscle Z-disks, we have investigated the protein composition of intermediate filaments in the hope that information on their protein composition would provide some clues on the physiological functioning of α-actinin, a protein located in skeletal muscle Z-disks (Schollmeyer, Stromer and Goll 1972; Schollmeyer et al. 1973, 1974a).

Our approach to determining the protein composition of intermediate filaments was to make a dense-body preparation by extracting minced chicken gizzard with 0.6 M KCl according to the method of Cooke and Chase (1971). This dense-body preparation was then extracted for 2 days at 2°C with 2 mM Tris pH 7.6, 1 mM dithiothreitol, which removes amorphous material

Figure 3
Semiadjacent sections of the same chicken gizzard cell as in Fig. 2 incubated with (*a*) anti-α-actinin antibodies or (*b*) normal rabbit serum, followed by sheep anti-rabbit IgG conjugated with peroxidase. Arrows indicate structures common to both micrographs. Anti-α-actinin clearly binds to intracellular dense bodies (DB) (*a*), whereas no obvious reaction product can be observed in the control (*b*). Magnification, 18,600×.

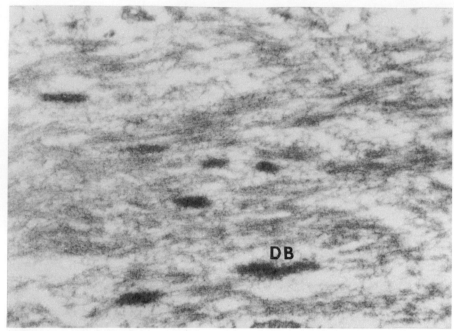

Figure 4
Chicken gizzard smooth muscle incubated with anti-tropomyosin followed by sheep anti-rabbit IgG conjugated with peroxidase (indirect method). Both intracellular dense bodies (DB) and a group of intracellular filaments, 6 to 9 nm in diameter, bind anti-tropomyosin, although the density of the stain associated with the filaments makes it difficult to see reaction product clearly. In some instances these filaments are associated with the dense bodies. Magnification, 20,000×.

from Z-disks of striated muscle (Stromer, Hartshorne and Rice 1967; Stromer et al. 1969). Low ionic strength extraction of the dense-body preparations was monitored by using SDS-polyacrylamide gel electrophoresis and electron microscope examination of dense-body preparations before and after low ionic strength extractions (Figs. 5, 6). Unextracted dense-body preparations contained numerous filaments arranged in bundles and coated with amorphous material together with individual 10-nm filaments that appear to extend into and course between separate dense bodies (Fig. 5a). No large, myosin-containing filaments are seen (Fig. 5a). SDS-polyacrylamide gel electrophoresis of unextracted dense-body preparations shows that these preparations are composed principally of five different polypeptide chains (Fig. 6a). These polypeptide chains have approximate molecular weights of 200,000 daltons (presumably from myosin that was not extracted and that is in the disaggregated form because no thick filaments are seen), 160,000 daltons (possibly M-protein; see next section), 100,000 daltons (α-actinin), 55,000 daltons, and 42,000 daltons (actin). The presence of α-actinin in these dense-body preparations was confirmed by showing that anti-α-actinin bound to the dense bodies and occasionally also to scattered regions along the filaments (Fig. 5b). Resolution obtainable in the peroxidase immunocytochemical procedure

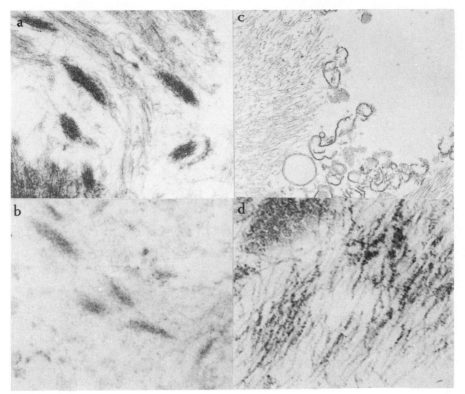

Figure 5
Dense-body preparations made from chicken gizzard by KCl extraction. (*a*) This
electron micrograph of a pelleted dense-body preparation shows the numerous
dense bodies and filaments that exist in this preparation. Magnification, 65,600×.
(*b*) Dense-body preparation after incubation with anti-α-actinin followed by sheep
anti-rabbit IgG conjugated with peroxidase. Reaction product is seen on dense
bodies and scattered along some filaments. Magnification, 49,200×. (*c*) Electron
micrograph of the residue remaining after low ionic strength extraction of a dense-
body preparation. Only 10-nm intermediate filaments remain. Magnification,
24,600×. (*d*) Electron micrograph showing binding of antibodies against the 55,-
000-dalton protein to the intermediate filament residue left after low ionic strength
extraction of a dense-body preparation. The presence of reaction product all along
the filaments in this micrograph indicates that 55,000-dalton component is located
throughout these filaments. Magnification, 24,600×.

permits no decision as to whether the infrequent, scattered anti-α-actinin
binding occurs to 6-nm or 10-nm filaments. It is possible that the scattered
anti-α-actinin binding seen along the filaments in Figure 5b originated from
rebinding of α-actinin extracted from the dense body since no anti-α-actinin
binding to filaments was observed in unextracted muscle (Fig. 2) and because we (Stromer and Goll 1972) have shown that added α-actinin binds
to filaments in skeletal muscle.

Low ionic strength extraction of dense-body preparations with 2 mM Tris
pH 7.6 removes the amorphous material from the dense bodies and leaves a

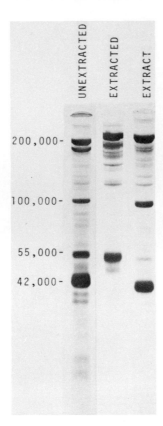

Figure 6

SDS (7.5%) polyacrylamide gels of dense-body preparations from chicken gizzard muscle before and after low ionic strength extraction. Unextracted: dense-body preparation before low ionic strength extraction; extracted: residue remaining after extraction with 2 mM Tris, 1 mM dithiothreitol, pH 7.6, for 2 days; extract: material extracted by the 2-day, 2 mM Tris, 1 mM dithiothreitol extraction. The insoluble dense-body preparations and residues after low ionic strength extraction of dense bodies were dissolved and electrophoresed according to the procedure described by Suzuki et al. (1976). Electrophoresis was done according to the procedure of Weber and Osborn (1969). Numbers at the left of the gels indicate approximate migration distances of polypeptides having the indicated molecular weights. Thirty-three μg protein was loaded onto each gel.

residue that seems to contain primarily 10-nm filaments (Fig. 5c). SDS-polyacrylamide gel electrophoresis shows that the extract contains three major polypeptide chains of 200,000, 100,000 and 42,000 daltons (Fig. 6) and minor polypeptide chains at 55,000, 115,000 and between 150,000 and 200,000 daltons (Fig. 6). These results agree with previous findings showing that α-actinin can be extracted almost completely from skeletal muscle fibrils at low ionic strength (Arakawa, Robson and Goll 1970). The extracted residue, on the other hand, contains three major polypeptide chains of 200,-000, 160,000 and 55,000 daltons (Fig. 6) and minor polypeptide chains of 115,000 and 130,000 daltons (Fig. 6). Together, these SDS-polyacrylamide gels and the electron micrographs of the extracted residue suggest that the intermediate filaments that constitute most of this insoluble residue must be composed of one or more of the 55,000-, 115,000-, 130,000-, 160,000- and 200,000-dalton polypeptides that exist in this residue. Because the extract solubilized by low ionic strength extraction of dense-body preparations contained a small amount of the 55,000-dalton polypeptide (Fig. 6), we elected to attempt to isolate this 55,000-dalton polypeptide from these extracts, make antibodies to it, and determine whether these antibodies would bind to the intermediate filaments remaining after 2 mM Tris extraction of the dense-body preparations.

Protein solubilized by the low ionic strength extraction was salted out at pH 6.8 between 0% and 15% ammonium sulfate saturation to produce a P_{0-15}

fraction that was dialyzed against 1 mM KHCO$_3$, 10 mM 2-mercaptoethanol, pH 7.2, overnight at 2°C. This salting-out procedure precipitated the 42,000- and 55,000-dalton components from the low ionic strength extracts. After dialysis and subsequent clarification, the 55,000-dalton protein was purified to homogeneity (Fig. 7b) by using preparative slab electrophoresis in SDS as described in the preceding section. SDS-polyacrylamide gel electrophoresis of the electrophoretically purified 55,000-dalton component was done in a Tris-glycine buffer at pH 8.3 (Bryan 1974) instead of the phosphate buffer used in the Weber and Osborn (1969) procedure because the Tris-glycine buffer system separates the α and β subunits of tubulin (Bryan 1974). β-Tubulin clearly migrates significantly faster than the 55,000-dalton component under these electrophoretic conditions (Fig. 7a), but electrophoretically purified 55,000-dalton component comigrates with α-tubulin (Fig. 7a). Other properties, such as electrophoretic mobility in urea, distinguish the 55,000-dalton component from α-tubulin, however. Antibodies prepared against electrophoretically purified 55,000-dalton component bind all along the intermediate filaments left after low ionic strength extraction of dense-body preparations (Fig. 5d). Consequently, the intermediate filaments in chicken gizzard must be at least partly composed of this 55,000-dalton protein. Properties of this protein are under investigation and will be reported elsewhere (N. Arakawa, T. Huiatt, R. Robson, D. Goll and M. Stromer, in prep.). These preliminary results obviously do not eliminate the possibility that intermediate filaments contain proteins in addition to the 55,000-dalton component, and additional studies will be necessary to establish the complete composition of intermediate filaments.

Although our studies on localization of binding of anti-α-actinin, anti-tropomyosin and antibody to the 55,000-dalton component show that dense bodies contain α-actinin, that the filaments associated with dense bodies as well as the dense bodies themselves contain tropomyosin, and that interme-

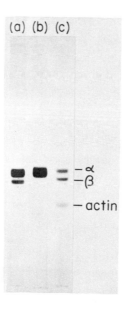

Figure 7
SDS (5%) polyacrylamide gels of the 55,000-dalton protein from chicken gizzard and of tubulin. (*a*) Tubulin and electrophoretically purified 55,000-dalton protein run on the same gel. (*b*) Electrophoretically purified 55,000-dalton protein alone. (*c*) Tubulin and actin. All gels were in 25 mM Tris-glycine pH 8.3 as described by Bryan (1974). Tubulin was prepared from porcine brain according to the procedure of Shelanski, Gaskin and Cantor (1973). Protein loads were: tubulin, 3 μg; actin, 2 μg; 55,000-dalton protein, 15 μg.

diate filaments are at least partly composed of a 55,000-dalton protein, the relatively poor resolution obtainable with peroxidase immunochemistry does not permit us to distinguish clearly what filament species course into the attachment plaques and the intracellular dense bodies. Previous studies (Rice and Brady 1973; Ashton, Somlyo and Somlyo 1975) have indicated that both 6-nm filaments, which are presumably composed of actin, and 10-nm intermediate filaments are associated with dense bodies in smooth muscle. Extraction of smooth muscle with 0.6 M KCl at a slightly alkaline pH leaves only dense bodies with 10-nm filaments, but this finding may simply indicate that both dense bodies and 10-nm filaments share a common insolubility in 0.6 M KCl. Our SDS-polyacrylamide gels of dense-body preparations prepared by 0.6 M KCl extraction of actomyosin show that these preparations contain considerable quantities of actin (cf. Fig. 6), which is presumably a constituent of the 6-nm filaments. Consequently, dense-body preparations probably contain 6-nm filaments or remnants of 6-nm filaments as well as dense bodies and intermediate filaments. Our results are therefore consistent with, but do not prove, the idea that both 6-nm actin-containing and 10-nm intermediate filaments containing the 55,000-dalton protein insert into dense bodies containing α-actinin. As we will discuss later, the common presence of α-actinin, actin and a 55,000-dalton protein in several motile systems suggests the possibility that a unique relationship exists among these three proteins.

Our results showing that dense bodies contain α-actinin substantiate several earlier suggestions (Prosser, Burnstock and Kahn 1960; Bois 1973; Somlyo et al. 1973) that dense bodies are smooth muscle analogs of the Z-disk in skeletal muscle and weaken some indirect arguments suggesting that dense bodies are not equivalent to Z-disks (Popescu and Ionescu 1970; Small and Squire 1972). Our assertion that dense bodies contain α-actinin is based on four lines of evidence: (1) anti-α-actinin binds exclusively to dense bodies and attachment plaques in sections of chicken gizzard muscle; (2) anti-α-actinin binds to dense bodies in dense-body preparations made by extracting actomyosin with 0.6 M KCl; (3) dense-body preparations made from chicken gizzard by extraction with 0.6 M KCl and containing only dense bodies and associated 6-nm and 10-nm filaments contain a peptide with a molecular weight of 100,000 daltons, which is the same as the subunit molecular weight of α-actinin; and (4) the 100,000-dalton peptide in dense-body preparations is solubilized by low ionic strength extraction just as α-actinin is solubilized from skeletal muscle Z-disks by low ionic strength extraction. These four lines of independent evidence leave little doubt that dense bodies contain α-actinin. That dense bodies also bind anti-tropomyosin extends the similarity between dense bodies and skeletal muscle Z-disks since Z-disks have also recently been shown to contain tropomyosin (Schollmeyer et al. 1974a).

M-Protein in Smooth Muscle and Some Other Motile Systems

The M-line is a readily recognizable structure in striated muscle fibrils, where it has been suggested to function physiologically to keep thick filaments in the highly ordered, three-dimensional array characteristic of skeletal muscle myofibrils (Knappeis and Carlsen 1968; Pepe 1971; Cherney and Di Dio 1975). Thick filaments evidently are also arranged in ordered, three-dimensional arrays in smooth muscle (Ashton, Somlyo and Somlyo 1975), but no organized M-line structure has ever been observed in electron micrographs of smooth muscle. Somlyo et al. (1971, 1973), however, have observed that smooth muscle thick filaments contain amorphous material connecting some of the adjacent thick filaments and have suggested that this amorphous material may be analogous with "M-substance" in striated muscle. It is unclear whether myosin in non-muscle motile systems is even always aggregated in the form of filaments (Pollard and Weihing 1974; Niederman and Pollard 1975), so no information is available on whether any myosin filaments that may exist in non-muscle motile systems are ordered in three-dimensional arrays. No organized structures comparable to striated muscle M-lines, however, have ever been seen in non-muscle motile systems. We have therefore investigated the possible presence of M-protein in smooth muscle and several other motile systems by examining the binding of antibodies made against a protein that was extracted from chicken gizzard with the same procedures used to extract M-protein from chicken skeletal myofibrils.

A putative M-protein was extracted and partly purified from minced chicken gizzard tissue by using the same procedure that Masaki and Takaiti (1974) used to extract M-protein from chicken skeletal muscle. This putative M-protein was then purified to complete homogeneity (Fig. 8) by using the preparative SDS slab electrophoresis procedure described earlier in this paper. The existence of M-protein in different motile systems was determined by using immunocytochemistry at the light microscope level with fluorescently labeled antibodies and by using Ouchterlony gel diffusion plates and low ionic strength extracts of different cells. Antibodies against electrophoretically purified M-protein were obtained as described earlier in this paper for anti-α-actinin and anti-tropomyosin antibodies and were conjugated with fluorescein

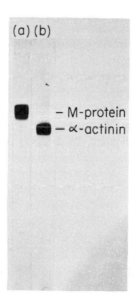

Figure 8
SDS (7.5%) polyacrylamide gels of (*a*) electrophoretically purified chicken gizzard M-protein and (*b*) chicken gizzard α-actinin. Each gel was loaded with 33 μg protein, and electrophoresis was done according to the procedure of Weber and Osborn (1969).

isothiocyanate according to the procedure of Wood, Thompson and Gold-stein (1965). Cultured cells were used for the fluorescence immunocyto-chemistry studies. A smooth muscle cell line, BC_3 HI, was generously furnished by Dr. David Schubert, Salk Institute, and Balb/c SV3T3 Cl_6 and 3T3 (A31) lines of fibroblast cells were obtained from Dr. George Todaro, National Cancer Institute.

The 3T3 cells were recloned, and clone f751a was used. All cells were grown for at least 3 days in Dulbecco's minimal essential media supplemented with 10% fetal calf serum, penicillin (100 μg/ml) and streptomycin (10 μg/ml) at 37°C in a 10% CO_2 atmosphere, as described by Furcht and Scott (1974). Cells were prepared for immunocytochemistry by removing them from growth media, washing them three times with 0.85% NaCl, 0.01 M potassium phosphate, pH 7.2, and then fixing them with a 50:50 v/v ethanol:ether solution for 20 minutes. After five more rinses with 0.85% NaCl, 0.01 M potassium phosphate, pH 7.2, the cells were incubated with an appropriately diluted antibody preparation in a humid atmosphere at 37°C for 60 minutes and were then examined in a Zeiss microscope equipped for epifluorescence.

Anti-M-protein-binding reveals a large number of M-protein-containing filaments that seem to span the interior of both smooth muscle (Fig. 9a) and fibroblast cells (Fig. 9b). The fluorescence pattern of anti-M-protein-binding shows striations or periodicity similar to the striations or interruptions seen in binding of fluorescently labeled anti-myosin to fibroblasts (Weber and Groeschel-Stewart 1974). A few areas of dispersed cytoplasmic staining similar to the "ground-plasma-like" areas of Weber and Groeschel-Stewart (1974) are also seen in the cells incubated with fluorescently labeled anti-M-protein (Fig. 9a,b). Because M-protein presumably binds to myosin, it would be expected that anti-M-protein would bind only to those parts of the cell that contain myosin. Nevertheless, the rather extensive distribution of M-protein in the smooth muscle cell and fibroblast shown in Figure 9a,b is surprising and suggests that cells may contain appreciable quantities of M-protein without displaying discrete M-lines. That the anti-M-protein antibodies used in this study also bound to M-lines in skeletal muscle myo-fibrils substantiates the conclusion that non-muscle cells may contain M-protein.

The presence of a putative M-protein in cells that contain no discernible M-line was also tested by using Ouchterlony diffusion plates. Gel diffusion was done in 1% agar (purified Difco) dissolved in 400 mM KCl, 0.01 M potassium phosphate, pH 7.4; 2.5 ml of this agar solution was pipetted onto precleaned microscope slides. Concentrations of the protein solutions used to produce clear precipitin lines were approximately 1.0 mg/ml. The diffusion was allowed to proceed for several days at 4°C, and the plates were then dried and stained with amido black to facilitate location of the precipitin lines. Anti-M-protein antibodies were placed in the center wells, and the peripheral wells contained crude P_{0-30} fractions prepared by salting out low ionic strength extracts of cells between 0% and 30% ammonium sulfate satu-ration. This is the same procedure used to prepare crude P_{0-30} α-actinin extracts (Arakawa, Robson and Goll 1970; Goll et al. 1972; Suzuki et al. 1976); we (J. Schollmeyer, unpubl. obs.) have previously noted that these crude

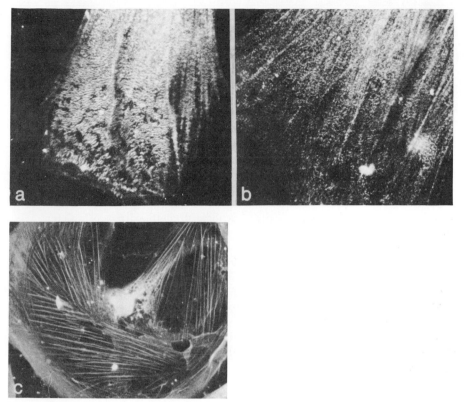

Figure 9

Micrographs showing cells viewed by fluorescence microscopy after fixation and incubation with fluorescein-labeled antibodies to either M-protein (*a* and *b*) or a sequential combination of anti-M, anti-α-actinin and anti-tropomyosin (*c*). (*a*) Fluorescence microscopy of anti-M-labeled smooth muscle cell (BC HI); (*b*) fluorescence microscopy of anti-M-labeled chick embryo fibroblast; (*c*) fluorescence microscopy of chick embryo fibroblasts labeled sequentially with anti-M, anti-α-actinin and anti-tropomyosin. Sequential labeling removes the striations observed when anti-M-protein alone is used for labeling but does not reveal two populations of filaments. Magnification, 394$\times$.

P_{0-30} α-actinin extracts contain M-protein. Double diffusion plates done with these crude P_{0-30} fractions showed that 3T3 fibroblasts, blood platelets and human uterus all contain a protein that is antigenically homologous with the putative M-protein prepared from chicken gizzard (Fig. 10). Presumably, M-protein functions in smooth muscle cells to keep myosin filaments in the three-dimensional array observed by Ashton, Somlyo and Somlyo (1975) in stereoelectron microscopy, but it is not yet known whether myosin in fibroblasts and blood platelets forms filaments that are arranged in a unique three-dimensional order (Pollard and Weihing 1974; Niederman and Pollard 1975).

Anti-tropomyosin also binds to filaments in fibroblasts to produce a striated pattern with fluorescent segments 1.2 μm long and nonfluorescent segments 0.4 μm long when viewed in fluorescence microscopy (Lazarides 1975a,b).

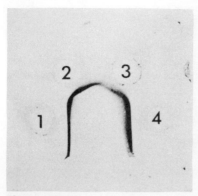

Figure 10
Ouchterlony double diffusion plates of low ionic strength extracts of different cells that were salted out between 0% and 30% ammonium sulfate saturation and then allowed to diffuse against anti-M-protein antibody in the central well. Central well has 500 μg anti-M-protein made against a putative M-protein isolated from chicken gizzard. The peripheral wells contain 15 μg each of crude P_{0-30} protein isolated from (1) chicken gizzard, (2) 3T3 fibroblasts, (3) human platelets and (4) human uterus. The single precipitin lines show that the M-protein antibody is homogeneous against a single muscle protein since the P_{0-30} extracts contain many different proteins. Continuity of the precipitin lines indicates all M-antigens from these four different tissues are homologous.

Anti-actin, on the other hand, seems to bind to the same filaments to produce continuous fluorescence (Lazarides and Weber 1974; Lazarides, 1975a,b). Lazarides (1975a) has proposed that the periodic absence of anti-tropomyosin binding to these filaments is due to the presence of a substance, "S," that binds periodically in segments 0.4 μm long to actin filaments and prevents binding of tropomyosin to these regions. Although Lazarides first suggested that the "S" substance might be myosin (Lazarides 1975a), he subsequently showed that anti-α-actinin sometimes bound periodically to actin filaments to produce fluorescent segments 0.4 μm long, and that the "S" substance might therefore be α-actinin. We incubated fibroblasts sequentially with fluorescently labeled anti-M-protein, anti-tropomyosin and anti-α-actinin to determine whether such sequential incubation would obscure the periodicity of anti-M-protein-binding or would reveal two populations of filaments, one having the periodicity of the anti-M-protein-binding, and hence presumably composed of myosin, and the other showing a continuous fluorescence such as might be expected if α-actinin were the "S" substance binding to actin filaments in 0.4 μm segments and preventing anti-tropomyosin-binding in these segments. The results indicate that only one set of filaments exhibiting continuous fluorescence can be observed after such sequential incubation (Fig. 9c). The complexity of this experiment and the limits of resolution of the light microscope prevent definite conclusions at this point, but these results suggest several possibilities. First, both myosin and α-actinin may act as the "S" substance proposed by Lazarides (1975a), and their complementary binding to actin filaments fills the segments left by anti-tropomyosin-binding to produce one population of filaments having continuous fluorescence. Weber and Groeschel-Stewart (1974) have suggested previously that myosin is located either in or very close to the microfilament system in fibroblasts, and it seems quite possible that this very close location of large myosin molecules could affect antibody binding to the microfilaments. Alternatively, the SDS-polyacrylamide gels of tropomyosin shown by Lazarides (1975a) seem to contain some minor contaminants with approximately the same molecular weight as M-protein. When considering the very high antigenicity that has been reported for M-protein (Masaki, Endo and Ebashi 1967) and the reported weak cross reactivity between antibodies against skeletal muscle tropomyosin, which Lazarides used as an antigen, and smooth muscle tropomyosin (Cummins and Perry 1974; J. Schollmeyer, unpubl.), it seems possible that some of Lazarides' results showing periodicity of anti-tropomyosin binding to fibroblasts may have originated from anti-M-protein-binding. Clearly, additional studies are necessary to determine the nature of the structures that bind antibodies against the putative M-protein isolated in the present study.

Localization of α-Actinin in Normal and Transformed Fibroblasts

As Goldman et al. (1975) have recently reported, significant numbers of densely staining substrate attachment sites can be observed in sections taken close and parallel to the adhesive side of flat-embedded 3T3 cells (Fig. 11a). If sections are taken in a slightly different plane, attachment plaques can be seen that have large numbers of microfilaments radiating from them (Fig. 11b), although the attachment plaques shown in Figure 11b are not stained as densely as those seen in Figure 11a. Fibroblasts that have been transformed with SV40-3T3 virus, on the other hand, contain few densely staining attachment sites (Fig. 12b) and, as has already been reported by others (McNutt, Culp and Black 1971, 1973; Gruenstein, Rich and Weihing 1975; Pollack, Osborn and Weber 1975; Robbins et al. 1975; Weber et al. 1975; Wickus et al. 1975), also contain no microfilaments attached to the cell membrane (Fig. 12b). Contact-inhibited control cells in our study have both densely staining attachment plaques and membrane-associated microfilament bundles (Fig. 12a). Although viral transformation is accompanied by almost total loss of membrane-associated actin filaments, it has been reported that the total amount of actin in these cells changes little after viral transformation (Robbins et al. 1975; Wickus et al. 1975). Myosin content, however, has been reported to be only 50% as great in transformed fibroblasts as in normal cells (Ostlund, Pastan and Adelstein 1974).

Intercalated disks in cardiac muscle, desmosomes in intestinal epithelial cells, Z-disks in skeletal and cardiac muscle and dense bodies and attachment plaques in smooth muscle all stain densely with ordinary electron microscope stains, and our antibody localization studies (Schollmeyer, Stromer and Goll 1972; Schollmeyer et al. 1973, 1974a,b) have shown that all these densely staining structures contain α-actinin. These findings and the morphological appearance of substrate attachment sites in fibroblasts suggested that substrate attachment sites might also contain α-actinin that is lost or dispersed during SV40-3T3 transformation of these cells. Use of the in-

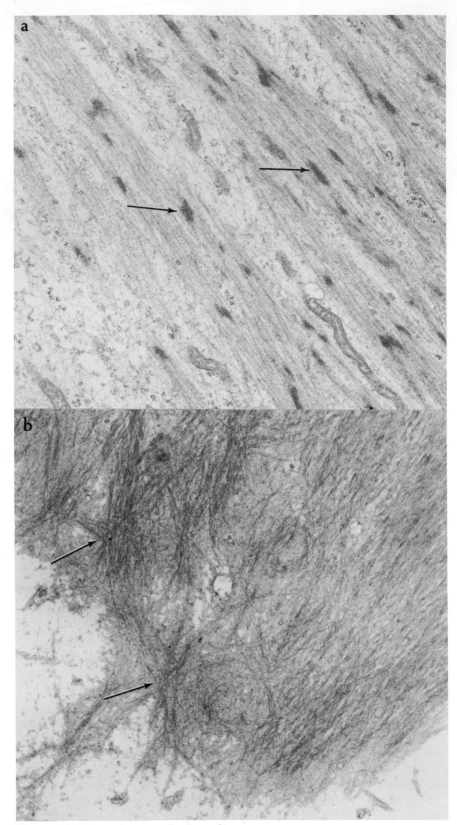

Figure 11 (*See facing page for legend*)

378

direct immunoperoxidase localization of anti-α-actinin-binding shows that anti-α-actinin binds to the substrate attachment sites in untransformed fibroblasts (Fig. 13). Application of a similar procedure to SV40-3T3-transformed cells showed that little or no anti-α-actinin binds to regions near the cell membrane, although some diffuse anti-α-actinin-binding could be detected in the perinuclear region (results not shown here). No peroxidase reaction products could be detected in control experiments using normal rabbit serum in place of the anti-α-actinin-containing serum. These results indicate that in addition to changes in cell shape, membrane transport, agglutination by plant lectins, contact inhibition of growth and membrane-associated microfilaments that occur during SV40 transformation of fibroblasts (Wickus et al. 1975), viral transformation is also associated with a loss or dispersal of α-actinin in fibroblasts.

Physiological Role of α-Actinin

Although it is now possible to obtain highly purified α-actinin in sizable quantities from skeletal (Suzuki et al. 1976), cardiac (Robson and Zeece 1973; I. Singh et al., unpubl.) and smooth (R. M. Robson, T. Huiatt, M. G. Zeece, D. E. Goll and M. H. Stromer, in prep.) muscle, the physiological role of α-actinin remains unclear (Temple and Goll 1970; Goll et al. 1972; Suzuki et al. 1973). The results of the present study together with results of four earlier studies on the localization (Schollmeyer, Stromer and Goll 1972; Schollmeyer et al. 1973, 1974a,b) and properties (Robson et al. 1970; Goll et al. 1972) of α-actinin suggest that at least three different physiological functions can be suggested for α-actinin (Suzuki et al. 1976). First, α-actinin may modify the structure of actin monomers while these monomers are in the aggregated, filamentous state. This possible role of α-actinin is based on α-actinin's ability to accelerate in vitro measures of contraction (Arakawa, Robson and Goll 1970; Robson et al. 1970; Temple and Goll 1970) and to alter the rate of exchange of the nucleotide bound to F-actin (Craig-Schmidt et al. 1975). Second, the ability of α-actinin to cross-link actin filaments in vitro (Maruyama and Ebashi 1965; Kawamura et al. 1970; Holmes, Goll and Suzuki 1971) may indicate that the physiological role of α-actinin is to bind or cross-link actin filaments to each other, as in Z-disks in striated muscle (Goll, Mommaerts and Seraydarian 1967; Masaki, Endo and Ebashi 1967; Goll et al. 1969; Schollmeyer, Stromer and Goll 1972; Schollmeyer et al. 1973, 1974a), intracellular dense bodies in smooth muscle (Schollmeyer et al. 1973) or microfilament bundles in acrosomal processes of *Limulus* sperm

Figure 11
Electron micrographs of thin sections through 3T3 fibroblasts at a location close to the substrate attachment surface. (*a*) Micrograph of a section parallel to the substrate attachment surface showing densely staining intracellular attachment sites (arrows) interspersed among bundles of microfilaments. Magnification, 30,000×. (*b*) Micrograph shows a slightly different orientation near the substrate attachment surface and permits observation of microfilaments as they appear to radiate from attachment sites (arrows) in large bundles. Magnification, 20,000×.

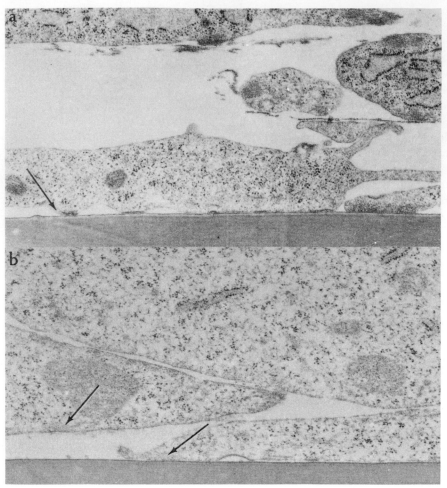

Figure 12

Electron micrographs of transverse sections through 3T3 fibroblasts that have been embedded on plastic cover slips. (*a*) Micrograph of a normal 3T3 fibroblast showing a cross section of a dense substrate attachment area (arrow) seen in longitudinal section in Fig. 11a. (*b*) No densely staining substrate attachment areas similar to those seen in Fig. 12a can be observed in SV40-3T3-transformed fibroblasts, although a few areas stain darkly enough (arrows) to suggest that they might be remnants of the substrate attachment sites seen in nontransformed cells. Magnifications, 20,000×.

(Schollmeyer et al. 1974b), or to bind actin filaments to membrane-associated attachment sites, as in fibroblasts, attachment plaques in smooth muscle (Schollmeyer et al. 1973) or the dense mat of desmosomes in intestinal epithelial cells (Schollmeyer 1974b). By selective binding to only one end or in only one orientation (Goll et al. 1972; Stromer and Goll 1972), α-actinin could control directionality of thin or microfilaments. Third, the ability of α-actinin to induce aggregation of actin monomers under conditions where these monomers would otherwise remain unaggregated (G. R. Holmes and D. E. Goll, unpubl. obs.) and the ability of α-actinin to accelerate the rate of

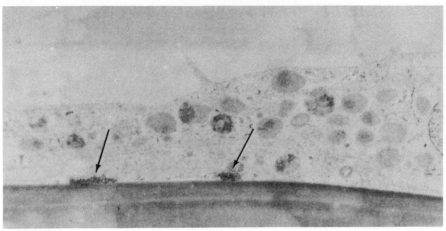

Figure 13

α-Actinin antibodies bind to the substrate attachment region (arrows) of 3T3 fibro-blasts after they have been incubated with anti-α-actinin antibodies followed by sheep anti-rabbit IgG conjugated with peroxidase. The electron-dense reaction products of the peroxidase-catalyzed oxidation of 3,3'-diaminobenzidene seem to cover almost the entire attachment site.

actin aggregation (Maruyama and Ebashi 1965) suggest that α-actinin may nucleate aggregation of actin monomers and thereby assist in regulating the proportion of aggregated to monomeric actin in cells (Lazarides and Burridge 1975; Suzuki et al. 1976).

A possible role for α-actinin in controlling the state of actin aggregation may account for the observations that the number of membrane-associated actin microfilaments decreases dramatically after viral transformations of some cells (McNutt, Culp and Black 1971, 1973; Pollack, Osborn and Weber 1975; Ambros, Chen and Buchanan 1975; Gruenstein, Rich and Weihing 1975; Robbins et al. 1975; Weber et al. 1975; Wickus et al. 1975) even though total actin content seems unaffected by such transformation (Robbins et al. 1975; Wickus et al. 1975). Our antibody localization studies show that the number of membrane-associated, densely staining attachment sites that contain α-actinin and that bind microfilaments are greatly decreased after viral transformation of fibroblasts just as the number of membrane-associated microfilaments are. After viral transformation, α-actinin seems dispersed through the cell rather than localized in attachment plaques. Although the simultaneous dispersion of α-actinin and loss of microfilaments does not necessarily denote a cause and effect relationship, it seems reasonable, in view of the known ability of α-actinin to cross-link actin filaments in vitro (Holmes, Goll and Suzuki 1971; Goll et al. 1972; Stromer and Goll 1972), to postulate that α-actinin attaches actin filaments to the cell membrane and that loss of α-actinin from the cell membrane leads to loss and disaggregation of membrane-associated actin filaments. We have shown previously (Goll et al. 1972; Stromer and Goll 1972) that α-actinin undergoes a temperature-induced conformational change that markedly affects its binding to actin, and it seems possible that this or a similar conformational change is involved in loss of α-actinin-containing attachment sites after viral transformation of fibroblasts. Such a role for α-actinin also indicates that α-actinin may be an extrinsic membrane protein under certain conditions, and that α-actinin undergoes functionally important changes in distribution during the cell cycle (Lazarides and Burridge 1975).

The three physiological roles proposed in the preceding paragraphs for α-actinin are not mutually exclusive, and α-actinin could function in any one, two, or all three of the proposed roles. Indeed, our proposal that α-actinin is involved in loss of membrane-associated microfilaments after viral transformation of fibroblasts requires that α-actinin both promote aggregation of actin monomers into microfilaments and bind actin filaments to the cell membrane. The nature of the binding of α-actinin to F-actin remains unknown, although we have shown in in vitro studies that α-actinin binds to F-actin filaments but not to G-actin, myosin, tropomyosin or the tropomyosin-troponin complex (Holmes et al. 1971). Specificity of in vitro binding studies involving actin has been questioned by the recent finding (Bray 1975) that actin binds to ten different proteins in vitro, but the close in vivo proximity of actin and α-actinin in all motile systems studied thus far suggests rather strongly that α-actinin binds to actin in vivo as well as in vitro. Several recent findings, however, when considered together, suggest that the α-actinin–F-actin binding may be influenced in a physiologically important way by the presence of a 55,000-dalton polypeptide. First, we (Dayton et al. 1974, 1975) have found that a proteolytic enzyme isolated from muscle releases α-actinin from skeletal muscle myofibrils, although it does not degrade either actin or α-actinin. The enzyme does not release purified α-actinin bound to purified F-actin in vitro. Release of α-actinin from skeletal muscle myofibrils by this enzyme is accompanied by destruction of a very small amount of 55,000-dalton component in these myofibrils (W. Dayton, D. Goll, J. Schollmeyer, R. Robson and M. Stromer, in prep.). These results imply that binding of α-actinin to myofibrils involves a 55,000-dalton component whose proteolytic destruction results in release of α-actinin from the myofibril. Second, a number of studies have shown that actin and α-actinin occur in association with a 55,000-dalton component in several different motile systems. For example, several studies (Rice and Brady 1973; Somlyo et al. 1973; Ashton, Somlyo and Somlyo 1975) have shown that smooth muscle dense bodies are associated with both actin-containing thin filaments and 10-nm intermediate filaments, and the present study has shown that smooth muscle dense bodies contain α-actinin and smooth muscle intermediate filaments contain a 55,000-dalton component. If 10-nm filaments in other tissues are also composed of this 55,000-dalton component, then desmosomes of intestinal epithelial cells also contain α-actinin and the 55,000-dalton component in association since we have shown that these desmosomes contain α-actinin (Schollmeyer et al. 1974b) and Ishikawa, Bischoff and Holtzer (1969) have shown that the 10-nm filaments coursing into these desmosomes do not bind heavy meromyosin and therefore do not contain actin. Consequently, desmosomes in intestinal epithelial cells may be an example of an association between α-actinin and the 55,000-dalton protein in the absence of actin, although the presence of actin filaments inserting into these desmosomes has not yet been definitely excluded. The acrosomal process of *Limulus* sperm may be the clearest example of an association among actin, α-actinin and the 55,000-dalton protein. Tilney (1975) has shown that this acrosomal process consists exclusively of 42,000-dalton (actin), 55,000-dalton and 100,000-dalton components, and our antibody studies (Schollmeyer et al. 1974b) have shown that the 100,000-dalton component is α-actinin. In other less well-characterized systems, Bullard, Dabrowska and Winkelman (1973) have shown that *Lethocerus* (water bug) flight muscle contains a 55,000-dalton polypeptide that co-purifies with actin, and Hammond and Goll (1975) have shown that *Lethocerus* flight muscle contains α-actinin. Similarly, the 8- to 9-nm filaments in axons from bovine brain contain a 51,000-dalton protein (Shelanski, Gaskin and Cantor 1973), and Chang and Goldman (1973) have shown that axons in cultured neuroblastoma cells contain actinlike filaments. Although the evidence thus far is only circumstantial, the repeated occurrence of α-actinin, actin and a 55,000-dalton protein in association may warrant future study.

Acknowledgments

We thank Sue Perry and Joan Andersen for expert and indefatigable assistance with the manuscript. We are also grateful to Dr. Ellis Benson for his support of this work.

Journal Paper No. J-8413 of the Iowa Agriculture and Home Economics Experiment Station, Project Nos. 1795, 1796, and 2025. This research was supported in part by grants from the National Institutes of Health (AM-12654, HL-15679, 1PO1-CA-16228), by a grant from the American Heart Association (71-679), and by grants from the Iowa Heart Association and the Muscular Dystrophy Associations of America.

REFERENCES

Ambros, V. R., L. B. Chen and J. M. Buchanan. 1975. Surface ruffles as markers for studies of cell transformation by Rous sarcoma virus. *Proc. Nat. Acad. Sci.* **72**.3144.

Anderson, W. A. and R. A. Ellis. 1965. Ultrastructure of *Trypanosoma lewisi:* Flagellum, microtubules and the kinetoplast, *J. Protozool.* **12**:483.

Arakawa, N., R. M. Robson and D. E. Goll. 1970. An improved method for the preparation of α-actinin from rabbit striated muscle. *Biochim. Biophys. Acta* **200**:284.

Ashton, F. T., A. V. Somlyo and A. P. Somlyo. 1975. The contractile apparatus of vascular smooth muscle: Intermediate high voltage stereo electron microscopy. *J. Mol. Biol.* **98**:17.

Avrameas, S. 1971. Peroxidase labelled antibody and Fab conjugates with enhanced intracellular penetration. *Immunochemistry* **8**:1175.

Bois, R. M. 1973. The organization of the contractile apparatus of vertebrate smooth muscle. *Anat. Rec.* **177**:61.

Bray, D. 1975. Sticky actin. *Nature* **256**:616.

Bryan, J. 1974. Biochemical properties of microtubules. *Fed. Proc.* **33**:152.

Bullard, B., R. Dabrowska and L. Winkelman. 1973. The contractile and regulatory proteins of insect flight muscle. *Biochem. J.* **135**:277.

Chang, C.-M. and R. D. Goldman. 1973. The localization of actin-like fibers in cultured neuroblastoma cells as revealed by heavy meromyosin binding. *J. Cell Biol.* **57**:867.

Cherney, D. D. and L. J. A. Di Dio. 1975. N-lines and M-bands in cardiac muscle. *Experientia* **31**:445.

Cooke, P. 1975. A filamentous cytoskeleton in vertebrate smooth muscle fibers. *J. Cell Biol.* **67**:78a.

Cooke, P. H. and R. H. Chase. 1971. Potassium chloride-insoluble myofilaments in vertebrate smooth muscle cells. *Exp. Cell Res.* **66**:417.

Cooke, P. H. and F. S. Fay. 1972. Thick myofilaments in contracted and relaxed mammalian smooth muscle cells. *Exp. Cell Res.* **71**:265.

Craig-Schmidt, M. C., R. M. Robson and D. E. Goll. 1975. Effect of α-actinin on the bound nucleotide of F-actin. *Fed. Proc.* **34**:539.

Cummins, P. and S. V. Perry. 1974. Chemical and immunochemical characteristics of tropomyosins from striated and smooth muscle. *Biochem. J.* **141**:43.

Dayton, W. R., D. E. Goll, M. H. Stromer, W. J. Reville, M. G. Zeece and R. M. Robson. 1975. Some properties of a Ca^{2+}-activated protease that may be involved in myofibrillar protein turnover. In *Proteases and biological control* (ed. E. Reich, D. B. Rifkin and E. Shaw), pp. 551–577. Cold Spring Harbor Laboratory, Cold Spring Harbor, New York.

Dayton, W. R., I. Singh, J. E. Schollmeyer, D. E. Goll, M. H. Stromer, W. J. Reville and R. M. Robson. 1974. Use of an endogenous muscle protease to study the Z-disk and Z-disk proteins. *J. Cell Biol.* **63**:73a.

Driska, S. P. and D. J. Hartshorne. 1974. Regulatory proteins of smooth muscle. *Fed. Proc.* **33**:2019.

————. 1975. The contractile proteins of smooth muscle. Properties and components of a Ca^{2+}-sensitive actomyosin from chicken gizzard. *Arch. Biochem. Biophys.* **167**:203.

Ebashi, S., M. Endo and I. Ohtsuki. 1969. Control of muscle contraction. *Q. Rev. Biophys.* **2**:351.

Fay, F. S. and P. H. Cooke. 1972. Reversible disaggregation of myofilaments in vertebrate smooth muscle. *J. Cell Biol.* **56**:399.

Furcht, L. T. and R. E. Scott. 1974. Influence of cell cycle and cell movement on the distribution of intramembranous particles in contact inhibited and transformed cells. *Exp. Cell Res.* **88**:311.

Goldman, R. D., E. Lazarides, R. Pollack and K. Weber. 1975. The distribution of actin in nonmuscle cells. The use of actin antibody in the localization of actin within the microfilament bundles of mouse 3T3 cells. *Exp. Cell Res.* **90**:333.

Goll, D. E., W. F. H. M. Mommaerts and K. Seraydarian. 1967. Is α-actinin a constituent of the Z-band of the muscle fibril? *Fed. Proc.* **26**:499.

Goll, D. E., W. F. H. M. Mommaerts, M. K. Reedy and K. Seraydarian. 1969. Studies on α-actinin-like proteins released during trypsin digestion of α-actinin and of myofibrils. *Biochim. Biophys. Acta* **175**:174.

Goll, D. E., A. Suzuki, J. Temple and G. R. Holmes. 1972. Studies on purified α-actinin. I. Effect of temperature and tropomyosin on the α-actinin/F-actin interaction. *J. Mol. Biol.* **67**:469.

Greaser, M. L. and J. Gergely. 1971. Reconstitution of troponin activity from three protein components. *J. Biol. Chem.* **246**:4226.

Gruenstein, E., A. Rich and R. R. Weihing. 1975. Actin associated with membranes from 3T3 mouse fibroblasts and HeLa cells. *J. Cell Biol.* **64**:223.

Hammond, K. S. and D. E. Goll. 1975. Purification of insect myosin and α-actinin. *Biochem. J.* **151**:189.

Hanson, J. 1973. Evidence from electron microscope studies on actin paracrystals concerning the origin of the cross-striation in the thin filaments of vertebrate skeletal muscle. *Proc. Roy. Soc. B.* **183**:39.

Hanson, J. and J. Lowy. 1963. Contractile mechanism of mammalian smooth muscle. *Proc. Roy. Soc. B.* **160**:523.

Hanson, J., V. Lednev, E. J. O'Brien and P. Bennett. 1973. Structure of the actin-containing filaments in vertebrate skeletal muscle. *Cold Spring Harbor Symp. Quant. Biol.* **37**:311.

Hitchcock, S. E., H. E. Huxley and A. Szent-Györgyi. 1973. Calcium-sensitive binding of troponin to actin-tropomyosin: A two-site model for troponin action. *J. Mol. Biol.* **80**:825.

Holmes, G. R., D. E. Goll and A. Suzuki. 1971. Effect of α-actinin on actin viscosity. *Biochim. Biophys. Acta* **253**:240.

Huxley, H. E. 1963. Electron microscope studies on the structure of natural and synthetic protein filaments from striated muscle. *J. Mol. Biol.* **7**:281.

Ishikawa, H., R. Bischoff and H. Holtzer. 1969. Formation of arrowhead complexes with heavy meromyosin in a variety of cell types. *J. Cell Biol.* **43**:312.

Kawamura, M., T. Masaki, Y. Nonomura and K. Maruyama. 1970. An electron microscopic study of the action of the 6S component of α-actinin on F-actin. *J. Biochem.* **68**:577.

Kelly, R. E. and R. V. Rice. 1968. Localization of myosin filaments in smooth muscle. *J. Cell Biol.* **37**:105.

Knappeis, G. G. and F. Carlsen. 1968. The ultrastructure of the M-line in skeletal muscle. *J. Cell Biol.* **38**:202.

Laszt, L. and G. Hamoir. 1961. Étude par électrophoresce et ultracentrifugation de la composition protéinique de la couche musculaire des carotides de bovidé. *Biochim. Biophys. Acta* **50**:430.

Lazarides, E. 1975a. Tropomyosin antibody: The specific localization of tropomyosin in nonmuscle cells. *J. Cell Biol.* **65**:549.

———. 1975b. Immunofluorescence studies on the structure of actin filaments in tissue culture cells. *J. Histochem. Cytochem.* **23**:507.

Lazarides, E. and K. Burridge. 1975. α-Actinin: Immunofluorescent localization of a muscle structural protein in nonmuscle cells. *Cell* **6**:289.

Lazarides, E. and K. Weber. 1974. Actin antibody: The specific visualization of actin filaments in nonmuscle cells. *Proc. Nat. Acad. Sci.* **71**:2268.

McNutt, N. S., L. A. Culp and P. H. Black. 1971. Contact-inhibited revertant cell lines isolated from SV40-transformed cells. II. Ultrastructural study. *J. Cell Biol.* **50**:691.

———. 1973. Contact-inhibited revertant cell lines isolated from SV40-transformed cells. IV. Microfilament distribution and cell shape in untransformed, transformed, and revertant Balb/c 3T3 cells. *J. Cell Biol.* **56**:412.

Maruyama, K. and S. Ebashi. 1965. α-Actinin, a new structural protein from striated muscle. II. Action on actin. *J. Biochem.* **58**:13.

Masaki, T. and O. Takaiti. 1974. M-protein. *J. Biochem.* **75**:367.

Masaki, T., M. Endo and S. Ebashi. 1967. Localization of 6S component of α-actinin at Z-band. *J. Biochem.* **62**:630.

Morimoto, K. and W. F. Harrington. 1973. Isolation and composition of thick filaments from rabbit skeletal muscle. *J. Mol. Biol.* **77**:165.

Murphy, R. A., J. T. Herlihy and J. Megerman. 1974. Force-generating capacity and contractile protein content of arterial smooth muscle. *J. Gen. Physiol.* **64**:691.

Nakane, P. K. and G. B. Pierce, Jr. 1967. Enzyme-labeled antibodies for the light and electron microscopic localization of tissue antigens. *J. Cell Biol.* **33**:307.

Needham, D. M. and C. F. Schoenberg. 1967. The biochemistry of the myometrium. In *Cellular biology of uterus* (ed. R. M. Wynn), pp. 291–352. North-Holland, Amsterdam.

Niederman, R. and T. D. Pollard. 1975. Human platelet myosin. II. *In vitro* assembly and structure of myosin filaments. *J. Cell Biol.* **67**:72.

Nonomura, Y. 1968. Myofilaments in smooth muscle of guinea pig's taenia coli. *J. Cell Biol.* **39**:741.

Offer, G. 1973. C-protein and the periodicity in the thick filaments of vertebrate skeletal muscle. *Cold Spring Harbor Symp. Quant. Biol.* **37**:87.

Offer, G., C. Moos and R. Starr. 1973. A new protein of the thick filaments of vertebrate skeletal myofibrils. Extraction, purification, and characterization. *J. Mol. Biol.* **74**:653.

Ohtsuki, I. 1975. Distribution of troponin components in the thin filament studied by immunoelectron microscopy. *J. Biochem.* **77**:633.

Ostlund, R. E., I. Pastan and R. S. Adelstein. 1974. Myosin in cultured fibroblasts. *J. Biol. Chem.* **249**:3903.

Panner, B. J. and C. R. Honig. 1967. Filament ultrastructure and organization in vertebrate smooth muscle. Contraction hypothesis based on localization of actin and myosin. *J. Cell Biol.* **35**:303.

Parry, D. A. D. and J. M. Squire. 1973. Structural role of tropomyosin in muscle regulation: Analysis of the X-ray diffraction patterns from relaxed and contracting muscles. *J. Mol. Biol.* **75**:33.

Pepe, F. A. 1971. Structural components of the striated muscle fibril. In *Subunits in biological systems* (ed. S. N. Timasheff and G. D. Fasman), vol. 5, part A, pp. 323–353. Marcel Dekker, New York.

Pollack, R., M. Osborn and K. Weber. 1975. Patterns of organization of actin and myosin in normal and transformed cultured cells. *Proc. Nat. Acad. Sci.* **72**:994.

Pollard, T. D. and R. R. Weihing. 1974. Actin and myosin and cell movement. *Crit. Rev. Biochem.* **2**:1.

Popescu, L. M. and N. Ionescu. 1970. On the equivalence between dense bodies and Z-bands. *Experientia* **26**:642.

Prosser, C. L., G. Burnstock and J. Kahn. 1960. Conduction in smooth muscle: Comparative structural properties. *Amer. J. Physiol.* **199**:545.

Reynolds, E. S. 1963. The use of lead citrate at high pH as an electron opaque stain in electron microscopy. *J. Cell Biol.* **17**:208.

Rice, R. V. and A. C. Brady. 1973. Biochemical and ultrastructural studies on vertebrate smooth muscle. *Cold Spring Harbor Symp. Quant. Biol.* **37**:429.

Rice, R. V., A. C. Brady and J. C. Baldassore. 1974. Mesosin—A new fibrous protein which may be ubiquitous. *Fed. Proc.* **33**:1466.

Robbins, P. W., G. G. Wickus, P. E. Branton, B. J. Gaffney, C. B. Hirschberg, P. Fuchs and P. M. Blumberg. 1975. The chick fibroblast cell surface after transformation by Rous sarcoma virus. *Cold Spring Harbor Symp. Quant. Biol.* **39**:1173.

Robson, R. M. and M. G. Zeece. 1973. Comparative studies of α-actinin from porcine cardiac and skeletal muscle. *Biochim. Biophys. Acta* **295**:208.

Robson, R. M., D. E. Goll, N. Arakawa and M. H. Stromer. 1970. Purification and properties of α-actinin from rabbit skeletal muscle. *Biochim. Biophys. Acta* **200**:296.

Schoenberg, C. F. 1969. An electron microscope study of the influence of divalent ions on myosin filament formation in chicken gizzard extracts and homogenates. *Tissue and Cell* **1**:83.

Schollmeyer, J. E., M. H. Stromer and D. E. Goll. 1972. α-Actinin and tropomyosin localization in normal and diseased muscle. *Biophys. J.* **12**:280a.

Schollmeyer, J. E., D. E. Goll, R. M. Robson and M. H. Stromer. 1973. Localization of α-actinin and tropomyosin in different muscles. *J. Cell Biol.* **59**:306a.

Schollmeyer, J. V., D. E. Goll, M. H. Stromer, W. Dayton, I. Singh and R. M. Robson. 1974a. Studies on the composition of the Z-disk. *J. Cell Biol.* **63**:303a.

Schollmeyer, J. V., D. E. Goll, L. Tilney, M. Mooseker, R. Robson and M. Stromer. 1974b. Localization of α-actinin in nonmuscle material. *J. Cell Biol.* **63**:304a.

Shelanski, M. L., F. Gaskin and C. R. Cantor. 1973. Microtubule assembly in the absence of added nucleotides. *Proc. Nat. Acad. Sci.* **70**:765.

Small, J. V. and J. M. Squire. 1972. Structural basis of contraction in vertebrate smooth muscle. *J. Mol. Biol.* **67**:117.

Sobieszek, A. and R. D. Bremel. 1975. Preparation and properties of vertebrate smooth-muscle myofibrils and actomyosin. *Eur. J. Biochem.* **55**:49.

Sobieszek, A. and J. V. Small. 1976. Myosin-linked Ca^{2+}-regulation in vertebrate smooth muscle. *J. Mol. Biol.* **102**:75.

Somlyo, A. P., C. E. Devine, A. V. Somlyo and R. V. Rice. 1973. Filament organization in vertebrate smooth muscle. *Proc. Roy. Soc. B.* **265**:223.

Somlyo, A. P., A. V. Somlyo, C. E. Devine and R. V. Rice. 1971. Aggregation of thick filaments into ribbons in mammalian smooth muscle. *Nature* **231**:243.

Spudich, J. A., H. E. Huxley and J. T. Finch. 1972. Regulation of skeletal muscle contraction. II. Structural studies of the interaction of the tropomyosin-troponin complex with actin. *J. Mol. Biol.* **72**:619.

Stromer, M. H. and D. E. Goll. 1972. Studies on purified α-actinin. II. Electron microscopic studies on the competitive binding of α-actinin and tropomyosin to Z-line extracted myofibrils. *J. Mol. Biol.* **67**:489.

Stromer, M. H., D. J. Hartshorne and R. V. Rice. 1967. Removal and reconstitution of Z-line material in a striated muscle. *J. Cell Biol.* **35**:C23.

Stromer, M. H., D. J. Hartshorne, H. Mueller and R. V. Rice. 1969. The effect of various protein fractions on Z- and M-line reconstitution. *J. Cell Biol.* **40**:167.

Stromer, M. H., L. B. Tabatabai, R. M. Robson, D. E. Goll and M. G. Zeece. 1976. Nemaline myopathy: An integrated study. I. Selective extraction. *Exp. Neurol.* (in press).

Suzuki, A., D. E. Goll, M. H. Stromer, I. Singh and J. Temple. 1973. α-Actinin from red and white porcine muscle. *Biochim. Biophys. Acta* **295**:188.

Suzuki, A., D. E. Goll, I. Singh, R. E. Allen, R. M. Robson and M. H. Stromer. 1976. Some properties of purified skeletal muscle α-actinin. *J. Biol. Chem.* (in press).

Temple, J. and D. E. Goll. 1970. Nucleotide and cation specificity of the α-actinin-induced increase in actomyosin nucleoside triphosphate phosphohydrolase activity. *Biochim. Biophys. Acta* **205**:121.

Tilney, L. 1975. Actin filaments in the acrosomal reaction of *Limulus* sperm. Motion generated by alterations in the packing of the filaments. *J. Cell Biol.* **64**:289.

Wakabayashi, T., H. E. Huxley, L. A. Amos and A. Klug. 1975. Three-dimensional image reconstruction of actin-tropomyosin complex and actin-tropomyosin-troponin-T-troponin-I complex. *J. Mol. Biol.* **93**:477.

Weber, K. and U. Groeschel-Stewart. 1974. Antibody to myosin: The specific visualization of myosin-containing filaments in nonmuscle cells. *Proc. Nat. Acad. Sci.* **71**:4561.

Weber, K. and M. Osborn. 1969. The reliability of molecular weight determinations by dodecyl sulfate-polyacrylamide gel electrophoresis. *J. Biol. Chem.* **244**:4406.

Weber, K., E. Lazarides, R. D. Goldman, A. Vogel and R. Pollack. 1975. Localization and distribution of actin fibers in normal, transformed, and revertant cells. *Cold Spring Harbor Symp. Quant. Biol.* **39**:363.

Wickus, G., E. Gruenstein, P. W. Robbins and A. Rich. 1975. Decrease in membrane-associated actin of fibroblasts after transformation by Rous sarcoma virus. *Proc. Nat. Acad. Sci.* **72**:746.

Wood, B. T., S. H. Thompson and G. Goldstein. 1965. Fluorescent antibody staining. III. Preparation of fluorescein isothiocyanate labelled antibodies. *J. Immunol.* **95**:225.

Modification of Mammalian Cell Shape: Redistribution of Intracellular Actin by SV40 Virus, Proteases, Cytochalasin B and Dimethylsulfoxide

Robert Pollack

Microbiology Department
SUNY, Stony Brook, New York 11794

Daniel B. Rifkin

The Rockefeller University
New York, New York 10021

Variations in shape (Fig. 1) are the most obvious of all the many changes that occur in a mammalian cell as it proceeds through its replicative cycle. In order to cycle between the mitotic ball and the interphase disc, every proliferating cell must be able to carry out simultaneously many cyclic physiological changes involving the cell membrane and the cortical cytoplasm. For example, the cell must alter its adhesions to the substrate (Tobey, Anderson and Petersen 1967) by modifying its plasma membrane (Burger et al. 1972; Mannino and Burger 1975) so that the same substrate is "wet" by the membrane to a cyclically varying degree (Folkman and Greenspan 1976). At the same time it must assemble and dissemble intracytoplasmic structures containing polymers of actin (Lazarides 1975a) and tubulin (Fuller, Brinkley and Boughter 1975; Weber, Pollack and Bibring 1975).

Normal and malignant cells differ both in their ability to traverse the cell cycle under limiting conditions and in the manner in which they carry out these cyclic changes in shape. For a normal cell to complete successive cycles (i.e., for it to be able to form a colony), serum must be present in high amounts, cells must initially be far from each other so that cell–cell contact is minimal, and a wettable substrate such as glass or plastic must be provided for the cells to spread out in early G_1 (Fig. 1). Transformed cells have a diminished or absent requirement for these three environmental signals and therefore can grow well in low serum, high cell density or in the absence of an anchoring substrate (Risser and Pollack 1974).

Paralleling one of these variations in growth control, a major perturbation in cortical cytoplasmic structure has recently been found. Anchorage-independent cell lines have been shown to lack intracellular actin-containing cables (Pollack, Osborn and Weber 1975) and intracellular microtubules (Brinkley, Fuller and Highfield 1975) at all points in interphase. Modifications of cell structure that correlate specifically with changes in anchorage dependence are especially interesting insofar as anchorage-independence and tumorigenicity seem to be very well correlated (Shin et al. 1975; Freedman

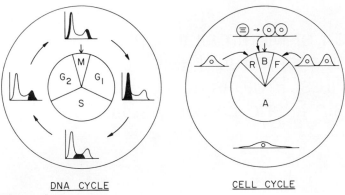

Figure 1
Shape changes in the cell cycle. On the left is the classic cycle, marked by the periods of DNA synthesis (S), mitosis (M) and the gaps between (G_1 and G_2). Histograms of DNA/cell show the shuttling of the population between 2C and 4C DNA content. On the right is an attempt to show the cell's shape changes as the cycle progresses. A indicates the adherent flattened cell, occurring in G_1, S and G_2; R, the rounding cell, occurring in late G_2 and early M; B, the ball or spherical cell, occurring in the middle of M; and F, the flattening cell, occurring in late M and early G_1. Actin-containing cables (see text) are absent throughout the cycle in anchorage-transformed cells and are present in normal cells only in A, disappearing in R and reappearing in F.

and Shin 1974; Evans and DiPaolo 1975). Clearly, experimental assembly and dissembly of the cortical actin-containing structures would contribute to our understanding both of the shape changes occurring in the cell cycle and of the mechanisms by which the oncogenic anchorage-independent phenotype is expressed.

With this in mind we attempted to convert the actin-containing cables within normal cells into the diffuse network characteristic of anchorage-transformed cells by use of a series of reagents known to alter cytoplasmic structure (Pollack and Rifkin 1975). Proteases were initially chosen because (1) one protease, plasmin, is a physiological concomitant of tumor growth in vivo (Reich, Rifkin and Shaw 1975), and (2) we had observed earlier that the amount of plasminogen activator produced by different SV40-transformed rat embryo fibroblast lines was well correlated with their ability to grow in the absence of anchorage (Pollack et al. 1974). More recently, the effects of cytochalasin B (CB) on the intracellular distribution of actin have been studied. In these studies we have found widely varying effects that are dependent on dosage (concentration × time of exposure).

METHODS

All studies reported here were carried out on an early passage population of cells isolated by trypsinization of 16-day-old rat embryos (Pollack and Rifkin 1975).

All experiments were performed at 36.5°C in an atmosphere of 10% CO_2, 90% air, with a relative humidity of 100%. Stock cultures were grown on plastic petri dishes (Falcon) in Dulbecco's modified Eagle's medium (DME; Gibco H21) with 10% fetal calf serum (Reheis) added. Transfers were made by trypsinization (0.25% trypsin, Gibco, in Ca^{++}- and Mg^{++}-free PBS), low-speed centrifugation, and resuspension in fresh DME plus 10% fetal calf serum. Cells were permitted to spread for 24 hours after transfer before any experimental protocol was initiated.

Proteases were made and used as described previously (Pollack and Rifkin 1975). Cytochalasin B was made as a stock of 1 mg/ml in DMSO. For all treatments of cells, cultures plated 24 hours earlier on 11-mm round glass cover slips at a density of $2 \times 10^4/cm^2$ were washed twice with prewarmed DME and held at 37°C for 1 hour. Proteases or other reagents were added

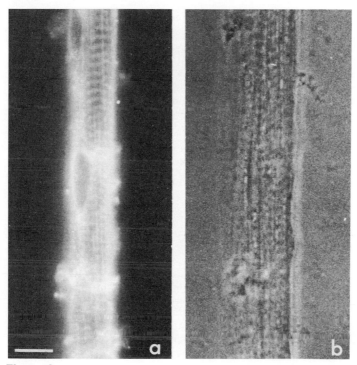

Figure 2

Immunofluorescence demonstration of specificity of antibody to actin. Antibody prepared against SDS-denatured actin (Lazarides 1975b) was layered on cover slips of cultured chick embryonic myotubes after formalin fixation-acetone postfixation (Pollack and Rifkin 1975). After washing in PBS, the cells were incubated with goat anti-rabbit IgG conjugated to fluorescein, washed, and mounted in Elvanol (Rodriguez and Deinhardt 1960). The same field is shown in *a* and *b:* (*a*) with UV-epi-illumination; (*b*) with phase contrast. Note the periodic variations in fluorescence and phase density characteristic of striated muscle. Presumably, bright fluorescent bands are the I-bands, and dense interbands are the A-bands. Note the absence of nuclear fluorescence and the presence of fine fluorescent fibers running perpendicular to the striations. Bar = 10 μ.

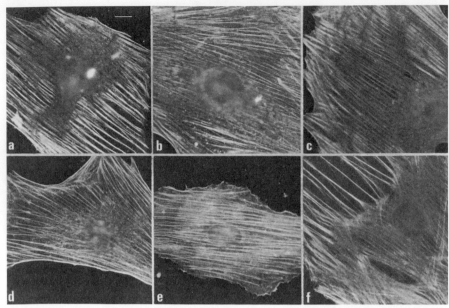

Figure 3
Actin cables remain in rat embryo cells in serum-free medium. Secondary rat
embryo cells were trypsinized and permitted to spread for 24 hours on glass cover
slips in DME plus 10% fetal calf serum. They were then washed twice with DME
and examined for actin cables (as in Methods) after (*a*) 15 min, (*b*) 1 hr,
(*c, d*) 2 hr, (*e*) 8 hr and (*f*) 24 hr in DME. These figures are typical of at least
90% of the rat embryo cells studied at any time. Bar = 10 μ. (Reprinted, with
permission, from Pollack and Rifkin 1975.)

in DME to bring final concentrations to the desired level, and cultures were
returned to 37°C until fixation.

Fixation and Staining

After exposure to reagents or control media, cover slips were fixed by immers-
ing them directly in formaldehyde fixative (3.8% formaldehyde in PBS).
Fixed cover slips were stored for periods of 1–5 days at 4°C before staining.
To open cell membranes to antibody, fixed cover slips were washed in PBS
and passaged for 5 minutes each in water:acetone (1:1), acetone, water:
acetone (1:1) and PBS. The postfixed cover slips were then put in a moist
chamber and incubated for 60 minutes at 37°C with 10 μl of a 1:40 PBS
dilution of rabbit antibody to actin. The cover slips were then washed three
times in PBS and incubated for 60 minutes with 10 μl of fluorescein isothiocy-
nate-conjugated goat anti-rabbit IgG (Miles), diluted 1:10 in PBS.

After this counterstain, cover slips were washed three times in PBS and once
in water and mounted cell-side down in Elvanol (Rodriguez and Deinhardt
1960). After drying for 24 hours, cells were examined at 830× with a Zeiss
photomicroscope II, by epi-illumination, using an FITC narrow-pass exciter
filter and a 500-nm barrier filter. Fields were photographed with Tri-X Kodak
film, exposed at DIN 32 and developed with Diafine developer. (See Fig. 2.)

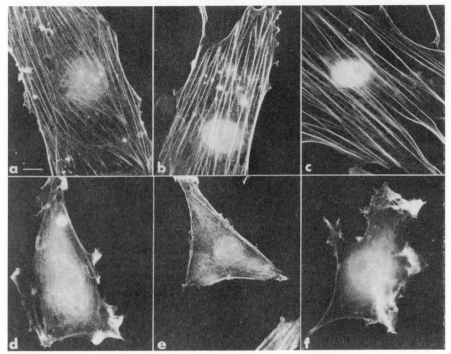

Figure 4

Effect of the plasmin system on actin cables in rat embryo cells. Cells were prepared as in Fig. 3. At 0 hr, DME was replaced by one of the following: (a) fresh DME; (b) DME plus 1.5 μg/ml plasminogen; (c) DME plus 25 μg/ml urokinase; (d, e, f) DME plus both 1.5 μg/ml plasminogen and 25 μg/ml urokinase. After 3 hr, cells were fixed and examined for actin cables as described in Methods. Bar = 10 μ. (Reprinted, with permission, from Pollack and Rifkin 1975.)

Scanning for Presence of Actin-containing Cables

All cover slips were labeled in an arbitrary fashion and scored by a person unable to identify the origin of the cover slip. At least 100 cells were scored on each cover slip. The microscope was focused on the edge of a cell and therefore on structures at the adherent side of the cell. If cables ran the length of the cell, the cell was scored as positive. Some cells had only finer structures, especially just behind ruffles. Such cells, and cells completely lacking cables, were scored as negative (Pollack and Rifkin 1975). (See Fig. 3.)

RESULTS AND DISCUSSION

Proteases

The results of protease treatment of REF cells are given in Table 1 (Pollack and Rifkin 1975). Trypsin and plasmin removed cables from within REF cells very efficiently, since 1–3 μg/ml of either protease converted more than

Table 1

Results of Protease Treatment of REF Cells

Protease[a]	RE cells[b] with cables (%) after protease exposure for			
	15 min	*1 hr*	*2 hr*	*4 hr*
None	97[b]	95	91	97
Plasmin				
plasminogen (1.5)	—	—	90	—
urokinase (25)	—	—	92	—
plasminogen (1.5) + urokinase (25)	—	—	23	—
plasminogen (7.5) + urokinase (50)	—	—	17	—
Trypsin				
(1.0)	83	64	—	—
(2.0)	100	35	13	0
(3.0)	56	30	0	—
(5.0)	67	16	0	—
Chymotrypsin				
(1.0)	—	79	—	47
(2.5)	—	—	51	—
(5.0)	76	—	65	26
(1.0)	—	—	37	14
Thrombin				
(5)	—	—	92	—
(10)	—	80	—	79
(50)	44	83	—	85
(100)	—	88	—	78

[a] Concentration in μg/ml given in parentheses.
[b] Average of at least two experiments for each point.

80% of the REF cells to a morphology resembling anchorage-transformed cells in 1–2 hours (Figs. 4, 5).

Not all proteases removed cables. For example, urokinase, whose proteolytic activity was necessary to activate the proenzyme plasminogen, was by itself unable to disrupt the cables (Fig. 5). Thrombin also had no effect on cables, even at very high doses (100 μg/ml for 4 hr). Chymotrypsin had an intermediate effect on cables: although no dose of chymotrypsin removed all cables, sufficiently high doses had some effect on them (Table 1). Although we have some evidence that proteases bound to beads also remove cables, we have not yet established with certainty whether the effects shown here are all the result of proteolysis from without or are instead the result of protease uptake (Hodges, Livingston and Franks 1973).

In these experiments, dosages of enzyme were low enough so that the cells were not dislodged from the cover slip as a result of exposure to the

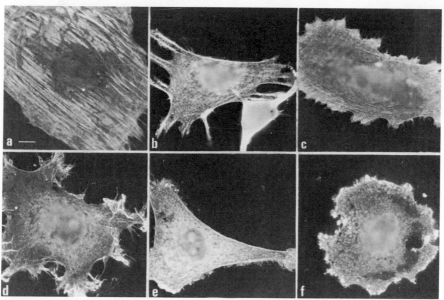

Figure 5

Time course of effect of trypsin on actin cables in rat embryo cells. Rat embryo cells were plated and examined for actin cables as described in Methods. Cells were fixed at various times after addition of DME plus 5 μg/ml trypsin: (*a*) 15 min; (*b*) 1 hr; (*c, d*) 2 hr; (*e*) 4 hr; (*f*) 8 hr. By 8 hr, approximately half of the rat embryo cells have detached from the glass. No cells detach after 2 hr in this concentration of trypsin. Bar = 10 μ. (Reprinted, with permission, from Pollack and Rifkin 1975.)

protease. Of course if a protease such as trypsin is presented to cells at high concentrations (> 100 μg/ml), cell detachment will occur rapidly. These results suggest, therefore, that dispersion of intracellular actin cables is necessary, but not sufficient, for rounding and detachment.

Cytochalasin B and DMSO

Cytochalasin B (CB) has many interesting effects on cell morphology and motility. The direct site of CB's activity is unknown. It is clear, however, that both the plasma membrane and the actin-containing microfilaments are affected by the drug. For this reason we examined the distribution of actin-containing structures by immunofluorescence in RE cells after treatment with different doses of CB. Because of the complexity of CB's effects on RE cells, treated cells were classified as spread, retracted or detached (Table 2), as well as being scored for the fraction of spread cells retaining cables (Table 3).

Retraction and detachment were minimal after 15 minutes exposure to CB concentrations up to 3 μg/ml (Fig. 6a,b,c). At 10 μg/ml, a minority of spread cells showed a new distribution of actin: a disoriented, unraveled dispersion (Fig. 6d). At low CB concentration, there was some conversion of cables to this diffuse distribution among the spread cells (Table 3).

Table 2
Effects of Cytochalasin B in DMSO on Adherence of RE Cells

Treatment			Distribution of cells (%)		
CB (µg/ml)	DMSO (%)	exposure (hr)	spread[a]	retracted[b]	detached[c]
0	0	0.25	90[a]	0	10
		1	91	0	9
		24	80	0	20
3.0	0.3	0.25	94	1	5
		1	26	58	34
		24	0	0	100
10	1.0	0.25	37	57	5
		1	1	84	15
		24	0	90	10

[a] See Fig. 6a for actin distribution in typical spread cells.
[b] See Fig. 7c for actin distribution in typical retracted cells.
[c] See Fig. 8c for actin distribution in typical detached cell sites; shows dots remaining.

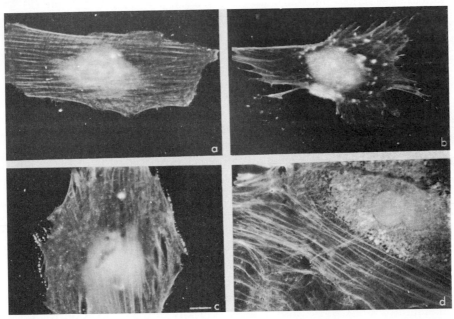

Figure 6
Intracellular distribution of actin in RE cells after 15 min exposure to cytochalasin B in serum-free medium. (*a*) Control, no CB; (*b*) 1 µg/ml CB, the cell has begun to retract, and actin-containing cables are less sharply defined; (*c*) µg/ml CB, retraction at the cell boundary has deposited pads of actin-containing material; (*d*) 10 µg/ml CB, the cytoplasm seems to have come apart, leaving unraveling, bent cables. All pictures at same magnification; bar = 10 µ.

Table 3

Effects of CB and DMSO on Intracellular Cables within Spread Cells

Treatment			Spread cells
CB (μg/ml)	DMSO (%)	exposure (hr)	retaining cables (%)
0	0	0.25	86
		1	90
		24	77
0	0.3	1	86
	1	1	71
	3	1	57
	10	1	43
1	0.1	0.25	60
		1	70
		24	89
3	0.3	0.25	80
		1	99
		24	NSC[a]
10	1.0	0.25	93
		1	NSC
		24	NSC

[a]NSC — no spread cells.

By 1 hour of exposure to concentrations of CB greater than 1 μg/ml (Fig. 7), cell retraction had proceeded extensively. The majority of cells exposed to 3 or 10 μg CB also retracted (Table 2). However, among the minority of spread cells, cables were still present (Table 3). The aberrant dispersion of cables was maintained at 6 and 10 μg/ml. While no fully spread cells remained after 1 hour in 6 or 10 μg of CB, retraction was less than in cells exposed to 3 μg/ml for 1 hour (Fig. 7, Tables 2 and 3).

After 24 hours of exposure to CB (Fig. 8), populations of RE cells became homogeneous, showing actin distributions that were dependent on drug dose (Table 3). In 3 μg/ml, all cells detached, leaving behind sets of cytoplasmic adherent pads that outlined the shapes of spread cells (Fig. 8c, Table 2). In 10 μg/ml CB, very few cells detached. Rather, almost all cells contained the same partially retracted dispersion of disoriented cables (Fig. 8d, Table 2). There were no fully spread cells after 1 hour in 10 μg/ml CB.

Since CB was given to RE cells as dilutions of a 1 mg/ml stock solution in 100% dimethylsulfoxide (DMSO), we tested the effect of DMSO itself. DMSO alone caused retraction of RE cells at concentrations of 3% and 10% (Table 4). These concentrations are three and ten times higher, respectively, than the concentration of DMSO present in a culture receiving 10 μg/ml of CB. These high concentrations of DMSO also disrupted actin cables in about half of the

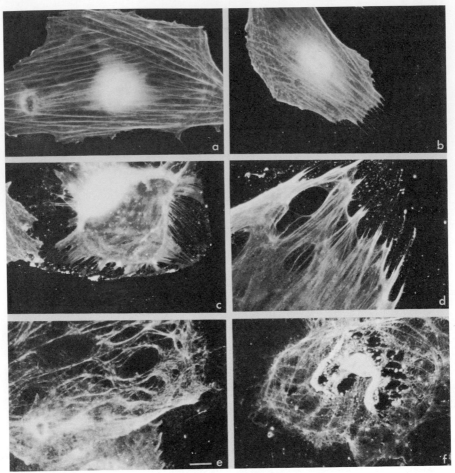

Figure 7

Intracellular distribution of actin in RE cells after 1 hr exposure to cytochalasin B in serum-free medium. (*a*) Control, no CB; (*b*) 0.3 μg/ml CB, slight retraction at edge of cell; (*c*) 1 μg/ml CB, extensive retraction; (*d*) 3 μg/ml CB, detail of large cell, retraction depositing pads; (*e*) 6 μg/ml CB, unraveling of cables as well as retraction; (*f*) 10 μg/ml CB, retraction, unraveling and disintegration of cytoplasm. All pictures at same magnification; bar = 10 μ.

cells (Table 5). However, 10% DMSO by itself for an hour did not cause appreciable cell detachment nor the appearance of unraveled cables. Thus these changes are apparently CB-specific.

The effects of CB and/or DMSO on the cables of spread and retracted RE cells were reversible within an hour after the cells were washed in fresh serum-free medium (Table 5). The material left behind by detached cells did not show any change in shape upon reversal, however.

CB-induced loss of actin cables within spread cells and detachment of cells from glass (Tables 3 and 4) did not proceed in a coordinated manner. Although cable loss in cells remaining attached during CB treatment was not markedly dosage-dependent, cell detachment was most effective at interme-

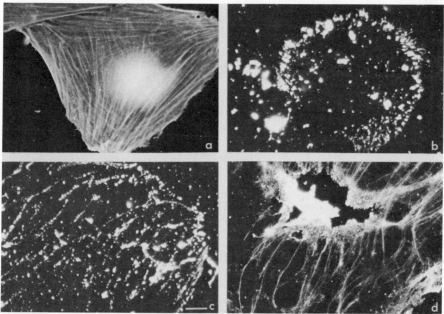

Figure 8

Intracellular distribution of actin in RE cells after 24 hr exposure to cytochalasin B in serum-free medium. (*a*) Control, no CB; (*b*) 1 μg/ml CB, complete detachment from glass, leaving pads outlining shape of spread cell; (*c*) 3 μg/ml CB, complete detachment of cell; (*d*) 10 μg/ml CB, unraveled cytoplasm does not detach. All pictures at same magnification; bar = 10 μ.

Table 4

Effect of DMSO Alone on Adherence of RE Cells

Treatment		Adherence (% of cells)		
DMSO (%)	exposure (hr)	spread	retracted	detached
0.3	0.25	90	8	2
	1	90	4	6
1.0	0.25	98	0	2
	1	87	8	5
3.0	0.25	46	32	12
10	0.25	16	73	11

diate concentrations of CB. At the optimal concentration of 3 μg/ml CB, cell loss exceeded 95%. After CB-induced detachment of the cell body, adherent bits of cytoplasm remained behind. These stained very brightly with anti-actin. Studies are underway to determine which other cell proteins are associated with these adherent bits of cell membrane and cytoplasm and to compare them with substrate-attached glycoproteins (Terry and Culp 1974; Culp 1974, 1975).

Table 5
Recovery from CB and DMSO

Treatment		Time (hr)		Adherence (% cells)			Spread cells with cables[a]
CB (μg/ml)	DMSO (%)	exposure	recovery	spread[a]	retracted	detached	(%)
0	0.3	1	—	90	8	2	86
		1	1	90	4	6	94
0	3	1	—	46	32	12	57
		1	1	70	16	14	91
0	10	1	—	16	73	11	43
		1	1	74	14	12	97
3	0.3	1	—	26	58	34	99
		1	1	67	10	23	88
10	1.0	1	—	1	84	15	NSC
		1	1	69	9	12	75

[a] Only the spread cells were scored for the presence of actin-containing cables.

Acknowledgments

This work was supported by grants from the Human Cell Program of the National Science Foundation; the National Cancer Institute, NIH; and the American Cancer Society. We thank Joel Gordon for chick embryo myotube cultures.

REFERENCES

Brinkley, B., G. Fuller and D. Highfield. 1975. Cytoplasmic microtubules in normal and transformed cells in culture. Analysis by tubulin antibody immunofluorescence. *Proc. Nat. Acad. Sci.* **72**:4981.

Burger, M. M., B. M. Bombik, B. M. Breckenridge and J. R. Sheppard. 1972. Growth control and cyclic alterations of cyclic AMP in the cell cycle. *Nature New Biol.* **239**:161.

Culp, L. A. 1974. Substrate-attached glycoproteins mediating adhesion of normal and virus-transformed mouse fibroblasts. *J. Cell Biol.* **63**:71.

Culp, L. A. 1975. Topography of substrate-attached glycoproteins from normal and virus-transformed cells. *Exp. Cell Res.* **92**:467.

Evans, C. and J. DiPaolo. 1975. Neoplastic transformation of guinea pig fetal cells in culture induced by chemical carcinogens. *Cancer Res.* **35**:1035.

Folkman, J. and H. P. Greenspan. 1976. Influence of geometry on control of cell growth. *Biochim. Biophys. Acta* **417**:211.

Freedman, V. H. and S. Shin. 1974. Cellular tumorigenicity in *nude* mice: Correlation with cell growth in semi-solid medium. *Cell* **3**:355.

Fuller, G. M., B. R. Brinkley and J. M. Boughter. 1975. Immunofluorescence of mitotic spindles by using monospecific antibody against bovine brain tubulin. *Science* **187**:948.

Hodges, G., D. Livingston and L. Franks. 1973. The localization of trypsin in cultured mammalian cells. *J. Cell Sci.* **12**:887.

Lazarides, E. 1975a. Immunofluorescence studies on the structure of actin filaments in tissue culture cells. *J. Histochem. Cytochem.* **23**:507.

———. 1975b. Tropomyosin antibody: The specific localization of tropomyosin in nonmuscle cells. *J. Cell Biol.* **65**:549.

Mannino, R. and M. M. Burger. 1975. Growth inhibition of animal cells by succinylated concanavalin A. *Nature* **256**:19.

Pollack, R. and D. Rifkin. 1975. Actin-containing cables within anchorage-dependent rat embryo cells are dissociated by plasmin and trypsin. *Cell* **6**:495.

Pollack, R., M. Osborn and K. Weber. 1975. Patterns of organization of actin and myosin in normal and transformed cultured cells. *Proc. Nat. Acad. Sci.* **72**:994.

Pollack, R., R. Risser, S. Conlon and D. Rifkin. 1974. Plasminogen activator production accompanies loss of anchorage regulation in transformation of primary rat embryo cells by simian virus 40. *Proc. Nat. Acad. Sci.* **71**:4792.

Reich, E., D. Rifkin and E. Shaw, eds. 1975. *Proteases and biological control. Cold Spring Harbor Conferences on Cell Proliferation,* vol. 2. Cold Spring Harbor Laboratory, Cold Spring Harbor, New York.

Risser, R. and R. Pollack. 1974. A non-selective analysis of SV40 transformation of mouse 3T3 cells. *Virology* **59**:477.

Rodriguez, J. and F. Deinhardt. 1960. Preparation of a semipermanent mounting medium for fluorescent Ab studies. *Virology* **12**:316.

Shin, D., V. Freedman, R. Risser, and R. Pollack. 1975. Tumorigenicity of virus-transformed cells in *nude* mice is correlated specifically with anchorage-independent growth *in vitro*. *Proc. Nat. Acad. Sci.* **72**:4435.

Terry, A. H. and L. A. Culp. 1974. Substrate-attached glycoproteins from normal and virus-transformed cells. *Biochemistry* **13**:414.

Tobey, R. A., E. C. Anderson and D. F. Petersen. 1967. Properties of mitotic cells prepared by mechanically shaking monolayer cultures of Chinese hamster cells. *J. Cell. Physiol.* **70**:63.

Weber, K., R. Pollack and T. Bibring. 1975. Microtubular antibody: The specific visualization of cytoplasmic microtubules in tissue culture cells. *Proc. Nat. Acad. Sci.* **72**:459.

Visualization of Tubulin-containing Structures by Immunofluorescence Microscopy: Cytoplasmic Microtubules, Mitotic Figures and Vinblastine-induced Paracrystals

Klaus Weber

Max-Planck-Institute for Biophysical Chemistry
D-3400 Göttingen, West Germany

Electron microscopy of animal cells grown in culture has established that three major fibrous systems can be detected in a wide variety of cells. These are the microfilaments (diameter of the individual fiber 6 nm), the filaments (diameter of the individual fiber 10 nm) and the microtubules (diameter most commonly 25 nm) (Goldman and Knipe 1973). Ultrastructural studies on tissues from different organisms are in agreement with these results, and at least the microfilament and microtubules can be considered to be structures characteristic of all eukaryotic organisms.

Antibodies against the structural proteins of these fibrous systems should be of help in elucidating the "proteinchemical anatomy" of these systems, since they can be used in immunological techniques such as immunoferritin labeling and immunofluorescence microscopy. The latter procedure was used initially to characterize the microfilament system. Thus actin, myosin and tropomyosin were shown to be localized in the microfilament bundles of non-muscle cells (Lazarides and Weber 1974; Weber and Groeschel-Stewart 1974; Lazarides 1975).

Here we report the characterization of the microtubular system by a similar technique. Originally an antibody prepared against tubulin from the outer doublets of sea urchin sperm flagella (*Strongylocentrotus pupuratus*) was used. This antibody cross-reacted with tubulin from mammalian cells, indicating the strong conservation of the tubulin amino acid sequence during evolution. Using indirect immunofluorescence we showed that the antibody decorated the well-defined structures within mammalian cells that are known to contain tubulin. These included cytoplasmic microtubules, vinblastine-induced paracrystals and the full spectrum of mitotic figures. As expected from electron microscopic studies, the expression of cytoplasmic microtubules was abolished by exposure of the cells to colchicine or to low temperature. The antibodies against tubulin and actin decorated different fibrous systems within the cells, and therefore indirect immunofluorescence could be used to distinguish the microfilament system from the microtubular organization (Weber, Pollack and Bibring 1975; Weber, Bibring and Osborn 1975).

Recently, antibodies against highly purified 6S brain tubulin have also become available (Fuller, Brinkley and Boughter 1975; Weber, Wehland and Herzog 1976). These antibodies have been used after affinity chromatography as monospecific antibodies against tubulin. Use of these monospecific antibodies in immunofluorescence microscopy yielded precisely the same microtubular organization in tissue culture cells as described previously. (Weber, Pollack and Bibring 1975; Weber, Bibring and Osborn 1975; see also Brinkley, Fuller and Highfield, this volume).

MATERIALS AND METHODS

Cell cultures and the procedures for indirect immunofluorescence microscopy have been described in detail elsewhere (Weber, Pollack and Bibring 1975; Weber, Bibring and Osborn 1975). The preparation of monospecific tubulin antibody is described in Results. The actin antibody has been described previously (Lazarides and Weber 1974).

RESULTS

Antibody Preparations against Tubulin

We have used two types of antibody preparation in our studies on microtubular organization in tissue culture cells. The first preparation was obtained in rabbits, using tubulin from the outer doublets of sea urchin sperm flagella (species *Strongylocentrotus purpuratus*) as an antigen. Originally this preparation was used for the visualization of cytoplasmic microtubules (Weber, Pollack and Bibring 1975) and vinblastine-induced paracrystals of tubulin, as well as to study mitotic figures (Weber, Bibring and Osborn 1975). The γ-globulins from this serum were used as described previously (Weber, Pollack and Bibring 1975). An aliquot of the γ-globulin fraction was used to isolate monospecific antibodies against tubulin, using the procedure of Fuller, Brinkley and Boughter (1975). The monospecific antibodies gave results in immunofluoresence microscopy that were identical to those obtained using the original preparation of total γ-globulins.

The second antibody preparation was obtained using highly purified 6S pig brain tubulin as antigen. Tubulin was purified by three cycles of depolymerization-polymerization according to a procedure described previously (Shelanski, Gaskin and Cantor 1973). The depolymerized tubulin was freed of high molecular weight components by passage through Sepharose-4B. The 6S tubulin was further purified by chromatography on DEAE-cellulose and phosphocellulose. Tubulin purified by this procedure is essentially free of other proteins, as judged by polyacrylamide gel electrophoresis in the presence of sodium dodecyl sulfate. Antibodies were raised in rabbits against this highly purified 6S tubulin, which had been cross-linked with 2% glutaraldehyde in the cold. The total protein injected per animal was 2 mg. The γ-globulin fraction of the serum was subjected to affinity chromatography on tubulin covalently bound to CNBr-activated Sepharose-4B, using essentially the procedure of Fuller, Brinkley and Boughter (1975). The monospecific anti-tubulin γ-globulins from the column (Fig. 1) were dialyzed into phosphate-buffered saline and adjusted to a concentration of 400 μg/ml. This preparation was used in immunofluorescence microscopy after a tenfold dilution into phosphate-buffered saline. The tubulin antibodies gave a precipitin line in immunodiffusion analysis using purified tubulin (Fig. 2).

Cytoplasmic Microtubules

Cells grown on cover slips were fixed in dilute formaldehyde for optimal preservation of structural organization. Exposure to cold methanol and/or acetone at −20°C was used to make the plasma membrane permeable to macromolecules. Indirect rather than direct immunofluorescence microscopy was used in order to obtain the amplification effect provided by the first technique.

Mouse 3T3 cells show within the cytoplasm a multitude of very thin, separate fluorescent fibers of apparently constant diameter (Figs. 3, 4).

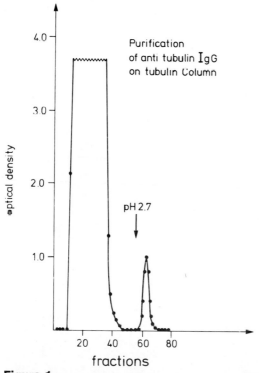

Figure 1

Absorbancy at 280 nm of a chromatography experiment of the tubulin γ-globulin fraction on the immunosorbent column to which highly purified tubulin was covalently linked. The tubulin-specific antibodies were eluted after the change of the elution buffer (arrow).

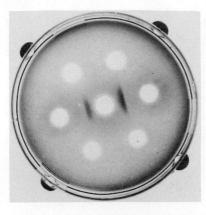

Figure 2
Immunodiffusion analysis of pure 6S tubulin against antibody eluted from the affinity column.

This elaborate system of fragile fibers was first described in a variety of fibroblastic cells of human, monkey, rat, mouse, hamster and chicken origin (Weber, Pollack and Bibring 1975) and was subsequently shown by Brinkley, Fuller and Highfield (this volume) in mouse and rat kangaroo cells. The cytoplasmic microtubules often extend radially from the nucleus. They seem to terminate near the cell surface. Some of the fibers are straight and some are bent and curved.

Some more or less pronounced nuclear fluorescence is sometimes apparent even in our studies with monospecific tubulin antibodies. The intensity can vary considerable from cell to cell and can be exaggerated when the nucleus is slightly out of the plane of focus. This happens often in photomicrographs, since one focuses on the filaments within the cytoplasmic processes. Colchicine did not diminish the nuclear fluorescence (Weber, Pollack and Bibring 1975).

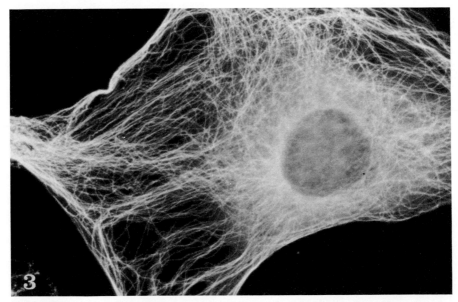

Figure 3
Mouse 3T3 cells. Indirect immunofluorescence using tubulin antibody.

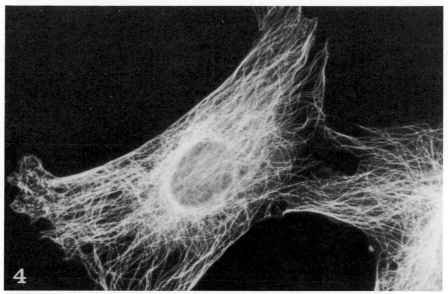

Figure 4
Mouse 3T3 cells. Indirect immunofluorescence using tubulin antibody.

At present we assume that nuclear immunofluorescence is not due to assembled microtubules. Further studies, however, are necessary to clarify the nature of this staining.

A series of experiments have been begun in the hope of determining how microtubules are assembled in vivo (Osborn and Weber 1976). Respreading of trypsinized cells, treatment of cells with colcemid and recovery from exposure to the drug were studied by immunofluorescence. The results of these experiments can be summarized briefly as follows: After exposure to colcemid for 1 hour, most of the 3T3 cells lose their cytoplasmic microtubules. Against a weak background fluorescence one can frequently see one or more bright cytoplasmic fluorescent spots close to the cell nucleus. These structures (Fig. 5) are colcemid-resistant and probably represent centriolar structures. They frequently have residual microtubular structures radiating from them and they could be the starting points for microtubular assembly in vivo. After removal of the drug, microtubules seem to regrow from these spots. With increasing time after removal of the drug, the microtubules become so numerous around the nucleus that it is difficult to determine their points of origin. Figure 6 indicates how microtubules have partially regrown from The perinuclear area and are "trying" to reach the plasma membrane during regrowth. Figures 7 and 8 show the expression of microtubules during replating experiments. Figure 7 is an early time point (60 min after replating). Figure 8 shows a later time point, at which time the cells are trying to reassume fibroblastic character and have already developed numerous microtubules. Replating experiments give essentially the same impression. Again microtubules are trying to reach from the perinuclear space to the plasma membrane. More experiments are necessary to prove that microtubules are coming from a centrospherelike structure and radiating to the margin of

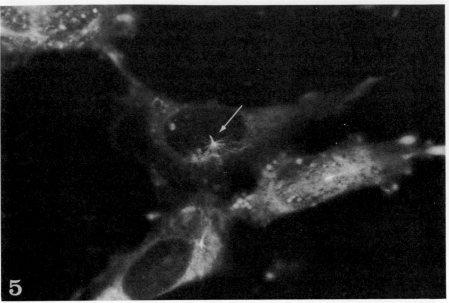

Figure 5
Mouse 3T3 cells treated for 1 hr with 0.7 μg/ml colchicine, stained for tubulin. Note loss of cytoplasmic microtubules. Note the colcemid-resistant fluorescent structure in all three cells (arrow). This structure is presumed to be the centrosphere.

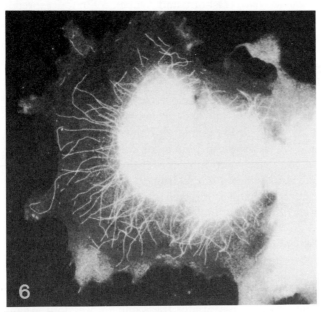

Figure 6
3T3 cells after partial recovery (20 hr) from treatment with colchicine (1 μg/ml for 1 hr). Note how microtubules "try to reach the plasma membrane." Staining for tubulin.

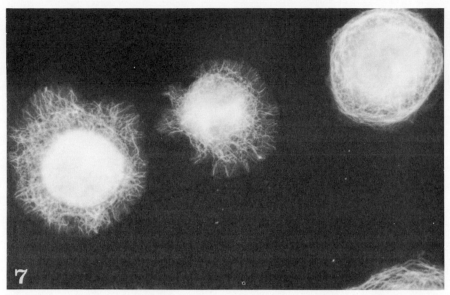

Figure 7
3T3 cells stained for tubulin 60 min after replating.

the cell. However it already seems apparent that their expression determines the shape of the cell body (Osborn and Weber 1976).

Comparison of the Cytoplasmic Microtubules and the Microfilament System

The pattern of organization of actin in a variety of cells of different origin has been described in detail previously (Lazarides and Weber 1974; Goldman et al. 1975). The existence of a complex array of fibrous structures in living 3T3 cells with phase-contrast, Normarski and polarized light optics has also been demonstrated (Goldman et al. 1975). When these cells are stained in indirect immunofluorescence with actin antibody, the same fibers show intense fluorescence, indicating that they contain actin. Electron microscopic studies revealed that these fibrous structures are submembraneous bundles of microfilaments located predominantly on the attached side of the cells inside the plasma membrane (Goldman et al. 1975).

The expression of microfilament bundles and microtubular structures in interphase mouse 3T3 cells have been studied in parallel using antibodies against actin and tubulin (Weber, Pollack and Bibring 1975) (Fig. 9). Figure 9b shows a well-spread 3T3 cell stained with actin antibody as originally described by Lazarides and Weber (1974). Figure 9a shows a 3T3 cell from a cover slip processed in parallel with antibody against tubulin. The actin and microtubular fibers in the cytoplasm are remarkably different in appearance and arrangement. The actin fibers vary in diameter (up to 1 μ) and length and are concentrated as submembraneous structures towards the adhesive side of the cell. The majority of actin fibers are arranged in parallel

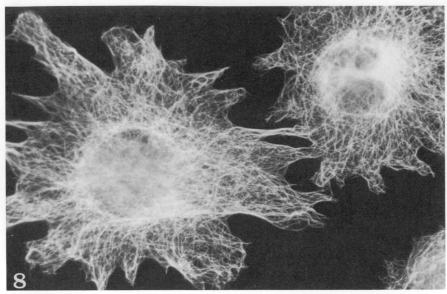

Figure 8
3T3 cells 120 min after replating. Staining for tubulin. Note the abundance of microtubules in both cells.

structures, although occasional focal points can be found. The margins of the cells often show strong actin fibers. In contrast to the cytoplasmic organization of the actin fibers, the microtubular fibers are very fragile structures, are separate from each other, and are very thin and probably of constant diameter.

Effects of Colchicine on Cytoplasmic Microtubules and Microfilaments

Extensive electron microscopic studies have proven that cytoplasmic microtubules disappear in the presence of the alkaloid colchicine. This in vivo result is in accord with the known in vitro properties of microtubules (see, for example, Olmsted and Borisy 1973). Therefore mouse 3T3 cells were plated into medium containing colchicine (15 μg/ml). Although somewhat reduced in size as a consequence of colchicine treatment, the cells were well spread over the surface of the cover slip. After processing for immunofluorescence, no organized fibers appeared in the cytoplasm when antibody against tubulin was used (Fig. 9c). Instead, there was a weak, diffuse fluorescence in the cytoplasm. When cover slips were processed in parallel with actin antibody (Fig. 9d), the cells displayed the actin-containing microfilament bundles (Weber, Pollack and Bibring 1975). This result is in excellent agreement with the known colchicine sensitivity of microtubules and the known colchicine insensitivity of microfilament bundles (see Goldman and Knipe 1973; Olmsted and Borisy 1973).

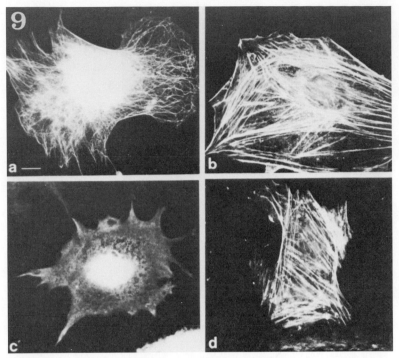

Figure 9

3T3 cells stained with tubulin antibody (*a, c*) or with actin antibody (*b, d*). The cells shown in *c* and *d* have been treated with colchicine (15 μg/ml) for 24 hr. Note the colchicine sensitivity of the microtubules (*a* vs. *c*) and the colchicine insensitivity of the microfilament bundles. Bar = 10 μm.

Tubulin Antibody Decorates Vinblastine-induced Paracrystals

Vinblastine is known to induce the reorganization of microtubular structures in living cells. In the presence of a sufficient concentration of this alkaloid, spindle microtubules and cytoplasmic microtubules disappear and crystalloid inclusion bodies or paracrystals appear. These paracrystals have been studied by different microscopic techniques and by proteinchemical studies (Bensch and Malawista 1968; Bryan 1972; Dales 1972; Nagayama and Dales 1973). The experiments have indicated that the vinblastine-induced paracrystals are most likely formed from homogeneous tubulin. Following the procedure of Dales (1972), growing mouse 3T3 and human HeLa cells were treated with vinblastine (at 5×10^{-5} M) for 12 to 16 hours prior to processing for immunofluorescence. All cells were found to be devoid of cytoplasmic microtubules. Instead, the cells contained several fluorescent paracrystals in their cytoplasm (Fig. 10). It is worthwhile to note that vinblastine-treated cells did not reveal crystalline inclusion bodies when stained with either antibodies to actin or antibodies to myosin (data not shown). Instead they revealed throughout the cytoplasm a strong fluorescence indicative of disorganized actin- and myosin-containing structures, although some of the typical actin-

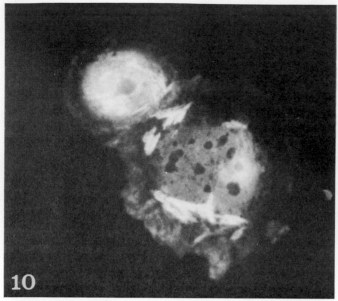

Figure 10
Vinblastine-induced paracrystals in 3T3 cells stained with tubulin antibody.

containing fibers were still present. Although actin and a variety of other Ca^{++} binding proteins can be precipitated in vitro by vinblastine (Wilson et al. 1970), the in vivo formation of vinblastine-induced paracrystals seems to occur in a highly specific manner, avoiding the incorporation of proteins other than tubulin into the crystalline matrix. This conclusion is in good agreement with the proteinchemical studies of Bryan (1972), which showed that paracrystals induced in sea urchin eggs are essentially pure tubulin.

Mitotic Figures

HeLa cells and cells of the established rat kangaroo line PtK1 were studied during mitosis using immunofluorescence microscopy (Weber, Bibring and Osborn 1975; Fuller, Brinkley and Boughter 1975). Figure 11 shows a well-developed metaphase structure in a PtK1 cell. Figure 12 shows a composite of mitotic HeLa cells, using antibody against tubulin. The different frames are arranged according to the well-known phases of mitosis. The different stages of prophase can be seen, and the orientation of the two centriole-connected structures is visualized outside the nuclear membrane. In pro-metaphase cells, a multitude of dark chromatids and two extended fluorescent "asters" are seen. The nuclear membrane is absent in these cells. During the early part of prophase, a weak background fluorescence can be seen around the condensing chromosomes. Metaphase cells show the nonfluorescent equatorial plate and the expression of a well-developed fluorescent spindle. In a variety of metaphase and anaphase cells, both chromosomal (kinetochore-to-pole) and interpolar (pole-to-pole) fibers are apparant. With the onset of anaphase, there is a loss in contrast of the fluorescent structures, although the region of

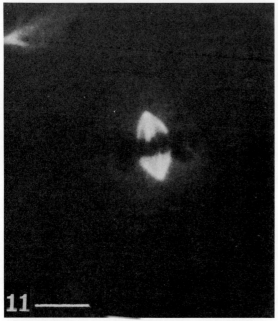

Figure 11
A rat kangaroo PtK1 cell during metaphase, stained with tubulin antibody.
Note the absence of cytoplasmic fluorescence outside the spindle. Bar =
10 μm.

the interzonal fibers is still clearly visible. This effect is most likely due to an
increase in cytoplasmic tubulin as a result of the partial disassembly of the
spindle. After the cleavage furrow has been well developed, the two daughter cells are still connected by the highly fluorescent intercellular bridge. The

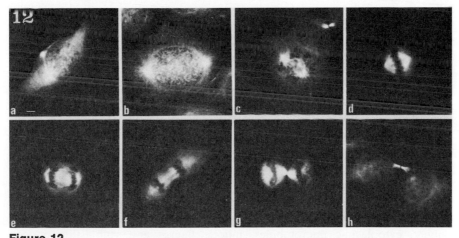

Figure 12
A composite of HeLa cells during different stages of mitosis, stained in direct
immunofluorescence with tubulin antibody. Bar in *a* = 10 μm.

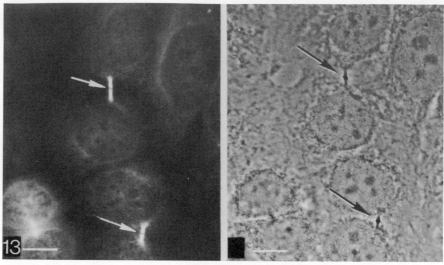

Figure 13
The intercellular bridge between daughter HeLa cells visualized by immuno-fluorescence with tubulin antibody (*left*) and by phase contrast (*right*). The arrows point to the midbodies, which show decreased fluorescence. Bars = 10 μm.

position of the midbody in the bridge is generally weaker in fluorescence than the two side parts of the bridge (Fig. 13). There are two possible explanations for this effect (Weber, Bibring and Osborn 1975). First, it is possible that in the midbody, the antibody has less access to the microtubular structures, and therefore a lower fluorescence level is obtained. An alternative explanation is the assumption that parts of the midbody have a lower concentration of tubulin than the remainder of the intercellular bridge. This seems unlikely in view of the electron micrographs obtained which show that microtubules are present throughout the intercellular bridge. Results on HeLa cells (Weber, Bibring and Osborn 1975) are similar to those obtained by Fuller, Brinkley and Boughter (1975) (see also Brinkley, Fuller and Highfield, this volume) using PtK1 cells. The PtK1 cell line is extremely suitable for such studies because cells do not round up during mitosis. However, both sets of results show that all mitotic stages can be clearly visualized by immunofluorescence microscopy.

DISCUSSION

Several results (Weber, Pollack and Bibring 1975; Weber, Bibring and Osborn 1975; Fuller, Brinkley and Boughter 1975; Brinkley, Fuller and Highfield, this volume) have now clearly established that the tubulin-specific antibodies specifically visualize those cellular structures known to contain tubulin: namely, cytoplasmic microtubules, mitotic figures and vinblastine-induced crystals. Furthermore, the expression of cytoplasmic microtubules is sensitive to both cold and colchicine (Weber, Pollack and Bibring 1975; Brinkley,

Fuller and Highfield, this volume). Cytoplasmic microtubules can also be readily distinguished from bundles of microfilaments (Weber, Pollack and Bibring 1975) that contain actin, myosin, tropomyosin and α-actinin (Lazarides and Weber 1974; Weber and Groeschel-Stewart 1974; Lazarides 1975; Lazarides, this volume) but not tubulin (Weber, Pollack and Bibring 1975). The tubulin antibodies have now been shown to be tubulin-specific by affinity chromatography, immunodiffusion and immunoelectrophoresis (Fuller, Brinkley and Boughter 1975; Weber, Bibring and Osborn 1975). The structures revealed by these antibodies in immunofluorescence microscopy clearly contain tubulin.

What is the relationship between the results obtained with tubulin-specific antibodies and those obtained previously by electron microscopy? It is possible that the results obtained with the antibodies may at times be astonishing to classical electron microscopists. However, it should be remembered that conventional transmission electron microscopy is restricted by small sample size and by requirements for fixations of the sample and for tedious serial section analysis to gain information on the three-dimensional distribution of microtubules. Immunofluorescence microscopy offers unique advantages in that whole cells can be observed and the antibody specifically visualizes the corresponding antigen distribution without the complicated details of the remainder of the cell structure. Furthermore, large numbers of cells can be viewed under a variety of experimental conditions, thus assuring a statistically valid observation. Therefore it is worth noting that the display and organization of microtubules during interphase as visualized by immunofluorescence microscopy is rather similar to that provided by high-voltage electron microscopy (Fonte and Porter 1974). Both these procedures provide an overview of the cell rather than an in-depth analysis of a small part of it. There is one further important advantage of immunofluorescence microscopy. Provided that a series of antibodies against different structural proteins is available, it becomes possible to obtain a "proteinchemical anatomy" of a certain cellular structure. This has already been done to some extent in the case of the microfilamentous bundles of certain tissue-culture cells. They were first shown to contain F-actin-like proteins by Ishikawa et al. (Ishikawa, Bischoff and Holtzer 1969). Immunofluorescence microscopy has not only verified this assignment (Lazarides and Weber 1974), but has also for the first time shown the presence of myosin (Weber and Groeschel-Stewart 1974) (thus providing the basis for contractility) and subsequently the presence of tropomyosin (Lazarides 1975) and α-actinin (Lazarides, this volume) as integral proteins of this cell structure. The assignment of the these three proteins was solely the result of immunofluorescence microscopy and not a result of classical transmission electron microscopy.

Is indirect immunofluorescence able to resolve individual cytoplasmic microtubules? Since microtubules measure 250 Å in diameter, they are impossible to resolve by normal light optical procedures, which have a limiting resolution of 200 nm. However, in indirect immunofluorescence microscopy we have two different contributions for potential visualization of individual microtubules. First, dark-field ultraviolet epi-illumination should permit a fluorescent fiber of any thickness less than 200 nm to be visualized at the limit of resolution since the fiber is emitting rather than absorbing light. Second, the tubule is coated by tubulin-specific antibody and also by the fluorescein-tagged second antibody. The coats of two antibodies can be assumed to increase the diameter of the individual tubule to 1000 Å or above. Therefore microtubules in the cytoplasm that are separated by a distance greater than 1000 Å should be resolved as individual thin fibers. Parallel microtubules that are closer together than this distance should be resolved as thicker filaments. The latter case would probably apply to most spindle microtubules since they are packed at distances of approximately 400 Å. Thus immunofluorescence microscopy reveals only thick filaments and in some cases a solid fluorescent mass.

One of the most striking results of immunofluorescence microscopy using tubulin antibody is the visualization of an elaborate and complex system of cytoplasmic microtubules in a variety of cultured cells. This was first reported by Weber, Pollack and Bibring (1975) and subsequently verified by Brinkley, Fuller and Highfield (this volume). In this context, the question of cytoplasmic organization of microtubules becomes an important problem. Preliminary data obtained from colcemid recovery and respreading of cells are in agreement with a proposal in which microtubules assemble from a defined origin (centrioles, centrospheres) in the perinuclear space and run towards the plasma membrane. However, more experiments are necessary to verify this assumption.

As pointed out previously (Weber, Bibring and Osborn 1975), the elaborate cytoplasmic microtubular system typical for interphase cells completely disappears during the onset of mitosis. By the end of prophase it becomes impossible to detect cytoplasmic microtubules by immunofluorescence microscopy. With the beginning of G_1 phase, cytoplasmic microtubules again become readily detectable by the antibody against tubulin. Thus it seems very likely that most of the tubulin from interphase cytoplasm, which is assembled into microtubules, becomes solubilized prior to the assembly of the spindle tubules and is used to assemble the spindle. This interpretation does not exclude the existence of a tubulin pool in cultured cells (Rubin and Weiss 1975) from which spindle assembly can also draw tubulin protein.

REFERENCES

Bensch, K. G. and S. E. Malawista. 1968. Microtubule crystals. A new biophysical phenomenon induced by vinca alkaloids. *Nature* **218**:1176.

Bryan, J. J. 1972. Vinblastine and microtubules. *J. Mol. Biol.* **66**:157.

Dales, S. J. 1972. Concerning the universality of a microtubule antigen in animal cells. *J. Cell Biol.* **52**:748.

Fonte, V. and K. R. Porter. 1974. Topographic changes associated with viral transformation of normal cells to tumorigenicity. In *8th Congress on Electron Microscopy*, p. 334. Australian Academy of Science, Canberra.

Fuller, G. M., B. R. Brinkley and J. M. Boughter. 1975. Immunofluorescence of mitotic spindles by using monospecific antibody against bovine brain tubulin. *Science* **187**:948.

Goldman, R. D. and D. M. Knipe. 1973. Functions of cytoplasmic fibers in non-muscle cell motility. *Cold Spring Harbor Symp. Quant. Biol.* **37**:523.

Goldman, R. D., E. Lazarides, R. Pollack and K. Weber. 1975. The distribution of actin in non-muscle cells. *Exp. Cell Res.* **90**:333.

Ishikawa, H., R. Bischoff and H. Holtzer. 1969. Formation of arrowhead complexes with heavy meromyosin in a variety of cell types. *J. Cell Biol.* **43**:312.

Lazarides, E. 1975. Tropomyosin antibody: The specific localization of tropomyosin in non-muscle cells. *J. Cell Biol.* **65**:549.

Lazarides, E. and K. Weber. 1974. Actin antibody: The specific visualization of actin filaments in non-muscle cells. *Proc. Nat. Acad. Sci.* **71**:2268.

Nagayama, A. and S. J. Dales. 1970. Rapid purification and the immunological specificity of mammalian microtubular paracrystals possessing an ATP-ase activity. *Proc. Nat. Acad. Sci.* **66**:464.

Olmsted, J. B. and G. G. Borisy. 1973. Microtubules. *Annu. Rev. Biochem.* **42**:507.

Osborn, M. and K. Weber. 1976. Cytoplasmic microtubules in tissue culture cells appear to grow from an organizing structure towards the cell membrane. *Proc. Nat. Acad. Sci.* (in press).

Rubin, R. W. and G. D. Weiss. 1975. Direct biochemical measurements in microtubule assembly and disassembly in Chinese hamster ovary cells. *J. Cell Biol.* **64**:42.

Shelanski, M. L., F. Gaskin and C. R. Cantor. 1973. Microtubule assembly in the absence of added nucleotides. *Proc. Nat. Acad. Sci.* **70**:765.

Weber, K. and U. Groeschel-Stewart. 1974. Antibody to myosin: The specific visualization of myosin containing filaments in non-muscle cells. *Proc. Nat. Acad. Sci.* **71**:4561.

Weber, K., R. Pollack and T. Bibring. 1975. Antibody against tubulin: The specific visualizations of cytoplasmic microtubules in tissue culture cells. *Proc. Nat. Acad. Sci.* **72**:459.

Weber, K., T. Bibring and M. Osborn. 1975. Specific visualization of tubulin-containing structures in tissue culture cells by immunofluorescence. *Exp. Cell Res.* **95**:111.

Weber, K., J. Wehland and W. Herzog. 1976. Griseofulvin interacts with microtubules both in vivo and in vitro. *J. Mol. Biol.* (in press).

Wilson, L., J. Bryan, A. Ruby and D. Mazia. 1970. Precipitation of proteins by vinblastine and calcium-ions. *Proc. Nat. Acad. Sci.* **66**:807.

Immunofluorescent Anti-tubulin Staining of Spindle Microtubules and Critique for the Technique

Hidemi Sato

Department of Biology, University of Pennsylvania
Philadelphia, Pennsylvania 19174

Yasushi Ohnuki

Pasadena Foundation for Medical Research
Pasadena, California 91109

Keigi Fujiwara

Department of Anatomy, Harvard Medical School
Boston, Massachusetts 02115

The structure of the in vivo mitotic spindle is rather difficult to visualize with phase-contrast or differential interference microscopy because of the minute difference of refractive indices encountered. However, owing to the anisotropy of spindle fine structure we can visualize the fibrous components of the spindle under a sensitive polarizing microscope. We have argued that the form birefringence reflects the amount of orderly aligned microtubules within the spindle (Sato, Ellis and Inoué 1975). Measuring spindle birefringence allows one to quantitate the amount of oriented microtubules present in the mitotic spindle that is controlled by various physiological parameters. From a series of observations and experiments we postulated that the labile spindle microtubules are formed by a reversible association of polymerizable tubulin molecules, and we determined the nature of the molecular equilibrium (Inoué 1964; Inoué and Sato 1967; Inoué et al. 1975; Sato 1975). Unfortunately, measurement of spindle birefringence does not allow us to determine the distribution of the unpolymerized tubulin pool.

To overcome this difficulty we tried indirect immunofluorescent staining on fixed specimens to visualize tubulin in both its polymerized form in mitotic microtubules and in an alternate unstructured form. Vinblastine-induced, mitotic tubulin paracrystals (VB-crystals) isolated from sea urchin gametes were chosen as the antigen. Tubulin-specific antibody was obtained from rabbit by injecting undenatured, pure VB-crystals. Primary culture of salamander lung epithelial cells is used as the experimental material because of their size, flatness and the clarity of their cytoplasm. Isolated spindles of mature oocytes of *Pisaster ochraceous* were used as test specimens to examine the fine structural artifacts.

In this article we describe (1) an improved method of cell fixation for fluorescent antibody localization of tubulin, (2) the pattern of tubulin distribution change in mitosis, (3) the fine structure of mitotic microtubules stained with fluorescent antibody, and (4) the appearance of actinlike filaments in isolated spindles with glycerol treatment.

MATERIALS AND METHODS

Induction and Specificity of the Antiserum against Tubulin

In addition to its normal assembly into mitotic spindles, tubulin in mature echinoderm eggs can be crystallized in vivo into two alternate crystalline forms: SM-crystals and VB-crystals (see Sato 1975). The birefringent SM-crystal, which appears to be a labile liquid crystal, is induced in mature oocytes and gametes by raising the temperature or altering the magnesium ion concentration. SM-crystals can readily be transformed from, or to, functional mitotic spindles but they are difficult to stabilize and isolate. The VB-crystal, an irreversibly formed tubulin crystal, is obtained by incubating echinoderm oocytes or gametes with 10^{-4} M vinblastine sulfate (Malawista and Sato 1969). VB-crystals are produced with a much larger yield per gamete than SM-crystals. The crystal morphology is different and the fine structure is characteristically double-helical (Fujiwara and Tilney 1975). VB-crystals can be stabilized and isolated in quantity (Bryan 1971). Colcemid (10^{-4} M) mobilizes tubulin and increases the yield of VB-crystals in mature sea urchin gametes threefold over that of the control (Strahs and Sato 1973). These VB-crystals consist of tubulin and vinblastine, contain little contaminant, and are easy to isolate.

Therefore rabbit antiserum directed against tubulin was prepared using VB-crystals as immunogen. The VB-crystals were induced in sea urchin gametes, *Strongylocentrotus purpuratus,* by incubating with 10^{-4} M vinblastine and colcemid according to a method developed in our laboratory (Fujiwara and Sato 1972; Strahs and Sato 1973; Fujiwara 1974). The crystals were isolated by a method modified after Bryan (1971). SDS-polyacrylamide gel electrophoresis of the purified paracrystal preparation showed that more than 95% of the protein was tubulin (Fig. 1).

Figure 1
SDS-polyacrylamide gel electrophoresis of the vinblastine crystal isolated from sea urchin gametes, *Strongylocentrotus purpuratus,* showing that the major protein component is tubulin.

Six rabbits were immunized three times each with approximately 0.5 mg of isolated tubulin paracrystals per injection. The initial subcutaneous injection was in the complete Freund adjuvant and the two boosts given at intervals of 10–14 days were in the incomplete Freund adjuvant. The rabbits were bled from their ears a week after the second boosts.

The specificity of the antiserum was determined by double immunodiffusion and immunoelectrophoresis performed on 0.6% agarose beds containing 50 mM veronal buffer at pH 8.6. The antigens used for these tests were purified sea urchin VB-crystals and unpurified tubulin extracted from sea urchin sperm tail axoneme, *Tetrahymena* ciliary axoneme, Chinese hamster tissue-culture cells, frog brain and fish brain. A single precipitin band was detected by double immunodiffusion between the antiserum and tubulin paracrystal, sperm tail axoneme or embryonic chick brain tubulin. By immunoelectrophoresis, all six antigens mentioned above formed a single precipitin band. From these results it was determined that the antiserum was specific to tubulin.

Choice of Experimental Specimen

Salamander Lung Epithelial Cells

Primary cultures of amphibian tissue have definite advantages for the present studies because (1) the tissues are composed of optically clear cells which are much larger than those obtained from avian or mammalian sources; (2) they contain larger and more birefringent spindles with fewer chromosomes; (3) mitosis proceeds with moderate speed (for instance, anaphase lasts 25 min at 24°C), which allows us to follow the mitotic sequences in great detail in monolayer culture; and (4) successful cultures can be achieved at room temperature without utilizing a CO_2 incubator.

Primary cultures of salamander (*Taricha granulosa;* salamanders provided by Dr. J. Kezer, University of Oregon) lung cells were cultured in Rose chambers according to Seto and Rounds' (1968) method with some modifications. Three to five pieces of $1 \times 1 \times 0.2$-mm size lung explants were set up in Rose chambers utilizing the cellophane strip method (Rose et al. 1958) and kept in an incubator at 25°C or at room temperature (20 ± 4°C), Eagle's minimum essential medium (MEM) with 10% fetal calf serum was used. Epithelial-like cells began to migrate from the explants on the third or fourth day in culture and started to divide on the fifth to seventh day, forming a beautiful monolayer culture. Mitosis usually reached a maximum rate between 8–14 days at 25°C. Observation chambers were usually kept at room temperature (from 21–24°C).

Isolated Spindles of Pisaster ochraceous

Mature oocytes were obtained by treating isolated sea-star ovaries with 10^{-6} M L-methyladenine (Kanatani et al. 1969) in artificial seawater (Cavanaugh 1964). One mg/ml Pronase (ICN; K & K Laboratories, Inc., Plainview, New York) and 0.1% mercaptoethylgluconamide (MEGA; Cyclo Chemical, Division of Travenol Laboratories, Inc., Los Angeles, California) were used to activate and remove the fertilization membrane. Spindles were isolated in quantity from these demembranated oocytes at 13°C in 12% hexylene glycol, pH 6.3, following the method described by Bryan and Sato (1970). The isolated medium is essentially the same as that described for sea urchins by Kane (1965). Isolated spindles are stored in isolation medium at 4°C. Experiments were carried out on spindles within 3 hours from isolation.

Indirect Immunofluorescence

Despite the specificity and sensitivity achievable by immunofluorescence, artifacts may well appear during the procedure of immunofluorescent staining. For instance, prefixation with 10% formalin-phosphate buffer solution (currently widely used as a standard procedure) definitely causes the shrinkage of both astral and spindle fibers and also changes the distribution of spindle birefringence. Cell rehydration after lyophilization by cold acetone at $-20°C$ also distorts spindle and other fibrous cytoplasmic structures as judged by their birefringence. Using images of in vivo spindles observed under the polarizing microscope as the standard (Fig. 2), we arrived at the following procedure for indirect immunofluorescence staining which introduced minimum spindle distortion:

1. Treat the monolayer culture of lung epithelium with absolute acetone ($-5°C$) for 8–10 minutes.
2. Gently rehydrate with physiological saline solution (PBS: 8 g NaCl; 0.2 g KCl; 1.15 g NaHPO$_4$; 0.2 g KH$_2$PO$_4$ in 1 liter of glass-redistilled water) pH 7.4 at 22°C.
3. Fix with $1 \sim 1.5\%$ formalin for $6 \sim 8$ minutes at 22°C.
4. Wash with PBS for 5 minutes at 22°C.
5. Incubate specimen with tubulin antiserum prepared against sea urchin VB-crystals for 30 minutes at 37°C in moist chamber (serum dilution: 1 to 4).
6. Wash specimen with PBS for 45 minutes to 1 hour at 37°C; two changes.
7. Postincubate with fluorescein-conjugated sheep antiserum to rabbit globulin (GIBCO, A Life Science Division of Mogul Corporation, Grand Island, New York) for 30 minutes at 37°C in a dark, moist chamber (final protein concentration, 0.5 mg/ml in PBS).
8. Postwash with PBS for 2 hours at 37°C in a dark, moist chamber; three changes.
9. Mount specimen with 50% glycerol-PBS. Seal with sticky wax (Engelhand Industries, Carteret, New Jersey).

Microscopy

Dark-field

Specimens were examined under a Zeiss GFL microscope equipped with a 100X Plan Achromatic objective (NA 1.25; for recording NA reduced to 1.1) and with a dark-field condenser (NA. 1.4, Achromatic-aplanatic). The output of an Osram HBO 200 w L-2 mercury arc (Osram GmbH, Berlin

Munchen, West Germany) was filtered with at least one heat-absorbing filter (American Optical Co., Rochester, New York) followed by AO #695 (American Optical Co.) filter as UV exciter. Combinations of the AO #723 and AO #724 filters (Americal Optical) were used as the barrier filter as well as the absorber of secondary fluorescence. Kodak Tri-X film (Eastman Kodak Co., Rochester, New York) was used for the photomicrography and was processed by Acufine developer (Acufine Inc., Chicago, Illinois) at an elevated ASA rating of 1200. Exposure time was either 1 or 2 minutes depending on the objective lens aperture.

Other Microscopic Techniques

Conditions for polarization and electron microscopy were essentially as described by Sato, Ellis and Inoué (1975).

OBSERVATIONS

Mitosis In Vivo

Mitosis in a tissue-culture cell of salamander lung epithelium is shown in Figure 2. From time-lapse cinematographic records using rectified polarized light optics we found that the prometaphase spindle, which is composed mainly of continuous spindle fibers, elongates past the nuclear envelope into the interior of the nucleus and rapidly gains in birefringence and length. The speed of spindle growth was 1.4 to 4 μm/minute (Fig. 2A–C). The advancing pole occasionally undergoes saltation, tilts, and swings. Birefringence is high in the banana-shaped early prometaphase spindle but decreases in the wider and longer late prometaphase spindle (Fig. 2C). The prometaphase spindle is not a stable structure: it twists, jerks, swims and sometimes migrates in the cell. These movements resemble those we recorded in grasshopper spermatocytes (Sato and Izutsu 1974) and are an exaggerated form of spindle movements we have noted in many other cell types.

Occasionally poles of the weakly birefringent, over-elongated spindle split and form a smaller but more birefringent spindle with two tandem satellite spindles. The cell is probably in a "tubulin hungry" state since the addition of 45% D_2O or temperature elevation (which usually enhance spindle birefringence and size; Inoué and Sato 1967) do not raise the spindle birefringence, suggesting that the concentration of polymerizable tubulin monomers is rather low. In anaphase, chromosomes moved at constant velocities, independent of the absolute value of spindle birefringence (Fuseler 1975).

Immunofluorescent Staining

Tubulin-containing structures were identified through indirect immunofluorescence by treating fixed cells of salamander lung epithelium with antibody prepared against sea urchin tubulin and fluorescein-conjugated sheep antiserum to rabbit globulin.

As shown in Figure 3A, the cytoplasm of interphase cells contains a complex array or network of fluorescent filaments and amorphous material ac-

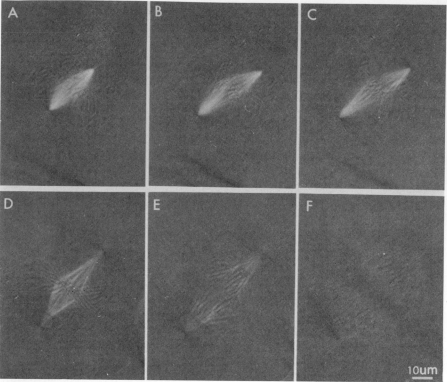

Figure 2

Mitosis observed in a tissue-culture cell of salamander (*Taricha granulosa*) lung epithelium. (*A,B,C*) The process of spindle formation in prometaphase. Note appearance of birefringent continuous spindle fibers and chromosomal spindle fibers in this elongating spindle. (*D*) Metaphase: spindle poles split and form two tandem but incomplete spindles. (*E*) Anaphase: spindle birefringence is reduced. (*F*) Late anaphase: birefringence has disappeared from the spindle region. Polarization microscopy with rectified optics.

cumulated around the nucleus. Most of these filaments are thought to be microtubules since they are greatly reduced in number by treatment with 10^{-5} M vinblastine or 10^{-5} M colcemid. These filament complexes are prominent in flat and elongated cells but have no affinity to anti-actin serum (courtesy of Dr. Frankel, School of Medicine, University of Pennsylvania). Immunofluorescence showed actin aggregates and actin cables with their own unique pattern of distribution, with little overlap with the microtubular pattern. With anti-tubulin, interphase nuclei did not stain except for the nucleoli. Nucleolar staining is presumably due to nonspecific adsorption of fluorescein-bound anti-IgG since staining also appears in the absence of tubulin antibody.

In dividing cells, the fluorescent image provides clear differentiation between tubulin organized into spindle fiber and astral ray microtubules and that present as unstructured amorphous material in the nonbirefringent area around the spindle (Fig. 3B–F).

The amount of amorphous material increased during prophase. The intense

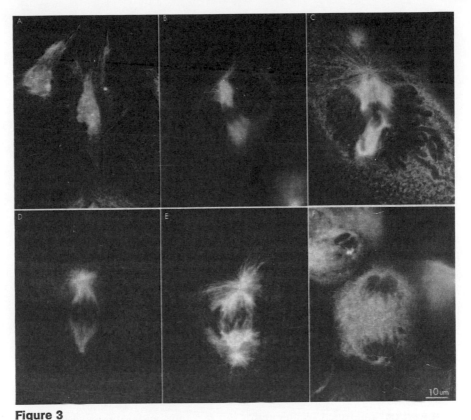

Figure 3

Indirect immunofluorescence observed in dividing tissue-culture cells of salamander lung epithelia. The fluorescent images clearly differentiate between tubulin organized into spindle fibers, astral rays and amorphous material. (*A*) Flat interphase cell, showing filamentous network throughout the cytoplasm with stronger fluorescence near the nucleus. (*B,C*) Growing prometaphase spindles composed mainly of continuous spindle fibers. (*D*) Metaphase: minimum amount of amorphous material. (*E*) Anaphase: chromosomal spindle fibers are clearly seen. Note the well-developed astral rays fanning out from centrosomes. (*F*) Late anaphase: fibrous spindle structures are decreasing in late anaphase and fluorescence of amorphous material is again increasing at this stage. (*B–F*) Same magnification; (*A*) half the magnification of *F*.

fluorescence surrounding the prophase nucleus probably reflects the accumulation of tubulin in the perinuclear "clear zone." In Figure 3, B and C show the growing prometaphase spindles at stages comparable to A and B in Figure 2. Figure 3D shows metaphase. For the chromosomal and continuous spindle fibers, these fluorescent images are similar to the in vivo polarized light images of the birefringent spindles. However, astral rays, which occasionally reach the cell membrane, appear more distinct than in the polarized light image. Chromosomes show no affinity for either anti-tubulin or fluorescein-bound sheep anti-IgG. In the micrographs, fluorescence of kinetochores is difficult to distinguish at metaphase due to the intense fluorescence of the midregion of the spindle. However, by careful focusing of the microscope we can definitely detect fluorescence of kinetochores and the chromosomal spindle fibers.

In anaphase, asters and chromosomal spindle fibers stain more intensely than the interzonal spindle region (Fig. 3E). Pole-to-pole distance increases, chromosomal spindle fibers become shorter, thinner and less fluorescent, and extended astral rays start to bend and disintegrate. Finally, the spindle structure becomes unrecognizable in late anaphase due to the dissociation of spindle fibers and the increase of amorphous material. During cytokinesis the fluorescence is intense in the stem body and the polar regions of the daughter cells. In general, these observations are in agreement with the report presented by Brinkley, Fuller and Highfield (this volume).

The fluorescence of amorphous material, which presumably reflects the amount of unstructured tubulin, increases in prophase, decreases during prometaphase, reaches minimum at metaphase, starts increasing during anaphase, and finally becomes maximum at late anaphase to telophase. This observation goes hand in hand with the in vivo finding that D_2O and temperature elevation have minimal influence on spindle birefringence in these cells. Unlike echinoderm gametes or zygotes (Inoué and Sato 1967; Sato 1975), these salamander lung epithelial cells must possess a rather small pool of polymerizable tubulin and use most of the available tubulin for spindle assembly during mitosis.

Artifact Due to the Staining Procedure

Even with our improved technique for indirect immunofluorescence, the birefringence of spindle structures decreases during the procedure of staining. This suggests the possibility of artifactual alterations in spindle fine structure during the procedure.

To examine this possibility we chose isolated metaphase spindles of *Pisaster ochraceous* oocytes as the experimental material because (1) they have good physical stability and a rather strong birefringence (3.4–3.8 nm retardation), and (2) there is a technique available for achieving a rather clean mass isolation of spindles without disturbing their initial size on birefringence (Bryan and Sato 1970; Sato, Ellis and Inoué 1975).

Fresh isolates are placed onto a strain-free cover glass (flamed and previously coated with a thin pellicle of gelatin solution). The spindles are then covered with another pellicle of gelatin and mounted in a modified "mini" Rose-type chamber. Fixative and warm PBS solution are perfused through this chamber with constant velocity. The whole assembly is maintained at 37°C during incubation with antisera. As controls, mass isolates of *Pisaster* spindles are also maintained in a centrifuge tube, and their birefringence is examined at each step of the staining procedure.

Figure 4A shows an isolated spindle of a *Pisaster* oocyte observed with polarized light and Figure 4D shows the image obtained from indirect immunofluorescent staining. Preserved spindles in both cases appear similar, but in fact, spindle birefringence is greatly reduced by the stage shown in Figure 4D.

Figure 4B shows a spindle immediately after rehydration by PBS after

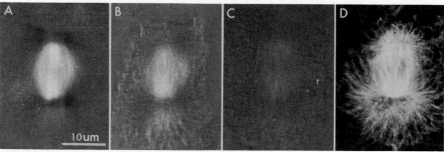

Figure 4

Spindle isolated from *Pisaster ochraceous* oocyte. (*A–C*) Shown in polarized light; (*D*) shows the image obtained by indirect immunofluorescent staining. Preserved spindle structures in *A* and *D* appear similar, but the spindle birefringence is greatly reduced during the staining procedure. (*B*) After cold acetone treatment and rehydration by PBS, the spindle is slightly deformed and birefringence is reduced. Only 30% of total birefringence is preserved in the spindle after completion of staining procedure, as shown in *C*. Dissociation, disorientation and breakdown of microtubules may be the cause of birefringence reduction. PBS ($N_D^{20} = 1.338$) is employed as the immersion medium observed under polarized light.

the cold acetone treatment. Although the acetone treatment itself has little effect on overall spindle birefringence, an average of 30% of the birefringence is lost from the spindle during rehydration. Figure 4C shows a photograph obtained after the completion of immunofluorescent staining. By this stage, only 30% of the original birefringence remains in the spindle.

Electron Microscopy of Stained Spindle

After completion of the staining procedure for indirect immunofluorescence, isolated spindles of *P. ochraceous* are fixed with either 3% glutaraldehyde-PBS or 3% glutaraldehyde 1% tannic acid-PBS, postfixed with 1% OsO₄-PBS, dehydrated with ethanol series, embedded with Epon, sectioned, stained with uranyl acetate and lead citrate, and observed.

As shown in Figure 5, the distribution of spindle microtubule remnants is more or less similar to the microtubules found in normal isolated spindles, as shown in Figure 6. However, upon closer inspection, the microtubular distribution in the spindle is altered, and their binomial distribution, commonly observed in control spindles (Fig. 6B), hardly exists in treated spindles. The average density of microtubules across the spindle is calculated to be 106 microtubules/μm² in the control but 80 or less/μm² in the treated spindles.

The individual microtubules in the stained spindles are badly damaged, giving the appearance of having been "riddled." This state appears during rehydration after the acetone treatment. However, tubulin in these "riddled" microtubules nevertheless possesses the ability to react with fluorescein-bound colchicine (observed with Dr. J. Clark, Harvard Medical School). Further loss of spindle birefringence may be due to the extensive washing.

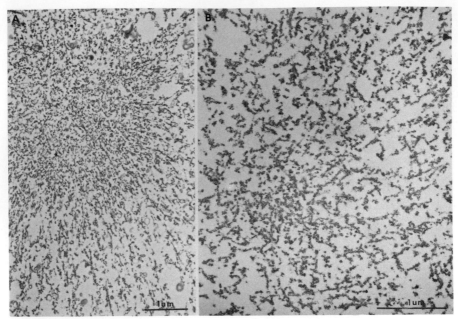

Figure 5
(*A*) Diagonal section of the polar region of an isolated spindle treated with immunofluorescent staining. The electron micrograph shows an image more or less
similar to the control (Fig. 6A), but the microtubules are badly damaged. (*B*)
Cross section shows "riddled" microtubules. The number of microtubules per unit
area is also reduced. The pattern of microtubule distribution in the spindle is somewhat altered.

Effect of Glycerination on Isolated Spindle

Clean isolated spindle of *Pisaster* is mainly composed of microtubules and
little else (see Fig. 6B). However, a bundle of filaments, about 60 Å in
diameter, progressively appears around the spindle microtubules when a large
quantity of clean isolates is preserved in 50% glycerol (Fig. 7). This suggests that actin was initially isolated along with the spindle as part of the
amorphous gel but was polymerized into filaments during storage in 50%
glycerol.

DISCUSSION

Antigenic similarity of tubulin obtained from various sources was tested by
several researchers (Dales 1972; Fuller, Brinkley and Boughter 1975; Brinkley, Fuller and Highfield, this volume). Their work supports the concept of
"universality of microtubule antigen." However, Fulton, Kane and Stephens
(1971) indicated that there is a distinctive difference among tubulin of
different origins. To avoid this potential problem we prepared antibody
against spindle tubulin (VB-crystal) obtained from unfertilized sea urchin
eggs. The antibody we obtained in fact did cross-react with tubulin from a

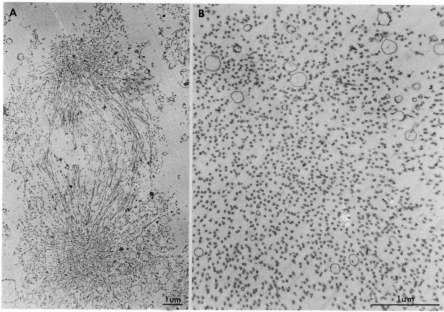

Figure 6

Tangential longitudinal section (*A*) and cross section (*B*) of isolated, untreated spindles of *Pisaster ochraceous* are mainly composed of microtubules and little else. The microtubules in the mid-spindle region are mostly aligned parallel to the spindle axis and exhibit a binomial distribution. Some of the vesicles in the photograph are mitochondria that have degenerated during the preparative procedure.

wide variety of sources such as cilia, flagella, brain tubulin, VB-crystals and spindles of a number of organisms. This indicates that the tubulin from various sources do share common antigenic sites. This immunological reaction is specific, since no staining occurs if nonspecific rabbit IgG (courtesy of Dr. F. A. Pepe, School of Medicine, University of Pennsylvania) is substituted for the immune serum. Also 100-Å diameter filaments induced in primary muscle cell culture (Holtzer et al. 1975) show no affinity for our antibody.

We attempted in situ indirect immunofluorescent staining of spindle microtubules and unpolymerized tubulin using cultured cells from salamander lung epithelium. Current techniques for cell fixation (Dales 1972; Lazarides and Weber 1974; Fuller, Brinkley and Boughter 1975; Lazarides 1975) were not adequate because of the rapid alterations in spindle birefringence and spindle fine structure. Using images of birefringent spindles observed in vivo as the critical standard we endeavored to improve the preservation and staining technique as described in this paper.

As shown in Figure 3, the pattern of fluorescence is similar to the birefringent in vivo spindle (Fig. 2). From prometaphase to mid-anaphase, both chromosomal spindle fibers and continuous spindle fibers were intensely stained. During these stages well-developed astral rays fanned out, many of them reaching the cell surface. In late anaphase, the chromosomal spindle

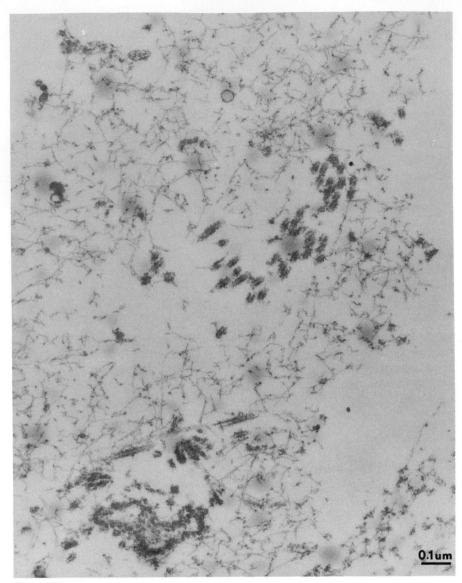

0.1um

Figure 7
Electron micrograph of glycerinated, isolated spindle, showing microtubules surrounded by many 60-Å diameter filaments. These actinlike filaments appear only when the mass of isolates is preserved in 50% glycerol. Compare this image with Fig. 6B.

fibers and astral rays become shorter and thinner, and finally disappeared.

In interphase, a filamentous network was commonly observed in the peripheral cytoplasm of flattened cells with a stronger amorphous fluorescence around the nucleus. The early apparent increase in perinuclear staining, however, could in part reflect an increase of cytoplasmic thickness. In prophase, the number of cytoplasmic filaments decreased and the perinuclear fluorescence increased. This amorphous fluorescence sharply decreased at the onset of mitosis and later increased during late anaphase. This suggests that most of the cellular tubulin accumulates around the nucleus in preparation for assembly into the mitotic spindle. The increase in amorphous fluorescence during anaphase clearly reflects the disassembly of spindle microtubules.

In general, our observations are in good agreement with the results described by Fuller, Brinkley and Boughter (1975), who used antibody to purified 6S tubulin isolated from bovine brain. The sensitivity of our technique was high enough to detect a cytoplasmic network whose filaments are believed to be made up of a few microtubules. In our optical system, the fluorescence can be preserved for at least 8 minutes under continuous excitation.

To analyze possible artifacts introduced by fixation and immunofluorescent staining we applied our technique to isolated meiosis I spindles obtained from mature oocytes of *Pisaster ochraceous*. We used this material because it has already been well characterized at the light and electron microscope levels in our laboratory (Sato 1975; Sato, Ellis and Inoué 1975).

Accordingly, $-5°C$ acetone treatment for 10 minutes has little effect on spindle birefringence or dimensions, in sharp contrast with the rather striking change observed after a $-20°C$ acetone treatment and rehydration. The rehydration of "half-dry" spindles by PBS considerably reduces the amount of spindle birefringence. "Riddled" spindle microtubules are the result of lyophylization by acetone and rehydration. It should be noted that spindles freshly isolated by hexylene glycol have no reactivity to antibody or to antimitotic poisons but gain reactivity to antibody as well as to colchicine after the cold acetone treatment. Postfixation by dilute formalin ($1 \sim 1.5\%$) is important for maintaining the spindle structure in vitro; otherwise, spindle fibers become wavy and more diffuse. Spindle birefringence is continuously lost during the staining and washing procedures. This suggests a further breakdown, disorientation and dissociation of microtubules. Electron microscope observations show that stained spindles contain less than 80% of the total microtubules contained in control spindles. Also, the distribution pattern of microtubules within the spindle is altered. The "riddled" state of individual microtubules and the alteration of microtubular orientation are consistent with dramatic loss of spindle birefringence. These observations demonstrate that the artifacts are induced during the fixation and staining procedure, and we would therefore like to point out that one should be extremely cautious about interpreting the structure and function of the in vivo spindle based on the images obtained by immunofluorescence alone. We believe, however, that this technique does provide important information on the mechanism of spindle assembly which complements data obtained by sensitive, high-resolution polarized light microscopy.

Finally, the accumulation of actinlike filaments in glycerinated isolated spindles can be explained as follows: Isolated spindles generally contain a significant amount of "gel," with some material remaining associated with the spindle fibers. Undoubtedly, this material contains actin that could polymerize during glycerination. The spindle is essentially a de novo structure formed at each mitosis in a cytoplasmic environment that contains a quantity of actin and some actin filaments. Thus it is not surprising to find some actinlike filaments both inside and outside the spindle. However, the accumulation of a mass of actinlike filaments (as shown in Fig. 7) should be regarded as an artifact induced by glycerination.

Acknowledgments

The authors wish to thank Dr. Shinya Inoué for his critical reading of the manuscript and helpful suggestions, as well as Mrs. D. Bush and Mrs. B. Woodward for technical assistance.

This work was supported by NIH Grant CA-10171, awarded by the National Cancer Institute, DHEW, and NSF Grant GB-5120, awarded by the National Science Foundation. Dr. K. Fujiwara is supported by a postdoctoral fellowship from the Muscular Dystrophy Association, Inc.

REFERENCES

Bryan, J. 1971. Vinblastine and microtubules. I. Induction and isolation of crystal from sea urchin oocytes. *Exp. Cell Res.* **66**:129.

Bryan, J. and H. Sato. 1970. The isolation of the meiosis I spindle from the mature oocyte of *Pisaster ochraceous*. *Exp. Cell Res.* **59**:371.

Cavanaugh, G. M. 1964. *Formulae and methods V*. Marine Biological Laboratories, Woods Hole, Massachusetts.

Dales, S. 1972. Concerning the universality of a microtubule antigen in animal cells. *J. Cell Biol.* **52**:748.

Fujiwara, K. 1974. Studies on the microtubule protein, tubulin. I. A fine structural analysis of a polymorphic form, the vinblastine induced crystal. II. A serological study of tubulin from various sources. Ph.D. thesis, University of Pennsylvania, Philadelphia.

Fujiwara, K. and H. Sato. 1972. VB-crystal formation in sea urchin eggs by short term VB-col incubation. *J. Cell Biol.* **55**:79a.

Fujiwara, K. and L. G. Tilney. 1975. Substructural analysis of the microtubule and its polymorphic forms. *Ann. N.Y. Acad. Sci.* **253**:27.

Fuller, G. M., B. R. Brinkley and J. M. Boughter. 1975. Immunofluorescence of mitotic spindles by using monospecific antibody against bovine brain tubulin. *Science* **187**:948.

Fulton, C., R. E. Kane and R. E. Stephens. 1971. Serological similarity of flagellar and mitotic microtubules. *J. Cell Biol.* **50**:762.

Fuseler, J. W. 1975. Temperature dependence of anaphase chromosome velocity and microtubule depolymerization. *J. Cell Biol.* **67**:789.

Holtzer, H., J. Groop, S. Dienstman, H. Ishikawa and A. P. Somlyo. 1975. Effect of cytochalasin B and colcemid on myogenic cultures. *Proc. Nat. Acad. Sci.* **72**:513.

Inoué, S. 1964. Organization and function of the mitotic spindle. In *Primitive motile systems in cell biology* (ed. R. D. Allen and N. Kamiya), pp. 549–598. Academic Press, New York.

Inoué, S. and H. Sato. 1967. Cell motility by labile association of molecules. The nature of mitotic spindle fibers and their role in chromosome movement. *J. Gen. Physiol.* **50**:259.

Inoué, S., J. Fuseler, E. D. Salmon and G. W. Ellis. 1975. Functional organization of mitotic microtubules: Physical chemistry of the *in vivo* equilibrium system. *Biophys. J.* **15**:725.

Kanatani, H., H. Shirai, K. Nakanishi and T. Kurokawa. 1969. Isolation and identification of meiosis inducing substance in starfish *Asterias amurensis. Nature* **221**:273.

Kane, R. E. 1965. The mitotic apparatus: Physical-chemical factors controlling stability. *J. Cell Biol.* **25**:137.

Lazarides, E. 1975. Immunofluorescence studies on the structure of actin filaments in tissue culture cells. *J. Histochem. Cytochem.* **23**:507.

Lazarides, E. and K. Weber. 1974. Actin antibody: The specific visualization of actin filaments in non-muscle cells. *Proc. Nat. Acad. Sci.* **71**:2268.

Malawista, S. E. and H. Sato. 1969. Vinblastine produces uniaxial birefringent crystals in starfish oocytes. *J. Cell Biol.* **42**:596.

Rose, G. G., C. M. Pomerat, T. O. Shindler and J. B. Trunnell. 1958. A cellophane-strip technique for culturing tissue in multipurpose culture chambers. *J. Biophys. Biochem. Cytol.* **4**:761.

Sato, H. 1975. The mitotic spindle: In *Aging gametes* (ed. R. D. Blandau), pp. 19–49. S. Karger, Basel, Switzerland.

Sato, H. and K. Izutsu. 1974. Birefringence in meiosis of grasshopper spermatocytes: *Chrysochraon japonicus* and *Trilophida annulata*. (time-lapse movie). Available from George W. Colburn Laboratory, Inc., Chicago, Illinois 60606.

Sato, H., G. W. Ellis and S. Inoué. 1975. Microtubular origin of mitotic spindle form birefringence Demonstration of the applicability of Wiener's equation. *J. Cell Biol.* **67**:501.

Seto, T. and D. E. Rounds. 1968. Cultivation of tissues and leukocytes from amphibians. In *Methods in cell physiology* (ed. D. M. Prescott), vol. 3, pp. 75–94. Academic Press, New York.

Strahs, K. R. and H. Sato. 1973. Potentiation of vinblastine crystal formation *in vivo* by puromycin and colcemid. *Exp. Cell Res.* **80**:10.

Tubulin Antibodies as Probes for Microtubules in Dividing and Nondividing Mammalian Cells

B. R. Brinkley, G. M. Fuller and D. P. Highfield

Divisions of Cell Biology and Human Genetics
Department of Human Biological Chemistry and Genetics
The University of Texas Medical Branch, Galveston, Texas 77550

Microtubules (MTs) in eukaryotic cells are associated with many important functions, including the maintenance of cell shape (Porter 1966), cell motility (Allison 1973), chromosome movement in mitosis and meiosis (Nicklas 1971), axonal transport in neurons (Smith, Jälfors and Cameron 1975), movements of pigments and other inclusions and organelles (Moellmann and McGuire 1975), cell membrane motility (Berlin 1975), mobilization and release of lysosomes (Malawista 1975), and transport and release of macromolecules including hormones (Wolff and Bhattacharyya 1975) and plasma proteins (Redman et al. 1975). Although MTs are most often identified by electron microscopy (EM), their dynamic nature and ubiquitous distribution are difficult to visualize by EM procedures. Light optical techniques, although limited to polarizing microscopy, have contributed much to the basic knowledge of MTs (Inoué 1964; Salmon 1975), and with the development of tubulin-specific antibodies (Fuller, Brinkley and Boughter 1975; Weber, Pollack and Bibring 1975), immunofluorescence procedures offer much promise as a tool for MT research.

The purpose of this report is to describe the use of monospecific antibodies made against 6S bovine brain tubulin as immunofluorescent probes for the detection of MTs in cultured mammalian cells. The experiments presented in this report attest to the reliability of this approach for studies of cytoplasmic and spindle MTs under a variety of experimental conditions.

MATERIALS AND METHODS

The cell culture conditions, as well as the procedures for isolating 6S tubulin and the production of tubulin-specific antibodies, have been described previously (Fuller, Brinkley and Boughter 1975).

435

Indirect Immunofluorescence

Cells grown as monolayers on cover slips were rinsed in phosphate-buffered saline (PBS) and fixed at room temperature for 20 minutes in 1.0–3.0% formaldehyde made up in PBS. Subsequently, the cover slips were rinsed thoroughly in PBS and fixed in absolute acetone at $-10°C$ for 7 minutes. After rinsing, the cover slips were incubated with antibody (0.20 mg/ml in borate-saline buffer) at 37°C for 30 minutes. In order to conserve antibody solution, the cover slips were inverted over a small drop of solution placed on the bottom of a 35-mm petri dish. The dish was placed into a 37°C incubator. After incubation with the antibody for 30 minutes, the cover slips were rinsed several times in PBS and incubated in a 1:1.5 dilution of fluorescein-tagged goat anti-rabbit immunoglobulin G (Meloy Laboratories, Springfield, Va.) in PBS for 30 minutes at 37°C. The cover slips were rinsed in PBS followed by distilled H_2O and mounted on glass slides in a drop of PBS: glycerol (1:9, pH 8.5) for viewing in a Leitz ultraviolet microscope equipped with a dark-field condenser. Photographs were recorded on Tri-X Pan film (Kodak). Control slides consisted of cells prepared as above but with rabbit albumin antiserum substituted for anti-tubulin.

Drugs and Chemicals

Colcemid (CIBA Pharmaceutical Products) was used at a final concentration of 0.06 μg/ml in complete medium. N^6,O^2-dibutyryl adenosine 3':5'-cyclic monophosphoric acid monosodium salt (Bt$_2$cAMP) was made up in warm medium at a final concentration of 3×10^{-4} M and was prepared on the day of the experiment. Testosterone propionate (Δ4-androstene-17β-propionate-3-one, Sigma Chemical) was made up in a stock solution at a concentration of 1.5×10^{-3} M in ethanol and diluted in warm Bt$_2$cAMP media to yield a final concentration of 1.5×10^{-5} M. Theophylline (Sigma Chemical) was used at a final concentration of 10^{-3} M in Bt$_2$cAMP medium. For controls, cultures treated with ethanol alone were always employed.

Cold Treatment

The conditions for cold treatment of cultured cells were precisely the same as those described by Brinkley and Cartwright (1975).

Electron Microscopy

Cells to be examined by electron microscopy were prepared according to the procedures described by Brinkley and Cartwright (1975).

RESULTS

Cytoplasmic Microtubular Complex in Interphase Cells

As shown in Figure 1, the cytoplasm of 3T3 cells contains an elaborate array of fine fluorescent filaments, which, by experiments to be described later, can be shown to be microtubules. In view of the results of numerous EM studies of cultured mammalian cells, the detection of cytoplasmic microtubules was not unexpected. However, we were surprised by the extent and array of the cytoplasmic microtubular complex (CMTC) revealed by tubulin immuno-fluorescence in all normal fibroblast cells. The fluorescent filaments often extended out radially from the nucleus and were aligned in bundles parallel to the major cell axes. Some filaments were seen to terminate near the cell surface, whereas others were bent to conform to the surface contours (Fig. 2). Some cells within a culture displayed a much more elaborate CMTC than others. Cells that were greatly elongated and anisotropic contained the most extensive CMTC, whereas the smaller and more pleomorphic cells (presumably G_1 cells) contained fewer and shorter cytoplasmic tubules. When the cultures became densely populated, individual cells achieved a more bipolar morphology and assumed a parallel alignment with the long axis of neighboring cells. In these cultures, the CMTC became more uni-formally organized, with very long bundles of MTs extending parallel to the long axis of each cell.

Although the above description is based upon observations of 3T3 cells, essentially the same immunofluorescent staining pattern was found in other nontransformed cells, including human skin fibroblasts, primary hamster embryo cells and rat kangaroo fibroblasts (strain PtK1).

Nuclear fluorescence in the form of discrete patches of more generalized staining throughout the nucleoplasm was apparent in many cells. The pattern and intensity of the stain within a given culture varied considerably from cell to cell. It was also found that the intensity of nuclear fluorescence could be exaggerated when the nucleus was slightly out of the plane of focus (cf. Figs. 9 and 10). This was often the case in our photomicrographs since we focused on the filaments within the thin cytoplasmic processes. Colcemid and cold treatment failed to diminish the nuclear fluorescence, and similar pat-terns were seen in both normal and transformed cells. Obviously, nuclear immunofluorescence is not due to assembled microtubules, and further investigations are needed before the nature of the nuclear material can be determined.

Effects of Microtubule Inhibitors

Although the antibody preparation used in this study was shown to be monospecific for soluble tubulin by double immunodiffusion and immuno-electrophoresis, it was essential to show that the fluorescent filaments seen in the cytoplasm of 3T3 cells were responsive to standard microtubule in-hibitors. When 3T3 cells were exposed for 1–3 hours to media containing colcemid, the CMTC disappeared or became greatly diminished (Fig. 4). In addition, the cells withdrew their long fibroblastic processes and became rounded and pleomorphic. The cytoplasm usually displayed a weak back-ground fluorescence, and not infrequently one or more bright fluorescent spots could be seen near the cell nucleus. Usually the cytoplasmic filaments disappeared after colcemid treatment, although the response varied somewhat from cell to cell. Occasionally, short fragments of filaments remained in the cytoplasm even after treatment in colcemid for 3 hours.

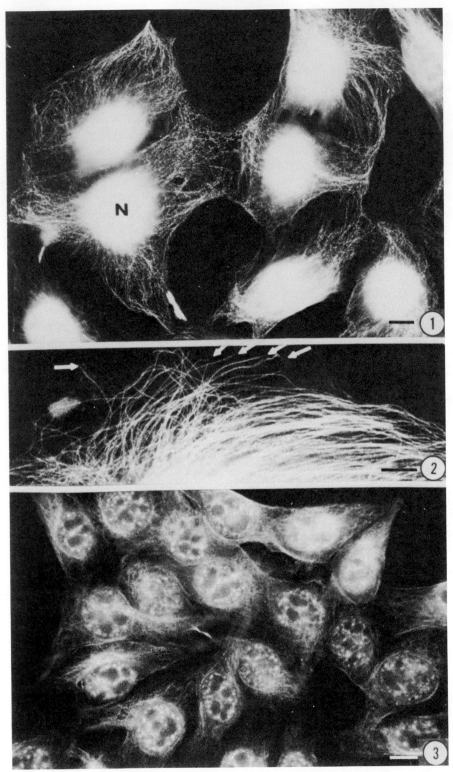

Figures 1–3 (*See facing page for legend*)

Reversal from Colcemid Inhibition

When the cultures were washed and placed in fresh media without colcemid for 15 minutes to 1 hour, the MTs reappeared in the cytoplasm. Within 15 minutes, MTs became associated with the fluorescent spots, which became brighter and displayed several elongated asterlike rays (Figs. 5, 6, 8). As shown in Figure 5, almost every cell displayed the asters, and subsequent examination of these regions by EM revealed two or more centrioles with associated microtubules (Fig. 7). Within 30 minutes after reversal, the microtubules became so numerous around the nucleus that it was difficult to determine their points of origin (Fig. 9). At this time, the cells began to achieve a more fibroblastic appearance, and bundles of MTs extended out into the developing process. By 60 minutes after reversal, the CMTC was essentially restored (Fig. 10), and the cells regained their normal fibroblastic appearance.

Essentially the same responses were obtained by chilling the cells to 0°C in ice bath for 30 minutes and then rewarming them in a 37°C water bath.

Cytoplasmic Microtubules in Transformed Cells

Malignant transformation of mammalian cells in tissue culture is characterized by several distinct morphological and growth-related features. Generally the cells lose their elongated anisotropic appearance and become rounded and polygonal in shape. Concomitantly, specific changes in cell surface properties occur, and the cultures lose density-dependent inhibition of growth (Stoker and Rubin 1967).

Recently we found that transformed cells could be identified in mixed cell populations by their tubulin antibody immunofluorescent patterns (Brinkley, Fuller and Highfield 1975). As shown in Figures 3 and 11, transformed SV3T3 cells have a greatly depleted CMTC when compared to their normal counterparts. Only a few fluorescent filaments are seen in the cytoplasm, and those that are present appear to be short, randomly distributed and have no apparent association with cell shape.

In addition to SV3T3, ten additional transformed cell lines were found to have similar patterns of cytoplasmic microtubules (Brinkley, Fuller and Highfield 1975). Thus our observations support and extend those of Porter and coworkers (Fonte and Porter 1974; Porter et al. 1974), who first identified altered cytoplasmic MTs in transformed cells using transmission EM.

Figures 1–3

(*Fig. 1*) 3T3 cells showing numerous fluorescent filaments in the cytoplasm. Nucleus (N) is brightly fluorescent because it is out of the plane of focus. Bar = 10 μ.

(*Fig. 2*) Higher magnification of cytoplasmic microtubules, some of which terminate at the cell surface (arrows), whereas others are bent to conform to the shape of the cell. Bar = 10 μ.

(*Fig. 3*) SV3T3 cells showing small, rounded appearance and paucity of CMTC. Bar = 10 μ.

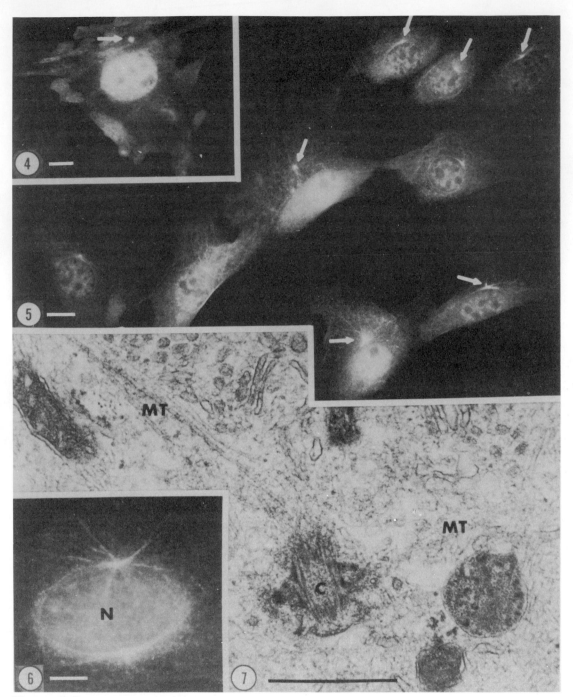

Figures 4–7

(*Fig. 4*) 3T3 cell treated with colcemid for 2 hr. Note the loss of fibroblastic shape and the absence of CMTC. Arrow indicates small fluorescent spot presumed to be the centrosphere. Bar = 10 μ. (*Fig. 5*) Survey of several 3T3 cells fixed 15 min after reversal from colcemid treatment. Note bright fluorescent spot in each cell (arrows). Bar = 5 μ. (*Fig. 6*) High magnification of asterlike body in cell 15 min after reversal from colcemid arrest. Bar = 5 μ. (*Fig. 7*) Electron micrograph of 3T3 cells, showing section through region corresponding to the asterlike body shown in Fig. 6. Centrioles (C) and associated microtubules (MTs) are apparent. Bar = 0.5 μ.

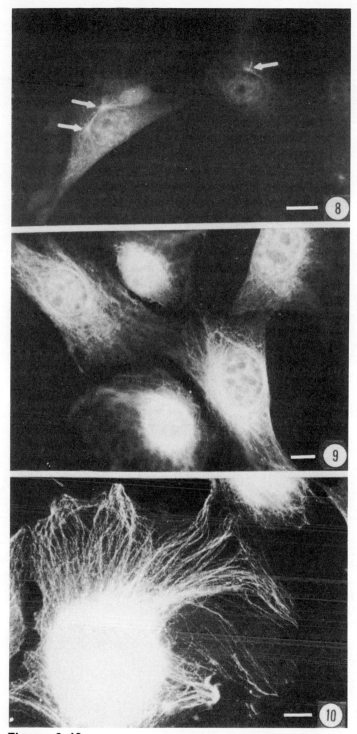

Figures 8–10
Stages in reappearance of CMTC after recovery from colcemid treatment: (*Fig. 8*) 15 min after reversal, note bright fluorescent centrospheres; (*Fig. 9*) 30 min after reversal, note numerous MTs around nucleus; (*Fig. 10*) 60 min after reversal, complete CMTC is present. Bars = 10 μ.

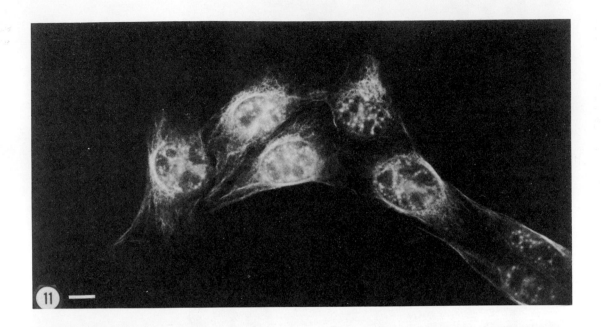

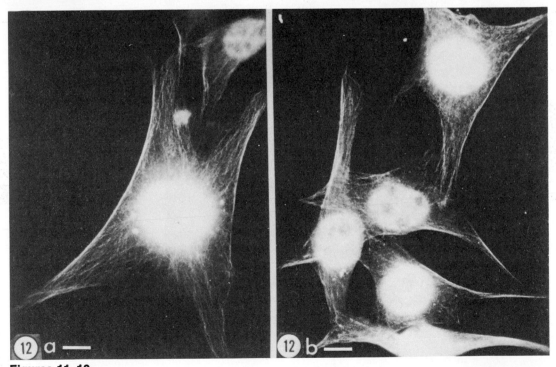

Figures 11–12

(*Fig. 11*) SV3T3 cells showing diminished cytoplasmic microtubules. Bar = 10 μ. (*Fig. 12*) SV3T3 cells after treatment with Bt$_2$cAMP for 6 hr. Note reappearance of fibroblastic shape and CMTC. Bar = 10 μ.

Effects of Reverse Transformation Agents

Several investigators (Hsie and Puck 1971; Johnson, Friedman and Pastan 1971; Sheppard 1971) have shown that derivatives of cyclic AMP alone or in synergistic combination with other compounds, such as testosterone or theophylline, could initiate a reversion of the morphology of transformed cells to more fibroblastic forms. We repeated these experiments to evaluate the disposition of cytoplasmic MTs during reverse transformation. As previously reported (Hsie and Puck 1971; Johnson, Friedman and Pastan 1971; Sheppard 1971), Chinese hamster ovary (CHO) and SV3T3 cells grown in the presence of dibutyryl 3′:5′-cyclic adenosine monophosphate (Bt$_2$cAMP) plus testosterone or theophylline underwent a dramatic reversion from the rounded, polygonal form to a more spindle-shaped, fibroblastic form (cf. Figs. 11 and 12). Although we did not evaluate the growth condition in Bt$_2$cAMP-treated cultures, the cells grew as monolayers, appeared to be more aligned along their long axes, and showed less tendency to pile up and become multilayered in culture.

As shown in Figure 12a,b, SV3T3 cells treated with Bt$_2$cAMP contained many more cytoplasmic MTs, which were less randomly oriented in the cytoplasm and were aligned into parallel bundles extending out into the cell processes. Essentially the same events were observed in CHO cells similarly treated with Bt$_2$cAMP. Thus the major morphological changes induced by Bt$_2$cAMP were accompanied by an assembly of cytoplasmic MTs which gave rise to a CMTC that was very similar to that seen in nontransformed cells. Moreover, tubulin antibody immunofluorescence permits one to evaluate microtubules in these cells without the need for EM.

Fate of CMTC during Mitosis

When cells progress from interphase into mitosis, much of the cytoplasmic tubulin presumably becomes available for spindle microtubule assembly (Hsie and Puck 1971). For this reason we were particularly interested in observing the fate of the CMTC as the cells progressed into mitosis. Although most of our observations are based on studies of rat kangaroo cells (strain PtK1), which are more ideally suited for light microscopic examination, the staining patterns to be described below were also seen in 3T3, SV3T3, human skin fibroblasts and HeLa cells.

Prophase

In all cells observed, the fluorescent filaments of the cytoplasm became greatly diminished during prophase and completely disappeared during subsequent stages of mitosis. Except for a slight detectable increase in background staining of the cytoplasm, most of the fluorescence became localized in the mitotic spindle. The disappearance of the CMTC was accompanied by a general rounding up of cells as they entered mitosis. As shown in Figures 13 and 14, the most obvious feature of prophase was the appearance of one or two brightly fluorescent spots near the cell nucleus. Radiating out from each spot (presumably the centrosphere) were a number of fine fluorescent filaments.

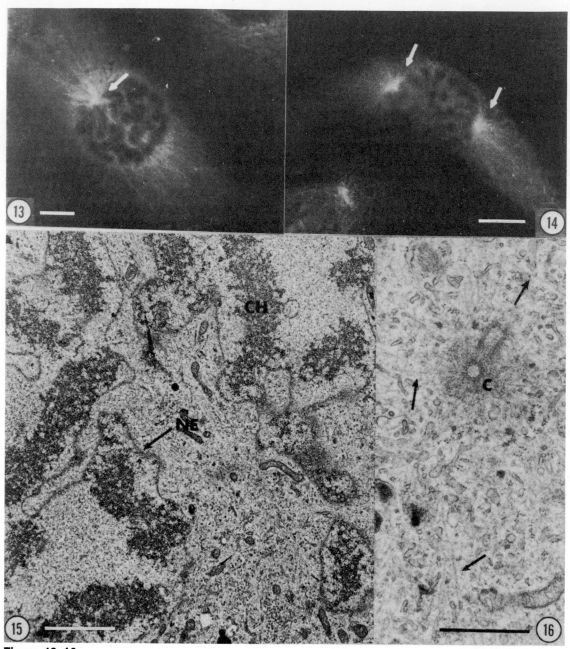

Figure 13–16

Stages of prophase in PtK1 cells, showing (*Fig. 13*) a single bright fluorescent spot at centrosphere (arrow) and (*Fig. 14*) fluorescent spots at opposite poles (arrows). (*Fig. 15*) Electron micrograph of prophase stage, showing chromosome (CH) and nuclear envelope (NE). (*Fig. 16*) Higher magnification of centriole (C) pair, showing microtubules (arrows) extending out in many directions. Bar = 5 μ in Figs. 13 and 14 and 1 μ in Figs. 15 and 16.

444

As described previously by Roos (1973) and J. B. Rattner and M. Berns (pers. comm.), prophase in rat kangaroo cells may be characterized by the early separation of centriole pairs to form two distinct poles (Fig. 14). Alternatively, a delay of centriole separation may occur, which results in the temporary formation of a single pole (Fig. 13).

The nuclear fluorescence observed in many interphase cells completely disappeared during prophase. The condensing chromosomes were nonfluorescent except for small spots at the position of the kinetochore. The latter were very faint and difficult to demonstrate photographically but were more apparent after colcemid treatment.

When prophase cells were examined by EM, the chromosomes appeared greatly condensed, the nuclear envelope remained intact, and a large number of MTs radiated out from an amorphous substance surrounding the centrioles (Figs. 15, 16).

Prometaphase

Prometaphase is characterized by the disruption of the nuclear envelope and the association of chromosomes with the mitotic spindle. As shown in Figures 17 and 18, the prometaphase spindle of PtK1 cells can assume two forms. If the centriole pairs separated prior to nuclear envelope disruption, an elongated bipolar spindle was formed, with chromosomes scattered throughout the spindle (Fig. 18). Examination of this stage by EM (Fig. 19) indicated that two distinct poles were formed, with centriole pairs present at each pole. Fragments of the nuclear envelope were present, as were numerous small, smooth-surfaced vesicles at the poles. The chromosomes displayed kinetochores from which bundles of MTs could be seen extending to the two poles. If the centrioles failed to separate at prophase, a unipolar spindle formed at prometaphase, with chromosomes arranged in a cluster around the pole (Fig. 17). In the latter case, a single bright fluorescent spot is apparent near the cell center. By this stage, all of the cytoplasmic microtubules have disappeared, and the cytoplasm assumes a structureless texture with a slightly increased level of background fluorescence. During late prometaphase, the spindle of PtK1 cells becomes considerably elongated, having a pole-to-pole distance of over 20 μ. By metaphase, however, the spindle shortens to approximately 10 μ in length.

Metaphase

The metaphase spindle was particularly well characterized by immunofluorescent staining (Figs. 20, 21). As the chromosomes became aligned on the metaphase plate, bundles of fluorescent filaments could be seen extending from the chromosomes to the poles. Often the poles were characterized by a bright fluorescent spot associated with a few asterial fibers. The kinetochores were not stained at metaphase unless the spindle was disrupted by colcemid as described below. By carefully focusing through the spindle, bundles of interpolar (pole-to-pole) fibers could also be observed extending across the metaphase plate.

The PtK1 strain of rat kangaroo fibroblast is aneuploid, having a modal chromosome number of $2N = 11$. The two cells shown in Figures 20 and 21

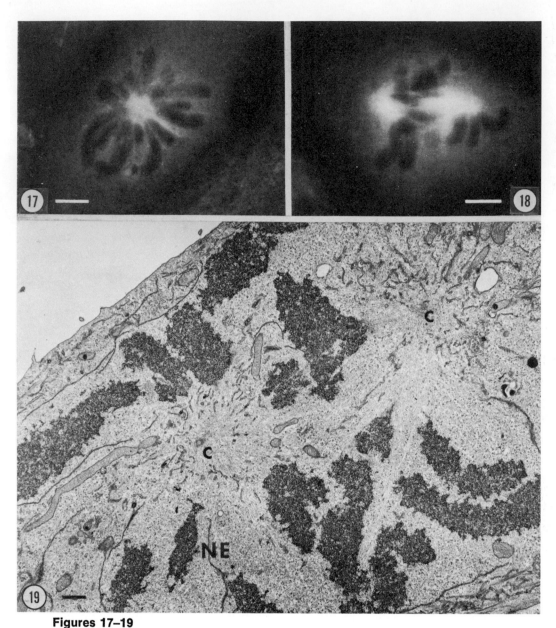

Figures 17–19
Prometaphase. (*Fig. 17*) Chromosomes grouped around a single bright spot. (*Fig. 18*) Spindle-shaped fluorescence where centrioles have separated to opposite poles. (*Fig. 19*) Prometaphase similar to Fig. 18 viewed in EM. Note fragments of nuclear envelope (NE) and centriole (C). Bar = 5 μ in Figs. 17 and 18 and 1 μ in Fig. 19.

are obviously from polyploid cells. This probably explains the tetrapolar spindle in one cell (Fig. 21) and the unusually large spindle (over 20 μ) in the other (Fig. 20).

The fluorescent images of metaphase cells were strikingly similar to images

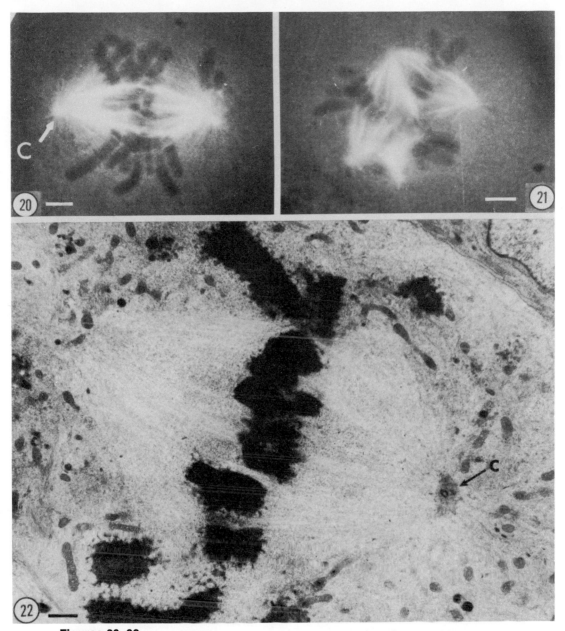

Figures 20–22

Metaphase. (*Fig. 20*) Metaphase in an unusually large polyploid cell. Arrow points to bright fluorescent spot at pole. (*Fig. 21*) Tetrapolar spindles. (*Fig. 22*) EM image of metaphase, showing centrioles (C) and bundles of microtubules from chromosomes. Bar = 5 μ for Figs. 20 and 21 and 1 μ for Fig. 22.

obtained by conventional transmission EM (Fig. 22) and by high-voltage EM (McIntosh et al. 1975). In thin sections viewed by conventional EM (Fig. 22), bundles of spindle MTs were found to extend from the kineto-chore to the poles. The latter consisted of pair of centrioles surrounded by

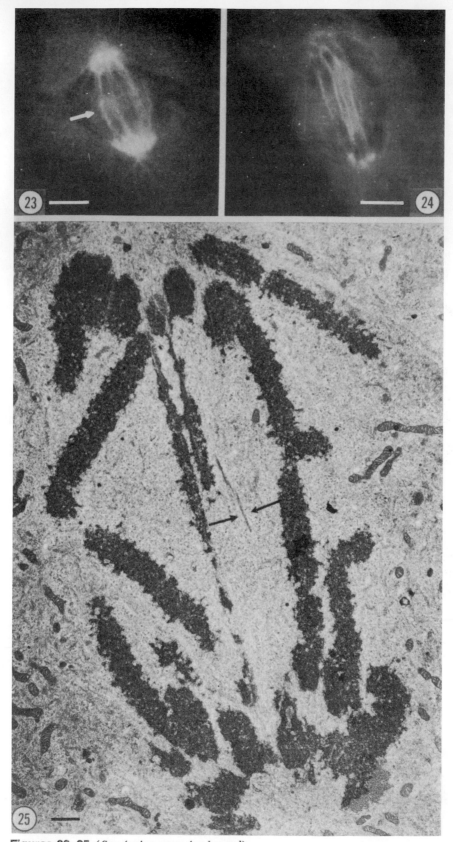

Figures 23–25 (*See facing page for legend*)

an amorphous, electron-dense matrix that probably corresponded to the bright fluorescent spots shown in Figure 20. As in the fluorescent preparations, the asters seen in electron micrographs consisted of only a few short MTs, which extended out radially from the centrioles. More detailed descriptions of the metaphase spindle of rat kangaroo cells are reported elsewhere (Brinkley and Cartwright 1970; Roos 1973; McIntosh et al. 1975).

Anaphase

The progression of cells into anaphase was characterized by the movement of chromosomes to the pole and the abrupt elongation of the entire spindle. In these preparations, the chromosome-to-pole fibers became shorter and bundles of interpolar fibers could be seen extending across the interzone (Fig. 23). The asters became more prominent during anaphase, suggesting that their microtubules may have become longer as the chromosome-to-pole fibers shortened. Later in anaphase, the interpolar microtubules became much thinner and more faintly fluorescent (Fig. 24). The cytoplasm surrounding the anaphase spindle displayed a somewhat brighter fluorescence during this stage, as might be expected if more free tubulin became dispensed throughout the cytoplasm as a result of spindle tubule disassembly. The cytoplasm, however, remained free of fluorescent filaments during this stage. Near the end of anaphase the interpolar fibers could not be detected in most cells.

Examination of comparable stages of anaphase by EM (Fig. 25) showed that the interzone contained microtubules, some of which appeared to overlap near the cell equator, forming the early Zwisenkörper (Fig. 25). An electron-dense, amorphous substance was found in close association with the overlapping MTs of the interzone.

Telophase

During this stage, a number of well-known events take place. The cells undergo cytokinesis, forming two daughter cells held together by a narrow, cytoplasmic bridge, as described in earlier studies (Robbins and Gonatas 1964; Byers and Abramson 1968). The bridge contains numerous microtubules, which extend from an electron-dense zone, the midbody, into the cytoplasm of each daughter cell. The midbody consists of numerous interdigitating MTs surrounded by osmophilic amorphous material (Fig. 27). When these regions were stained by tubulin antibody immunofluorescence, images such as those shown in Figures 26 and 28 were observed. In early telophase, most of the fluorescence was confined to the region of the bridge where MTs extend into daughter cells. A distinct region corresponding to the midbody remained unstained. Apparently the structural arrangement

Figures 23–25
Anaphase. (*Fig. 23*) Spindle has elongated and interpolar filaments are seen in the interzone (arrow). (*Fig. 24*) Later in anaphase the interpolar filaments became thinner. (*Fig. 25*) Electron micrograph of anaphase. Arrows point to Zwisekörper at the interzone. Bar = 5 μ in Figs. 23 and 24 and 1 μ in Fig. 25.

of MTs plus the surrounding electron-dense matrix interferes with antibody-binding. Later in telophase or early G_1 phase, the cytoplasmic bridge becomes thinner and more elongated (Fig. 28). At this stage, numerous filaments reappeared in the cytoplasm. Telophase cells resulting from multipolar mitoses displayed three or more cytoplasmic bridges (see inset, Fig. 28).

Effects of Microtubule Inhibition on Spindle Fluorescence

As described earlier, microtubule inhibitors reversibly disrupted the CMTC of interphase cells. When mitotic cells were examined following colcemid treatment, the chromosomes were seen arranged into a group near the cell center (Figs. 29, 30). The spindle was not present, but the cytoplasm exhibited a background fluorescence brighter than that usually seen in untreated mitotic cells. Of particular interest was the presence of small discrete spots of fluorescence associated with the chromosomes (arrows in Figs. 29 and 30). These were often paired and generally corresponded to the position of the kinetochore on each chromosome. Often two or more chromosomes appeared to be closely associated with a single paired spot, suggesting some type of kinetochore-to-kinetochore association. EM studies have shown that short segments of MTs may remain associated with the kinetochores in cells blocked with minimal doses of colcemid (Brinkley, Stubblefield and Hsu 1967). Thus the fluorescent spots may be due to remaining MT segments or to tubulin, which may coat the kinetochore under these conditions. Alternatively, the kinetochore itself may be composed, in part, of tubulin protein which binds the antibody. These observations obviously require further study.

Cold-treated Cells

As previously described (Brinkley and Cartwright 1975), treatment of mitotic cells with cold temperatures leads to differential loss of spindle MTs. The interpolar MTs are the most susceptible to cold and become disassembled, whereas the chromosome-to-pole fibers are the most resistant. This is particularly well demonstrated after immunofluorescent staining (Figs. 31–33). During prometaphase, metaphase and anaphase, bundles of chromosomal tubules were seen. The interpolar filaments, however, were clearly absent. Although a bright fluorescent spot persisted at each pole, asteral filaments were also missing after cold treatment. Thus it is possible using the immunofluorescence procedure to evaluate accurately spindle damage induced by either colcemid or cold temperature.

DISCUSSION

In view of the rapid and expanding pace of MT research today, more expedient methods for identifying MTs in intact cells are needed. Conventional transmission electron microscopy provides reliable images of MTs, but such methods are restricted by small sample size, expense of operation and requirements for tedious serial section analysis to gain information on the three-dimensional distribution of MTs. High-voltage EM eliminates the need

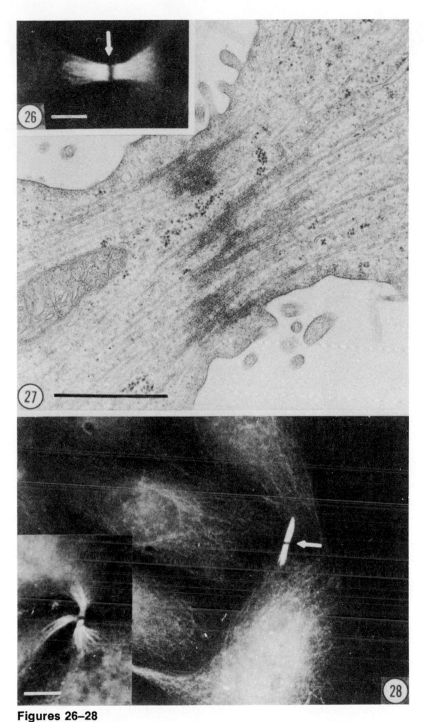

Figures 26–28

Telophase. (*Fig. 26*) Cytoplasmic bridge is brightly fluorescent except at midbody. (*Fig. 27*) Electron micrograph of bridge, showing numerous microtubules extending into electron-dense material at midbody. (*Fig. 28*) Late telophase, showing elongated cytoplasmic bridge (arrow). Inset shows bridge from tripolar cell. Bar = 5 μ in Figs. 26 and 28 and 1 μ in Fig. 27.

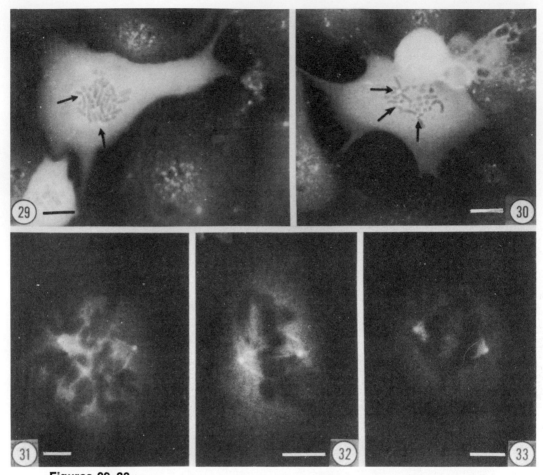

Figures 29–33

(*Figs. 29–30*) Colcemid inhibition of mitotic spindle. Note bright fluorescent spots on chromosomes (arrows) which correspond to kinetochores. Bars = 5 μ.

(*Figs. 31–33*) Effects of cold treatment on spindle. (*Fig. 31*) Prometaphase; (*Fig. 32*) metaphase; (*Fig. 33*) anaphase—note the absence of interpolar fibers and the reduced intensity of chromosomal fibers. Bar = 5 μ in Figs. 31–33.

for thin serial sections, but the cost and availability of these instruments are also limiting factors.

Results obtained in studies reported here using indirect immunofluorescence with tubulin-specific antibodies suggest that this procedure could be useful for many types of MT investigations. The specificity of the staining procedure for soluble tubulin and intact MTs has been well documented. The resolution of individual colcemid- and cold-sensitive filaments in interphase cells is possible with the dark-field ultraviolet microscope. The distribution of fluorescent filaments in the cytoplasm and the mitotic spindle closely corresponds to images obtained by high-voltage and conventional EM. For these reasons it is apparent that the fluorescent filaments described in this study represent intact microtubules.

Since MTs measure 240–270 Å in diameter, they are impossible to resolve by light optical procedures. However, if individual MTs are coated with tubulin-specific rabbit immunoglobulin molecules plus goat anti-rabbit immunoglobulin tagged with fluorescein, the diameter of the tubule is apparently increased to the point of resolution by ultraviolet microscope ($>$ 1000 Å). Therefore MTs within the cell that are separated by a distance greater than 1000 Å will be resolved as single thin filaments. Conversely, parallel MTs closer together than this distance will be resolved as thicker filaments. Since the spindle MTs are closely packed, with center-to-center distances of 400–500 Å, they often appear as coarse fluorescent filaments. In some instances, the entire spindle appears as a solid, stained mass. Cytoplasmic MTs, however, are usually much farther apart, and thus excellent resolution can be obtained by immunofluorescence.

These studies support and extend the observation of Porter and coworkers (Fonte and Porter 1974; Porter et al. 1974) that numerous microtubules exist within the cytoplasm of fibroblastic cells. Because of its elaborate organization, we have termed this system the "cytoplasmic microtubule complex" (CMTC) to distinguish it from the microtubular apparatus constituting the mitotic spindle. Colcemid and cold temperatures dissolve the CMTC and lead to the retraction of cell processes and loss of fibroblastic shape. Incubation of colcemid-treated cells in fresh medium leads to the rapid restoration of fibroblastic shape as well as the reappearance of the CMTC. These and similar observations (Porter 1966) strongly suggest that the CMTC is important in maintaining shape and symmetry in cultured mammalian cells. Obviously, the CMTC could be involved in other microtubular functions, such as intracellular movement of organelles and inclusions, maintenance of cell surface topography, and transport and release of substances for cells.

The immunofluorescent staining pattern observed in transformed cells indicates that the CMTC is greatly diminished in these cells. It follows from observations of normal cells that the altered shape of transformed cells could be directly related to the paucity and apparent randomness of MTs within the cytoplasm. What is the relationship of altered MT distribution to other well-known properties of transformed cells? From studies in several laboratories (Berlin 1975; Ichiro and Edelman 1975), MTs appear to play a role in regulating the distribution of cell surface receptors. Therefore depletion of the CMTC could affect surface properties in such a way as to impair contact and association with neighboring cells and lead to the loss of density-dependent inhibition of growth. Further studies are obviously needed before the total influence of microtubules on tumor cell properties is known.

Previous reports that reverse transformation agents, such as cAMP derivatives, could bring about major shape changes in transformed cells are confirmed by the present study. Moreover, the large-scale sampling of MT behavior by our precedures establishes unequivocally that reverse transformation accompanies the assembly and alignment of large numbers of MTs within the cytoplasm. Our results are also consistent with the observation that malignant cells have reduced levels of cAMP (Monahan, Fritz and Abell 1973, 1975). The precise role of cAMP in promoting assembly of MTs in cells remains unknown.

One of our most consistent findings was that the elaborate CMTC of inter-phase cells completely disappeared during mitosis. In thousands of cells examined, the CMTC was completely disassembled by the end of prophase and reappeared at the end of telophase or the beginning of G_1 phase. This cyclic response suggests that MT subunits of the CMTC may be recycled into the MTs of the mitotic spindle. Such reutilization of tubulin subunits is not unprecedented. Rosenbaum, Moulder and Ringo (1973) observed that in regenerating flagella of *Chlamydomonas*, MTs of the new flagellum were derived in part from subunits of the uninjured flagellum. Moreover, recycling is facilitated by the dynamic equilibrium model of Inoué (1964). For tubulin subunits to be taken from the CMTC and assembled into spindle MTs, how-ever, new MT organization centers (MTOCs) must be activated. Obviously, other organizational events would take place simultaneously. Since both soluble and polymerized tubulin pools exist in cultured cells (Rubin and Weiss 1975), some of the tubulin for the mitotic spindle could be derived from the existing soluble tubulin. More likely from our observation, how-ever, is that mostly all of the tubulin from the interphase cytoplasm becomes solubilized prior to the assembly of the spindle tubules.

Since transformed cells have a diminished CMTC but form an apparently normal, functional spindle, the control of assembly in the CMTC must be different from that in the mitotic spindle. Experiments reported elsewhere (Fuller et al. 1975) show that tubulin from SV3T3 is competent to assemble in vitro. Therefore it appears that in intact cells, the greatly limited initiation of assembly is possibly due to an aberration at the level of the MTOC. Addi-tional studies are underway in our laboratory to elucidate the role of cyto-plasmic MTs in normal and malignant cells.

Acknowledgments

We are grateful to Joan Ellison for excellent technical assistance in the prep-aration of tubulin antibodies. Appreciation is also extended to Linda Wible for assistance in electron microscopy and Timothy Stacey for photographic assistance. This work was supported in part by NCI Grant CA14675 and a grant from Dow Chemical Company.

REFERENCES

Allison, A. C. 1973. The role of microfilaments and microtubules in cell move-ment, endocytosis and exocytosis. In *Locomotion of tissue cells. CIBA Founda-tion Symposium* (ed. M. Abercrombie), vol. 14, p. 109. Associated Scientific Publishers, New York.

Berlin, R. D. 1975. Microtubules and the fluidity of the cell surface. *Ann. N.Y. Acad. Sci.* **253**:445.

Brinkley, B. R. and J. Cartwright, Jr. 1970. Organization of microtubules in mitotic spindle: Differential effects of cold shock on microtubule stability. *J. Cell Biol.* **47**:25a.

———. 1975. Cold labile and cold stable microtubules in the mitotic spindle of mammalian cells. *Ann. N.Y. Acad. Sci.* **253**:428.

Brinkley, B. R., G. M. Fuller and D. P. Highfield. 1975. Cytoplasmic microtubules in normal and transformed cells in culture: Analysis by tubulin antibody immunofluorescence. *Proc. Nat. Acad. Sci.* **72**:4981.

Brinkley, B. R., E. Stubblefield and T. C. Hsu. 1967. The effects of colcemid inhibition and reversal on the fine structure of the mitotic apparatus of Chinese hamster cells *in vitro. J. Ultrastruc. Res.* **19**:1.

Byers, B. and D. H. Abramson. 1968. Cytokinesis in HeLa: Post-telophase delay and microtubule-associated motility. *Protoplasm* **66**:413.

Fonte, V. and K. R. Porter. 1974. Topographic changes associated with viral transformation of normal cells to tumorigenicity. In *8th International Congress on Electron Microscopy,* p. 334. Australian Academy of Science, Canberra.

Fuller, G. M., B. R. Brinkley and J. M. Boughter. 1975. Immunofluorescence of mitotic spindles by using monospecific antibody against bovine brain tubulin. *Science* **187**:948.

Fuller, G. M., J. J. Ellison, M. McGill, L. Sordahl and B. R. Brinkley. 1975. Studies on the inhibitory role of calcium in the regulation of microtubule assembly *in vitro and in vivo.* Presented at the *International Symposium on Microtubules and Microtubular Inhibitors.* Beerse, Belgium, September, 1975.

Hsie, A. W. and T. T. Puck. 1971. Morphological transformation of Chinese hamster cells by dibutyryl adenosine cyclic 3′:5′-monophosphate and testosterone. *Proc. Nat. Acad. Sci.* **68**:358.

Ichiro, Y. and G. M. Edelman. 1975. Modulation of lymphocyte receptor mobility by concanavalin A and colchicine. *Ann. N.Y. Acad. Sci.* **253**:455.

Inoué, S. 1964. Organization and function of the mitotic spindle. In *Primitive motile systems in cell biology* (ed. R. D. Allen and N. Kamiya), p. 549. Academic Press, New York.

Johnson, G. E., R. M. Friedman and I. Pastan. 1971. Restoration of several morphological characteristics of normal fibroblasts in sarcoma treated with adenosine-3′:5′-cyclic monophosphate and its derivatives. *Proc. Nat. Acad. Sci.* **68**:425.

Malawista, S. E. 1975. Microtubules and the mobilization of lysosomes in phagocytizing human lymphocytes. *Ann. N.Y. Acad. Sci.* **253**:738.

McIntosh, J. R., Z. Cande, J. Snyder and K. Vanderslice. 1975. Studies on the mechanism of mitosis. *Ann. N.Y. Acad. Sci.* **253**:407.

Moellmann, G. and J. McGuire. 1975. Correlation of cytoplasmic microtubules and 10 nm filaments with the movement of pigment granules in cutaneous melanocytes of *Rana pipiens. Ann. N.Y. Acad. Sci.* **253**:711.

Monahan, J. M., R. R. Fritz and C. W. Abell. 1973. Levels of cyclic AMP in murine L5178Y lymphoblasts grown in different concentrations of serum. *Biochem. Biophys. Res. Comm.* **55**:642.

―――. 1975. Cyclic adenosine 3′:5′-phosphate levels and activities of related enzymes in normal and leukemic lymphocytes. *Cancer Res.* **35**:2540.

Nicklas, R. B. 1971. Mitosis. In *Advances in cell biology* (ed. D. M. Prescott, L. Goldstein and E. N. McConkey), vol. 2, pp. 225–298. Appleton-Century-Crofts, New York.

Porter, K. R. 1966. Cytoplasmic microtubules and their functions. In *Principles of biomolecular organization* (ed. G. E. Wolstenholm and M. O'Conner), p. 308. Little, Brown & Co., Boston.

Porter, K. R., T. T. Puck, A. W. Hsie and D. Kelly. 1974. An electron microscope study of the effects of dibutyryl cyclic AMP on Chinese hamster ovary cells. *Cell* **2**:145.

Redman, C. M., D. Banerjee, K. Howell and G. E. Palade. 1975. The step at which colchicine blocks the secretion of plasma proteins by rat liver. *Ann. N.Y. Acad. Sci.* **253**:780.

Robbins, E. and N. K. Gonatas. 1964. The ultrastructure of a mammalian cell during the mitotic cycle. *J. Cell Biol.* **21**:429.

Roos, V. P. 1973. Light and electron microscopy of rat kangaroo cells in mitosis. I. Formation and breakdown of the mitotic apparatus. *Chromosoma* **40**:43.

Rosenbaum, J. L., J. E. Moulder and D. L. Ringo. 1973. Flagellar elongation and shortening in *Chlamydomonas*. *J. Cell Biol.* **41**:600.

Rubin, R. W. and G. D. Weiss. 1975. Direct biochemical measurements of microtubule assembly and disassembly in Chinese hamster ovary cells. *J. Cell Biol.* **64**:42.

Salmon, E. D. 1975. Spindle microtubules. Thermodynamics of *in vivo* assembly and role in chromosome movement. *Ann. N.Y. Acad. Sci.* **253**:383.

Sheppard, J. R. 1971. Restoration of contact-inhibited growth to transformed cells by dibutyryl adenosine 3':5' cyclic monophosphate. *Proc. Nat. Acad. Sci.* **68**:1316.

Smith, D. S., V. Jälfors and B. F. Cameron. 1975. Morphological evidence for the participation of microtubules in axonal transport. *Ann. N.Y. Acad. Sci.* **253**:472.

Stoker, M. G. P. and H. Rubin. 1967. Density dependent inhibition of cell growth in culture. *Nature* **215**:171.

Weber, K., R. Pollack and T. Bibring. 1975. Antibody against tubulin: The specific visualization of cytoplasmic microtubules in tissue culture cells. *Proc. Nat. Acad. Sci.* **72**:459.

Wolff, J. and B. Bhattacharyya. 1975. Microtubules and thyroid hormone mobilization. *Ann. N.Y. Acad. Sci.* **253**:763.

Name Index

Italics indicate where full reference can be found; **boldface** type designates where author's article in this volume is located.

Subject Index

A-tubule. *See* Outer doublet tubules
Acanthamoeba castellani, 492, 624, 671, 689
   calcium regulation, 687
   cofactor, 690
   extracts, 716–718
Acrosomal reaction, 514
Actin. *See also* Thin filament; Microfilament
   antibody to, 348–351, 354, 356–357, 376, 399, 403, 499, 1263–1264
   in axonal transport, 1040, 1043–1045
   bound nucleotide, 495
   bundle formation, 580, 583
   concentration through cell cycle, 389, 673, 680–684
   in cytoplasmic movements, 1
   depolymerization, 477–478, 481
   in dividing cells, 265–276
   in enveloped viruses, 589–597
   in erythrocytes, 653
   filament network, 357–358
   in force production, 3, 4
   in HeLa cells, 270–273, 671
   HMM decoration of, 116, 118, 220–221, 225, 227, 233, 237, 267, 270–271, 495, 625, 716, 799, 807, 1262, 1269, 1273–1275, 1295–1314
   in lymphocytes, 312–314
   magnesium-ATPase, 495
   membrane association, 527, 623, 631, 671
   in mitotic apparatus, 269–276, 712, 1262–1265, 1268–1269, 1273–1290, 1302–1314
   in motile amoeba extracts, 809, 811, 814–816
   in muscle systems, 115–125
   -myosin interaction, role of calcium in, 142-143
   in non-muscle systems, 116, 1295–1314
   nuclear, 831–832
   paracrystals, 470
   in paramyxoviruses, 590
   in *Physarum,* 750–753
   in plasmodium, 500
   in platelets, 691
   polymerization, 477, 632, 690
   preservation, 697
   in RNA tumor viruses, 594
   in smooth muscle cells, 361–363, 372
   in streaming, 8, 809, 811, 814–816
   transformation, 809, 817–819
   in transformed fibroblasts, 377
   viscosity, 495
Actin-binding protein, 531, 535, 540, 626
Actin filament, 18, 530, 561. *See also* Microfilament; Thin filament
   amoeboid movement, 807, 809, 812–819
   in cultured cells, 347–358
   membrane attachment, 631, 632
   microvillar movement, 633
   polarity, 632, 635, 637, 779–782, 1269
   in smooth muscle, 172–176, 180
   transformation of, 812–815
α-Actinin, 18, 380–383
   antibody to, 364, 626, 643
     in fibroblasts, 351–354, 356–358, 376, 380